Adaptive Techniques for Dynamic Processor Optimization

Theory and Practice

Series on Integrated Circuits and Systems

Series Editor: Anantha Chandrakasan
Massachusetts Institute of Technology
Cambridge, Massachusetts

Continued after index

Alice Wang • Samuel Naffziger

Editors

Adaptive Techniques for Dynamic Processor Optimization

Theory and Practice

 Springer

Editors
Alice Wang
Texas Instruments, Inc.
Dallas, TX
USA
aliwang@ti.com

Samuel Naffziger
Advanced Micro Devices
Fort Collins, CO
USA
samuel.naffziger@amd.com

Series Editor
Anantha Chandrakasan
Department of Electrical Engineering
  and Computer Science
Massachusetts Institute of Technology
Cambridge, MA 02139
USA

ISBN: 978-0-387-76471-9   e-ISBN: 978-0-387-76472-6
DOI: 10.1007/978-0-387-76472-6

Library of Congress Control Number: 2007943527

Printed on acid-free paper

9 8 7 6 5 4 3 2 1

springer.com

Preface

The integrated circuit has evolved tremendously in recent years as Moore's Law has enabled exponentially more devices and functionality to be packed onto a single piece of silicon. In some ways however, these highly integrated circuits, of which microprocessors are the flagship example, have become victims of their own success. Despite dramatic reductions in the switching energy of the transistors, these reductions have kept pace neither with the increased integration levels nor with the higher switching frequencies. In addition, the atomic dimensions being utilized by these highly integrated processors have given rise to much higher levels of random and systematic variation which undercut the gains from process scaling that would otherwise be realized. So these factors—the increasing impact of variation and the struggle to control power consumption—have given rise to a tremendous amount of innovation in the area of adaptive techniques for dynamic processor optimization.

The fundamental premise behind adaptive processor design is the recognition that variations in manufacturing and environment cause a statically configured operating point to be far too inefficient. Inefficient designs waste power and performance and will quickly be surpassed by more adaptive designs, just as it happens in the biological realm. Organisms must adapt to survive, and a similar trend is seen with processors – those that are enabled to adapt to their environment, will be far more competitive. The adaptive processor needs to be made aware of its environment and operating conditions through the use of various sensors. It must then have some ability to usefully respond to the sensor stimulus. The focus of this book is not so much on a static configuration of each manufactured part that may be unique, but on *dynamic* adaptation, where the part optimizes itself on the fly.

Many different responses and adaptive approaches have been explored in recent years. These range from circuits that make voltage changes and set body biases to those that generate clock frequency adjustments on logic. New circuit techniques are needed to address the special challenges created by scaling embedded memories. Finally, system level techniques rely on self-correction in the processor logic or asynchronous techniques which remove the reliance on clocks. Each approach has unique challenges and benefits, and it adds value in particular situations, but regardless of the method, the challenge of reliably testing these adaptive approaches looms as one of the largest. Hence the subtitle the book: Theory and Practice. Ideas (not necessarily good ones) on adaptive designs are easy to come by, but putting these in working silicon that demonstrates the benefits is much harder. The final level of achievement is actually productizing the capability in a high-volume manufacturing flow.

In order for the book to do justice to such a broad and relatively new topic, we invited authors who have already been pioneers in this area to present data on the approaches they have explored. Many of the authors presented at ISSCC2007, either in the Microprocessor Forum, or in the conference sessions. We are humbled to have collected contributions from such an impressive group of experts on the subject, many of whom have been pioneers in the field and produced results that will be impacting the processor design world for years to come. We believe this topic of adaptive design will continue to be a fertile area for research and integrated circuit improvements for the foreseeable future.

Alice Wang   *Texas Instruments, Inc.*
Samuel Naffziger   *Advanced Micro Devices, Inc.*

Table of Contents

Chapter 8 Architectural Techniques for Adaptive Computing.........175
Shidhartha Das, David Roberts, David Blaauw, David Bull, Trevor Mudge

List of Contributors

Alan Drake	IBM
Alice Wang	Texas Instruments
Anantha Chandrakasan	Massachusetts Institute of Technology
Brian Cherkauer	Intel Corporation
David Blaauw	University of Michigan
David Bull	ARM Ltd.
David Roberts	University of Michigan
David Scott	Taiwan Semiconductor Manufacturing Company Ltd.
Diana Marculescu	Carnegie Mellon University
Eric Fetzer	Intel Corporation
Franco Ricci	Marvell Semiconductor Inc.
James Tschanz	Intel Corporation
Jason Stinson	Intel Corporation
Jim Garside	The University of Manchester
John J. Wuu	Advanced Micro Devices, Inc.
José Pineda de Gyvez	NXP Semiconductors, Eindhoven University of Technology
Joyce Kwong	Massachusetts Institute of Technology
Lawrence T. Clark	Arizona State University

Maurice Meijer	NXP Semiconductors
Naveen Verma	Massachusetts Institute of Technology
Sebastian Herbert	Carnegie Mellon University
Shidhartha Das	ARM Ltd., University of Michigan
Steve Furber	The University of Manchester
Steve Poehlman	Intel Corporation
Tadahiro Kuroda	Keio University
Takayasu Sakurai	University of Tokyo
Trevor Mudge	University of Michigan
William E. Brown	Ellutions, LLC
Yogesh K. Ramadass	Massachusetts Institute of Technology

Chapter 1 Technology Challenges Motivating Adaptive Techniques

David Scott,[1] Alice Wang[2]

[1]Taiwan Semiconductor Manufacturing Company Ltd., [2]Texas Instruments, Inc.

1.1 Introduction

In the design of an integrated circuit, the designer is faced with the challenge of having circuits and systems function over multiple operating points. From the point of view of performance, the circuit must meet its speed requirements over a range of voltages and temperatures that reflect the environment that the circuit is operating in. Also while the performance requirement must be met at a set of worst-case conditions for speed, the power requirement must be simultaneously met at another set of worst-case conditions for power.

Although each design is unique, the resulting instances of fabricated integrated circuits will number potentially in the billions. In addition, the number of components for each of the integrated circuits will also potentially number in the billions. Every single one of the billions of transistors in every one of the billions of circuits is unique. The success of an integrated circuit design is simply measured by the percentage of the fabricated integrated circuits with the transistors, as well as interconnections, meeting all the requirements.

The use of adaptive techniques allows for an integrated circuit to adapt for variations in the environment as reflected by both voltage and temperature and also for variations in the fabricated transistors. Adaptive techniques are intended to allow minimization of both dynamic and leakage power and also to increase the frequency of operation of the integrated circuit.

A. Wang, S. Naffziger (eds.), *Adaptive Techniques for Dynamic Processor Optimization*,
DOI: 10.1007/978-0-387-76472-6_1, © Springer Science+Business Media, LLC 2008

1.2 Motivation for Adaptive Techniques

1.2.1 Components of Power

The total power dissipation of an integrated circuit can be simply represented by the power equation below. The power is divided into three major components: the dynamic component, the subthreshold leakage component, and the parasitic leakage components. The dynamic component depends on the overall capacitance of the integrated circuit and the charge that must be displaced for each clock cycle. This is the power that is actually doing work to implement the function of the integrated circuit. Techniques such as clock gating [1] reduce power by gating the clock in unused parts of the integrated circuit, thereby reducing the effective capacitance of the integrated circuit:

$$P = CV^2f + V N (I_0 exp(-nV_{th}/(kT/q)) + I_{tox} + I_{gedl}) \quad (1.1)$$

The subthreshold leakage current is simply tied to the threshold voltage of the transistors in the integrated circuit. As the threshold voltage (V_{th}) increases, the subthreshold current decreases exponentially. However, while significant leakage savings can be achieved by increasing the threshold voltage, a high threshold voltage tends to force designers to operate the circuits at higher voltages in order to achieve the performance goals. Gate oxide leakage (I_{tox}) and gate edge diode leakage (I_{gedl}) relate to the characteristics of both the gate oxide and the silicon. These parasitic components of leakage will be discussed in a later section.

1.2.2 Relation Between Frequency and Voltage

As shown in Figure 1.1, operating frequency increases as the supply voltage of the integrated circuit increases [2]. Note that the straight line in this plot does not extrapolate back to zero but rather a larger value that depends on the threshold voltage of the transistors in the circuit. Hence, at a given supply voltage, the frequency of the integrated circuit can be changed by changing the threshold voltage. Threshold voltage can be controlled dynamically by changing the transistor body bias. Hence, supply voltage and body bias provide two degrees of freedom over which to control both frequency and power.

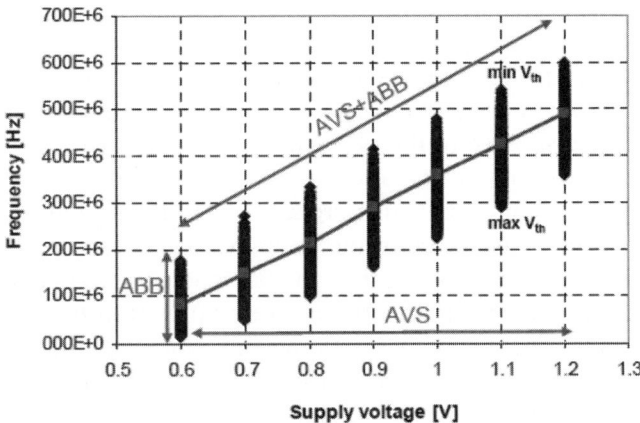

Figure 1.1 Frequency versus voltage [2]. (© 2005 IEEE)

A conceptual plot for the case where the frequency is constrained to be a constant is shown in Figure 1.2. This illustrates the tradeoffs that can be made between the choice of supply voltage and threshold voltage. If the transistors have a low threshold voltage, the leakage power is very high and the dynamic power is quite low. That is because the operating frequency can be achieved at a relatively lower supply voltage and yet the low threshold voltage results in a high leakage current. As the threshold voltage is increased, the supply voltage to maintain the operating frequency is also increased and hence dynamic power increases. At the same time, the increasing threshold voltage results in a lower leakage power. For a given integrated circuit, there is an optimum point where the power is minimized. This is the point where the increase in dynamic power is offset by the decrease in leakage power.

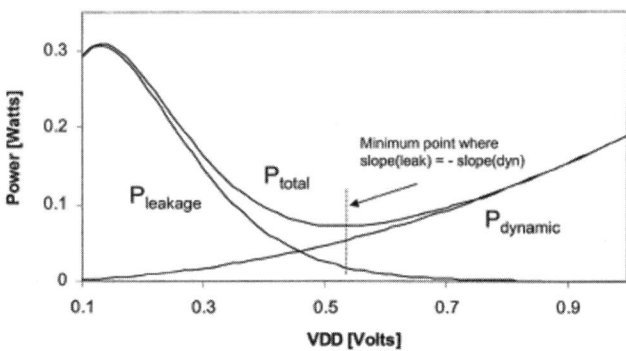

Figure 1.2 As VDD is increased, the body bias is adjusted to keep operating frequency constant [3]. (© 2002 IEEE)

1.2.3 Control Loop Implementation

An example control loop to control body bias is shown in Figure 1.3 [3]. A clock signal is input into a replica circuit and into a phase detector at the same time. The purpose of the phase detector is to detect whether the signal edge is able to pass through the replica circuit in a single clock cycle. Based on whether the signal edge precedes or follows a single clock cycle, the output of the phase detector increases or lowers the body bias accordingly. For this scheme to work, the replica circuit must be representative of the other circuits within the chip that are being controlled by the control loop. A similar scheme can be implemented to control the supply voltage of the replica line where in this case the supply voltage is either incremented or decremented in order to control the speed of the replica circuit.

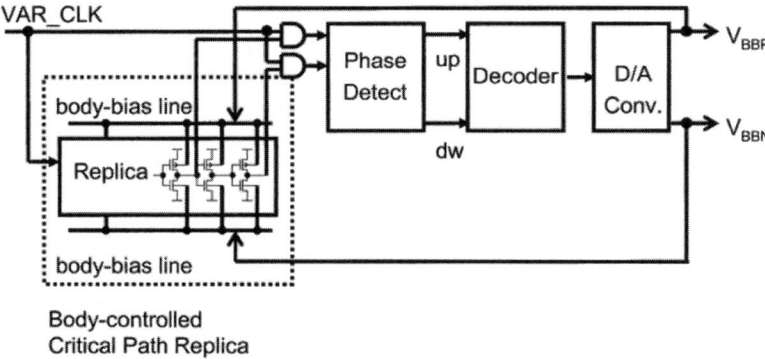

Figure 1.3 Illustration of replica path [3]. (© 2005 IEEE)

1.2.4 Practical Considerations

The key limitation of implementing an adaptive technique is the extent to which the replica circuit represents the integrated circuit. The replica is just one circuit while an integrated circuit has literally thousands of delay paths. This oversimplification is often resolved, assuming that the replica circuit represents the worst-case delay path.

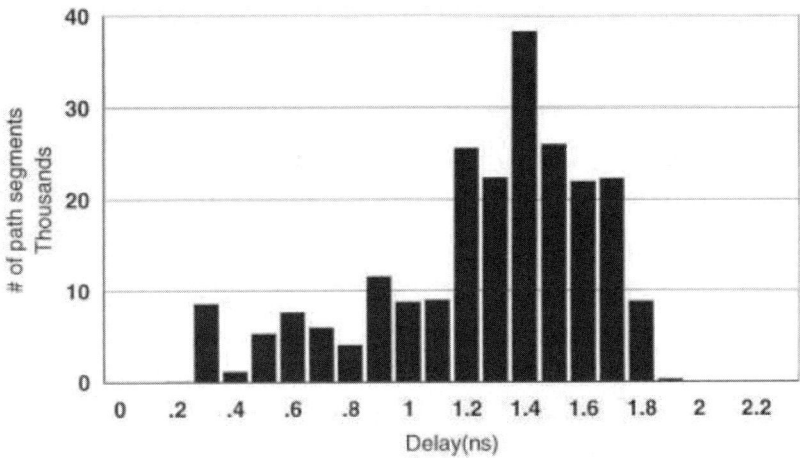

Figure 1.4 An illustration of critical paths in a design [4]. (© 2004 IEEE)

A typical histogram of delay path segments is shown in Figure 1.4 [4]. As seen from observing this histogram, many of the paths are much faster than the slowest path, and this variation represents a further opportunity to reduce power. The transistors in the faster paths can be substituted with transistors with lower leakage. One way to do this is by selective use of transistors with longer channel length. Due to the longer channel length, these transistors will be slower but they will also have reduced leakage.

An example of this has already been implemented in an integrated circuit [5] through the use of a library of circuits that were implemented with both long and short gate lengths. A slight area penalty was incurred to make each circuit in the library footprint and layout compatible as in Figure 1.5. Hence, these circuits can be freely interchanged at any point in the design cycle to minimize power at the expense of path delay. This algorithm can be similarly implemented using multiple threshold voltage transistors.

The use of the above algorithm for substitution of longer gate length transistors to reduce leakage can occur on a massive scale as is shown in Figure 1.6. One result of implementing this type of algorithm is that all delay path segments become more critical as the extra slack in the design is harvested in order to reduce leakage current. Making all these paths more critical will tend to make the design less tolerant of circuit variations or circuit modeling inaccuracies.

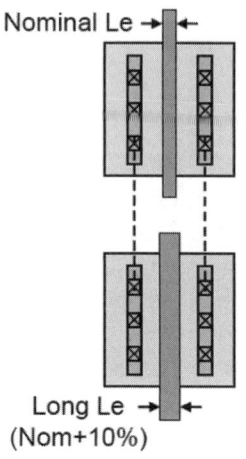

Nominal Le →| |←

Long Le →| |←
(Nom+10%)

- All transistors can be either nominal or long-Le
- Most library cells are available in both flavors
- Long-Le transistors are about 10% slower, but have 3x lower leakage
- All paths with timing slack use long-Le transistors
- Initial design uses only long channel devices

Figure 1.5 Two transistors with the same layout footprint. Layout area efficiency is sacrificed in order to make the shorter channel transistor replaceable by the longer channel transistor [5]. (© 2006 IEEE)

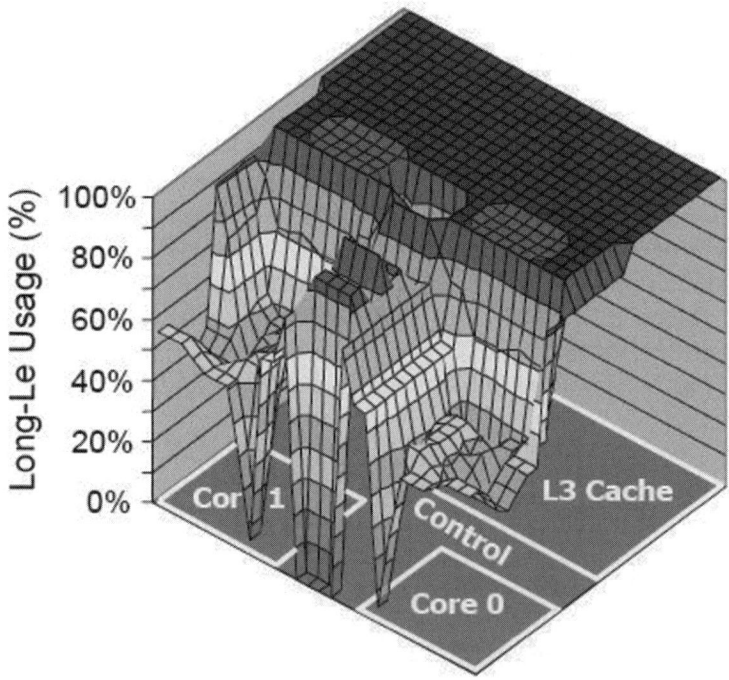

Figure 1.6 Usage of long channel transistors in a design showing that a shorter channel is required for a small fraction of the transistors in this design [5]. (© 2006 IEEE)

1.2.5 Impact of Temperature and Supply Voltage Variations

In the last section, we showed how the operating frequency varies with supply voltage. In this section, we also factor in the temperature dependence as well as across chip variations. The operating frequency as a function of supply voltage is shown for two different temperatures in Figure 1.7. At low temperatures, the mobility of the carriers is lower and hence the operating frequency is lower when the supply voltage is high. At high temperature, the lower threshold voltage favors low voltage operation. These two curves cross at what is normally considered the nominal operating voltage. Hence in the absence of adaptive techniques, modern integrated circuits show very little sensitivity of operating frequency to temperature.

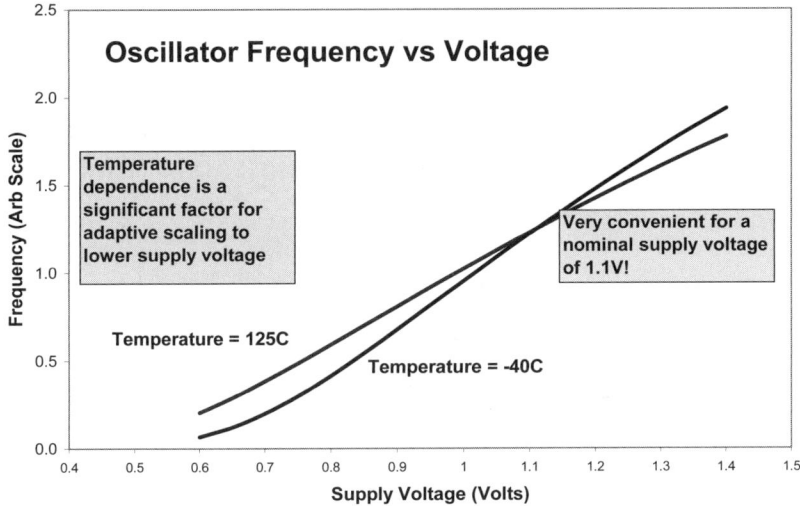

Figure 1.7 A plot of frequency versus voltage for a circuit at two different temperatures [6]. (© 2007 IEEE)

At low voltages, the type of behavior described in Figure 1.7 greatly favors high-temperature operation, and hence there is a great deal of sensitivity of operating frequency to temperature at low voltages. In general, there is a great deal of temperature variation across a chip [7]. At low voltages, the coldest parts of the chip will have the most problem, while at high temperature, the hottest parts of the chip will be slowest. Any adaptive scheme must account for the changing sensitivity to temperature at the low and high operating voltages.

An analysis of the impact of supply voltage variation on the chip has been previously done for two different cases [8]. This analysis was facilitated by recognizing that in steady state, the supply voltage across the chip must satisfy the following equation:

$$\nabla^2 V = R_s J_O \tag{1.2}$$

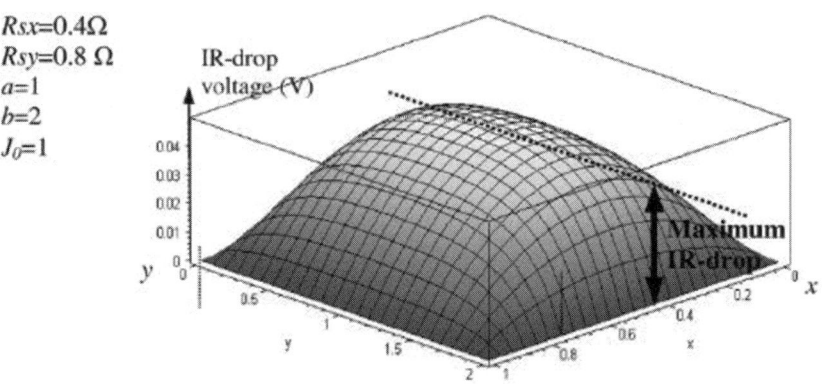

Figure 1.8a (Case 1) Supply voltage drop across a chip that has been wire bonded with supply pads on the edge of the chip [8]. (© 2005 IEEE)

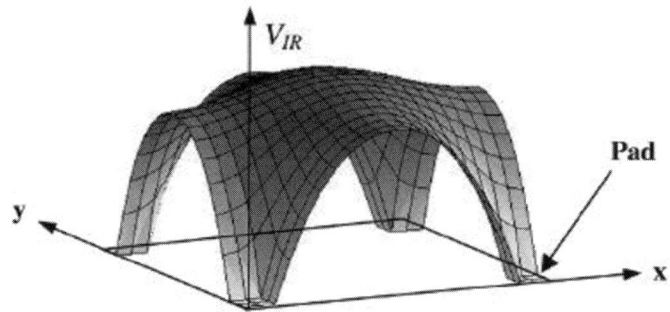

Figure 1.8b (Case 2) Supply voltage variations for flip chip where the power supply pads are arrayed over the complete chip area [8]. (© 2005 IEEE)

Case 1 is the wire bond case where the perimeter of the chip is pinned to the supply voltage and case 2, is the flip chip case where the supply and grounds pins are placed in a mesh across the chip. In Figure 1.8a, the wire-bound case is shown where clearly the maximum voltage supply loss is at the center of the chip. For the flip chip case shown in Figure 1.8b, each small part of the chip has the supply voltage pinned in the corners only, and this pattern is arrayed over the entire chip with the amplitude dependent on the local power supply current. The maximum operating frequency of each of the path segments will depend on the local supply voltage.

1.3 Technology Issues Relating to Performance-Enhancing Techniques

1.3.1 Threshold Voltage Variation

While in the previous section we dealt with variations in the supply voltage caused by on-chip voltage drops, in this section we discuss the variation in threshold voltage. The impact of threshold voltage variation is shown in Figure 1.9 involving circuits with two different threshold voltages. The sensitivity of the frequency to threshold voltage has the impact of shifting the curve to the right as the threshold voltage increases.

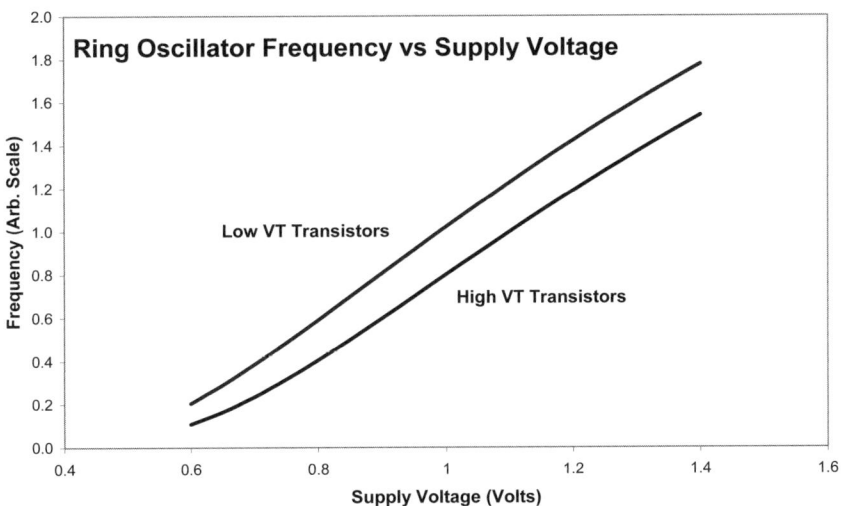

Figure 1.9 As the threshold voltage is increased, the frequency versus supply voltage curve is shifted to the right [6]. (© 2005 IEEE)

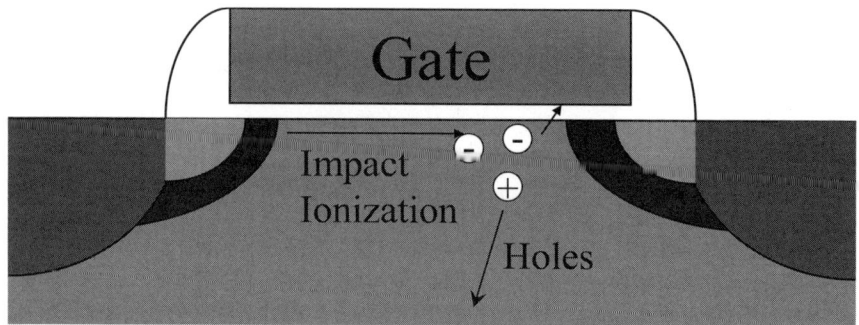

Figure 1.10a The mechanism of impact ionization is illustrated, with electron–hole pairs being generated at the drain end of the channel. Some of these generated carriers end up being trapped in the gate oxide [6]. (© 2005 IEEE)

The impact of hot carrier-induced threshold voltage shift is shown in Figure 1.10a. At high electric fields, carriers generated from impact ionization are trapped in the gate oxide. In general, the transistor lifetime decreases with the cube of the substrate current, and the gate voltage dependence of the substrate current is illustrated in Figure 1.10b. Also, there are separate degradation characteristics from both ac- and dc-related stress currents. It has also been found that the threshold voltage and other device parameters can shift over the lifetime of the product and not always in the same direction [8]. For example, the threshold voltage can recover after the stress is removed [9].

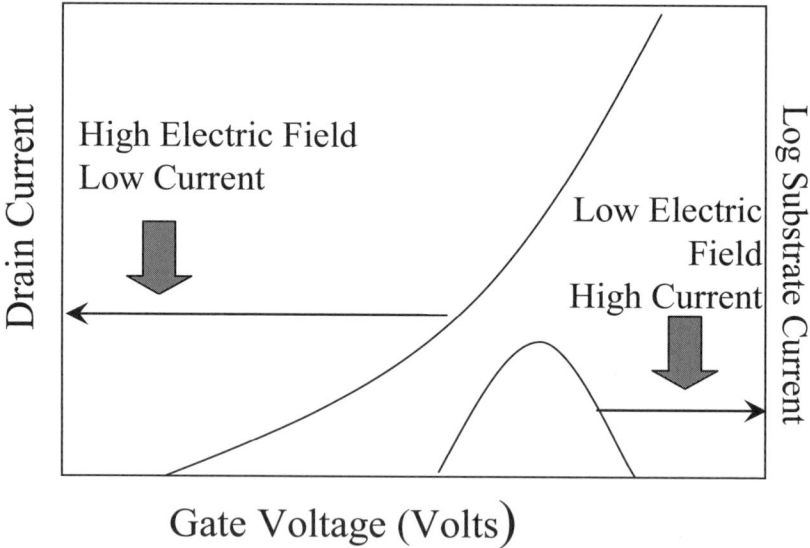

Figure 1.10b A peak in the substrate current occurs when high current flow and high electric field occur both at the same time [6]. (© 2005 IEEE)

Another source of threshold variation is negative bias temperature instability (NBTI). This phenomenon is commonly associated with p-channel transistors and is caused by the movement of charge in the gate oxide and at the interface. Of course in an integrated circuit, each transistor has a unique set of bias conditions over its lifetime and hence each transistor degrades differently. For a typical stress condition of negative bias, the variation of threshold voltage with time is given by the following equation [10]:

$$\Delta V_{th} = C \exp(-E_A / kT) \exp(\beta \mid V_G \mid) t^n \qquad (1.3)$$

where E_A is the activation energy, V_G is the gate voltage, and t is the stress time, and the other parameters are constants.

NBTI degradation has the biggest impact at lower supply voltage. This is due to the loss of headroom as the p-channel threshold voltage increases in absolute value. Previous work [10] has shown that as a ring oscillator is stressed, the low voltage frequency of operation is degraded due to the increase of p-channel threshold voltage. The impact of threshold voltage on low frequency operation can be observed in Figure 1.7. The circuit using transistors with the higher threshold voltage has a lower frequency of operation.

1.3.2 Random Dopant Fluctuations

As dimensions continue to shrink, the number of dopants in the channel has become discrete and measurable in discrete quantities The small number doping atoms in the channel means that the threshold voltage will be highly variable and will vary for transistors with otherwise identical characteristics. The variation in threshold voltage can be related to the average doping by the following equation [11]:

$$\sigma_{V_{th}} = \left(\frac{\sqrt[4]{4q^3 \varepsilon_{Si} \phi_B}}{2} \right) \cdot \frac{T_{ox}}{\varepsilon_{ox}} \cdot \frac{\sqrt[4]{N}}{\sqrt{W_{eff} L_{eff}}} \qquad (1.4)$$

Measured data shown in Figure 1.11 shows that the standard deviation for typical state of the art dimensions is quite significant, with standard deviation in the range of 30 mV–50 mV being quite feasible. For a chip with many millions of gates, transistors with threshold voltages more than 5 standard deviations from the mean are relatively common.

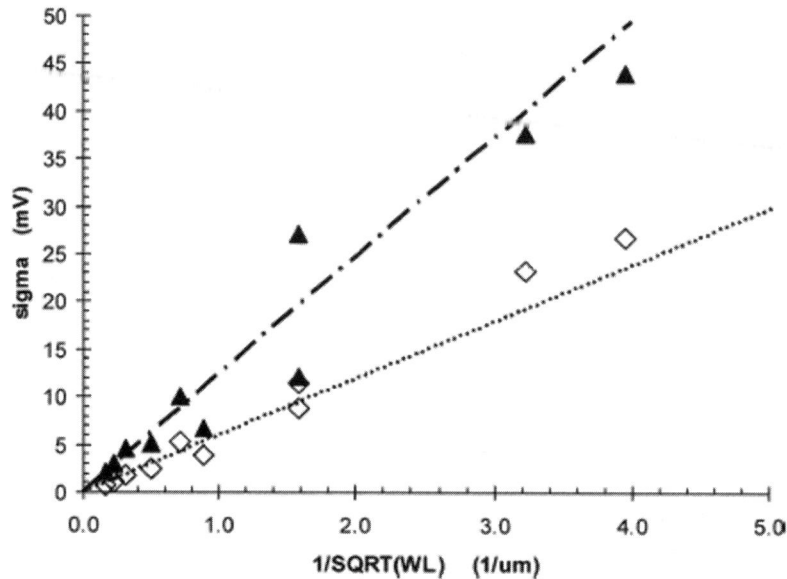

Figure 1.11 Measured data showing the increase in threshold voltage variation as the area of the transistor is decreased. Diamonds are for strong inversion while triangles are for subthreshold region [11]. (© 2005 IEEE)

In terms of applying adaptive techniques, the difficulty that the circuit designer faces is compounded. Identical transistors placed in different parts of the circuit tend to have randomly different values of threshold voltage as described by Equation 1.4 and Figure 1.11.

In addition, as body bias is applied, the randomness of the threshold voltage will tend to increase [12]. When body bias is increased, more dopants are incorporated into the depletion region and hence the randomness of these additional dopants is also incorporated into the transistor.

New transistor design techniques are continuously under development, and these scaled transistors offer new challenges to the designer. Taking advantage of these new transistor design techniques can be of great value to the circuit design. As shown in Figure 1.12, Yasuda [12] found that as body bias is applied to a collection of transistors, their threshold voltage distribution has a tendency to reorder. That is to say different transistors have different responses to body bias and hence the transistor with the lowest threshold voltage in a distribution may no longer be the lowest when body bias is applied. Certainly, this is a concern to a designer who is using body bias to control transistor performance or leakage.

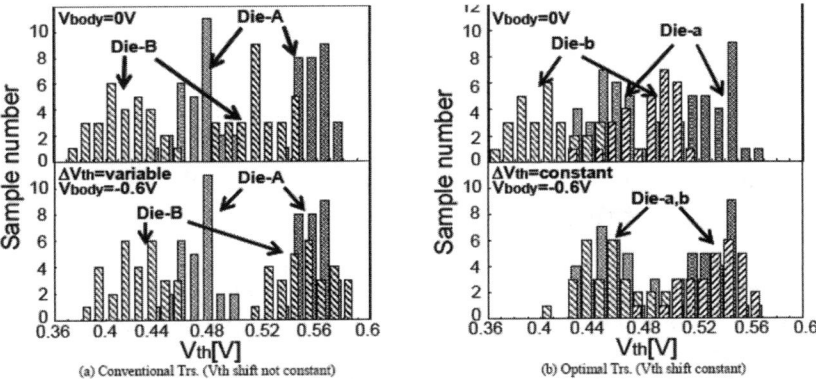

Figure 1.12 The benefit of an optimized transistor design is shown where the threshold voltage shift with body bias is constant [12]. (© 2005 IEEE)

It has been shown that new transistor design techniques, taking advantage of Fermi level pinning present in [13], offer an advantage in the transistor design. If the channel of the transistor is optimally designed, the response of the transistor to applied body bias can be made much more predictable.

1.3.3 Design in the Presence of Threshold Voltage Variation

When designing an adaptive system, the designers must contend with a number of sources of threshold voltage variation that are not under their direct control. Transistor characteristics and in particular threshold voltage can vary from wafer to wafer and also from die to die within a wafer. These variations are generally known as global variations. In addition, the designers must contend with the local variations that occur within a die.

Local variations can be due to random dopant fluctuations including the transistors having different sensitivities to back gate bias, line edge roughness of the gate material, and systematic changes in device behavior such as temperature and temperature gradients across the die.

Transistor characteristics can also change over the lifetime of the integrated circuit as a result of hot carrier effects or negative bias temperature-induced (NBTI) changes in the threshold voltage. In addition, new techniques to improve transistor performance by using mechanical stress [14] also will bring additional sources of variation.

As a result of all of the above, designers are generally not looking at a single line on the frequency versus voltage curve that can be modulated with back gate bias. They must think of this line as having considerable width, with the width of this line being defined by the sum total of all the variations that can occur in the transistor through fabrication, during its operation, and over its lifetime.

1.4 Technology Issues Associated with Leakage Reduction Techniques

A common technique to reduce subthreshold leakage is to merely increase the threshold voltage by applying back gate bias [15, 16]. In Figure 1.13, the waveforms show how the leakage of an integrated circuit can be reduced when the system is going into a lower power mode of operation. As discussed earlier, this type of scheme requires substrate terminals of the transistors to be available globally, and hence these two extra supply lines must be available to be globally routed. Also the substrate pump requires additional area and consumes current in order to operate. The operating current associated with the substrate pump offsets the leakage gains. The well bias generator function is often conveniently provided by the same supply as is used for the IO circuits, and hence an extra penalty for providing this extra supply is normally not incurred.

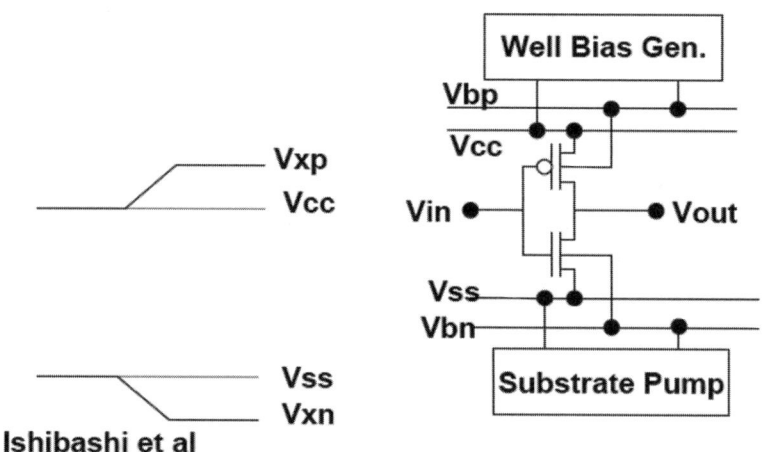

Figure 1.13 One scheme to reduce leakage is to merely apply back gate bias to all transistors [6]. (© 2005 IEEE)

The need for a substrate pump has been avoided by some designers by raising Vss rather than having Vbn negative [1]. This type of scheme has the added benefit of current being reduced due to both back gate bias and the lower operating voltage. However, switching of the Vss supply does require footer transistors that are large enough to conduct the entire active current of the circuit without having an excessive voltage drop. In addition, once added the footer transistors themselves become leakage sources.

1.4.1 Practical Considerations

The ability of a technology to support state-of-the-art integrated circuits and systems is conventionally judged by its leakage versus "on" current. Figure 1.14 shows such plots for three different technologies [14]. The individual data points for each different technology are achieved as a result of measuring transistors with different gate lengths and threshold voltage implants. In addition the normal process variations, discussed in the previous section, can play a role in smoothing out these curves. One important aspect of this tradeoff to note is that the "Ion" scale is linear while the leakage current scale is exponential.

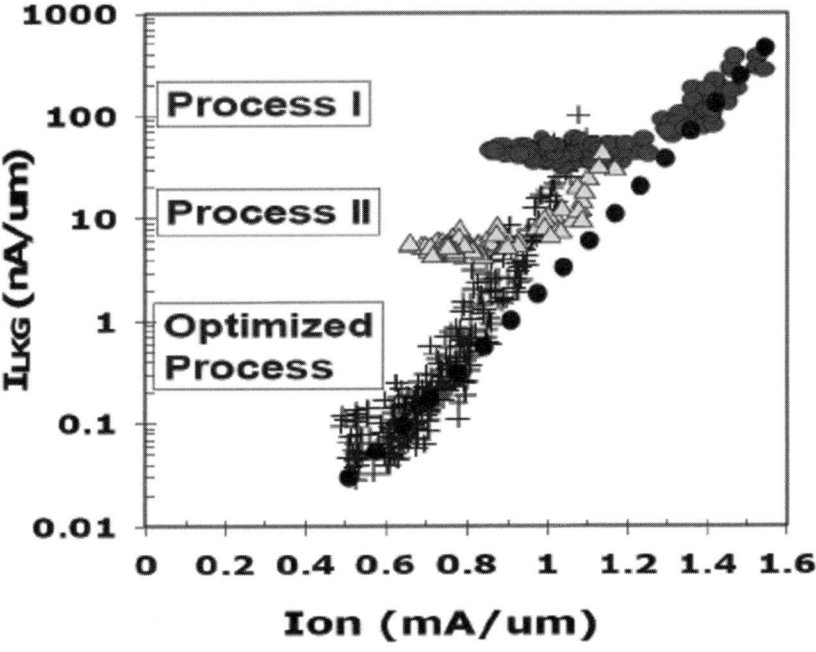

Figure 1.14 Leakage versus ion for three different technologies [14]. (© 2005 IEEE)

Although Process 1 has a higher "on" current and higher "off" current than the process optimized for mobile applications, both of the plots are asymptotic to the same line. In general, when the data points are near this line, the leakage within the transistor is dominated by subthreshold leakage. In this regime, the designer can trade "on" current for leakage current. Once the transistor is away from this line, the use of adaptive techniques to control leakage actually only results in lower performance with no savings in leakage.

1.4.2 Sources of Leakage Current

As seen in Figure 1.15, there are several sources of leakage current, and each of these has a different dependence on both voltage and temperature [17]. Understanding of the relation between leakage and both voltage and temperature requires consideration of each leakage mechanism separately.

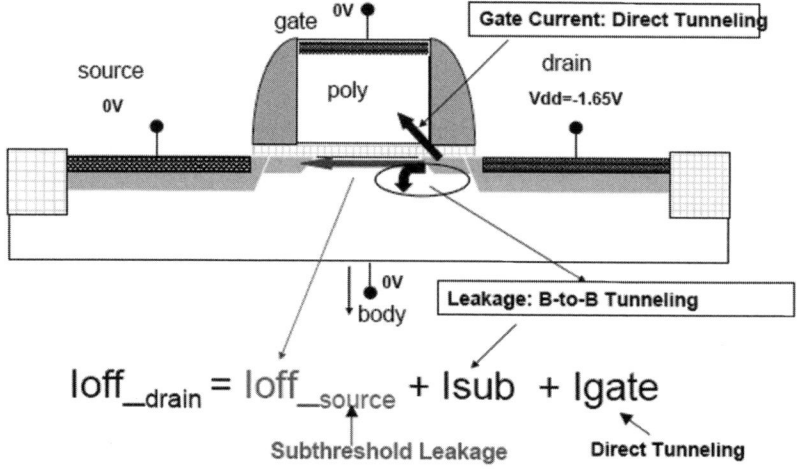

Figure 1.15 Sources of transistor leakage [17]. (© 2005 IEEE)

For different technologies, different leakage current sources dominate depending on how the requirements for the transistor design. In addition, within a given temperature, different components of leakage dominate as the voltage and temperature vary. In this section, we break down the different sources of leakage so that each can be discussed independently with the knowledge that in the end the components are summed together.

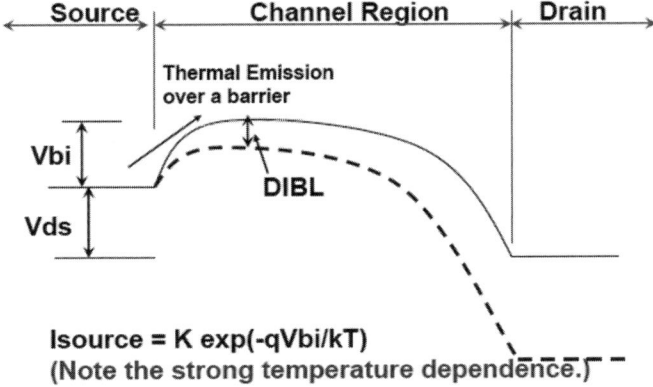

Figure 1.16 Source current or subthreshold leakage current [17].
(© 2005 IEEE)

As show in Figure 1.16, source current, referred to most often as sub-threshold leakage current, is due to thermal emission of the carriers over a barrier. The magnitude of the leakage current depends exponentially on the barrier height with temperature modulating this exponential dependence. Transistors with a high threshold voltage have a large barrier and hence have little subthreshold leakage. High-performance technologies tend to have lower threshold voltages. This in turn means a lower barrier height and hence high-performance technologies generally have a relatively high subthreshold leakage or source current. The height of the barrier in short channel devices can be modulated by the drain voltage, and this phenomenon is often referred to as drain-induced barrier lowering (DIBL) [18]. In a real transistor, the barrier height varies along the width of the transistor. Hence the current will always preferentially flow where the height of the barrier is lowest.

Carriers can also flow through the gate oxide by tunneling. This is a quantum mechanical effect, and the amount of current depends on the work function between the silicon and the insulator and the insulator thickness and the applied voltage. Since the amount of current depends on the oxide thickness but not the dielectric constant, technologist in modern technologies try to scale the dielectric constant rather than the oxide thickness in order to increase the transistor operating current.

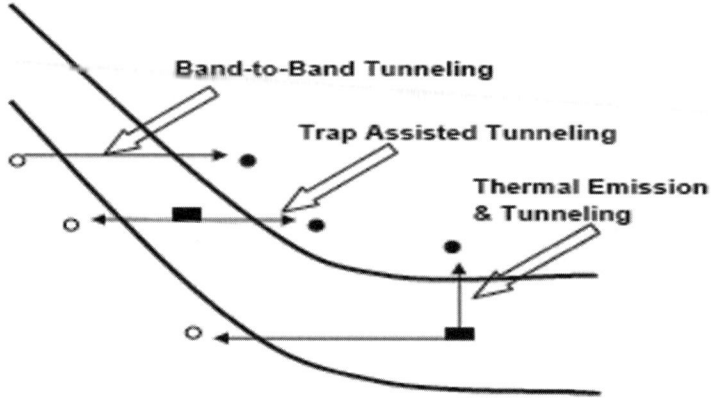

Figure 1.17 Sources of GEDL current are due to band-to-band tunneling that is often assisted by traps [19]. (© 2002 IEEE)

The source of leakage current that is the most challenging to control in scaled technologies is the gate edge diode leakage (GEDL) [19] as illustrated in Figure 1.17. This current is due to band-to-band tunneling in the presence of high electric field and traps in the band gap. If the electric field is high enough, carriers can simply tunnel across the band gap. However, most often traps in the silicon allow the tunneling to be trap assisted, and the current flow is increased significantly due to the smaller tunneling distance involved. An illustration of this mechanism is given in Figure 1.17. As seen in Figure 1.17, the presence of traps in the mid-gap can decrease the required tunneling distance by a factor of 2. Also as the number of traps increase, the tunneling current will increase. The tunnel current is given as [19]

$$J_{BTBT} \sim AL_{ov}E^2 \exp(-\beta / E) \tag{1.5}$$

where L_{ov} is the gate to drain overlap and E is the electric field.

Extensive simulations have been done to understand this mechanism [19]. GEDL is so heavily dependant on peak electric field that its characteristic behavior depends on the location of the peak electric field. Hence the profile of the doping species near the drain of the transistor design has a large role in determining this behavior. Figure 1.18 shows an NMOS transistor with the peak electric field at the semiconductor surface. The resulting current flow is often referred to gate-induced diode leakage (GIDL) current reflecting the proximity and sensitivity of the resulting electric field to the gate voltage.

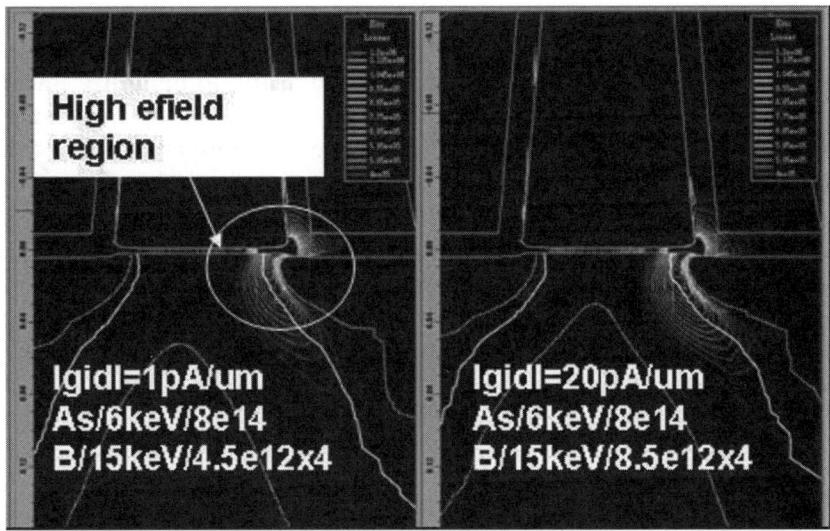

Figure 1.18 The peak electric field is at the surface and influenced by the gate voltage. (© 2005 IEEE) [19]

In Figure 1.19, the potential profile of a PMOS transistor is shown. In this transistor, the peak electric file is below the surface and hence the electric field is not sensitive to the gate voltage.

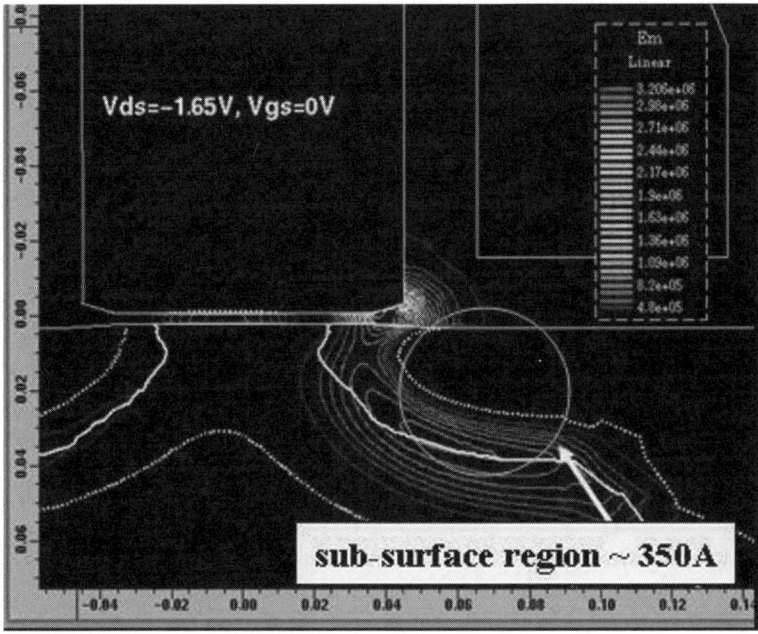

Figure 1.19 When the peak electric field is below the surface, the current is not sensitive to gate voltage [19]. (© 2005 IEEE)

1.4.3 Transistor Design for Low Leakage

As technologies scale, it is becoming more difficult to minimize transistor leakage. Reducing the subthreshold current is achieved by increasing the number of dopants in the channel in order to get more charge into the depletion region. However, the increase in dopants in the depletion region also results in a higher electric field which in turn increases the GIDL or GEDL current. In addition, if body bias is applied, the resulting electric field will be even higher. Often the total leakage will increase even though the subthreshold leakage is decreasing. An illustration of this effect is shown in Figure 1.20. These authors [20] show that for devices using a conventional dielectric the GIDL current increase with a body bias offsets the decrease in subthreshold leakage.

This same figure shows the promise of new transistor design techniques that can be applied as the result of employing an alternate dielectric such as HfSiON. Due to the Fermi level pinning associated with the resulting dielectric interface, the number of dopants required and the peak electric field are substantially reduced. It is, however, necessary to design the transistor profile to be able to stand off the short channel effects in modern semiconductor devices. As can be seen from observing Figure 1.20, the GIDL is substantially reduced, and thus, body bias is a viable technique for reducing total leakage in this technology.

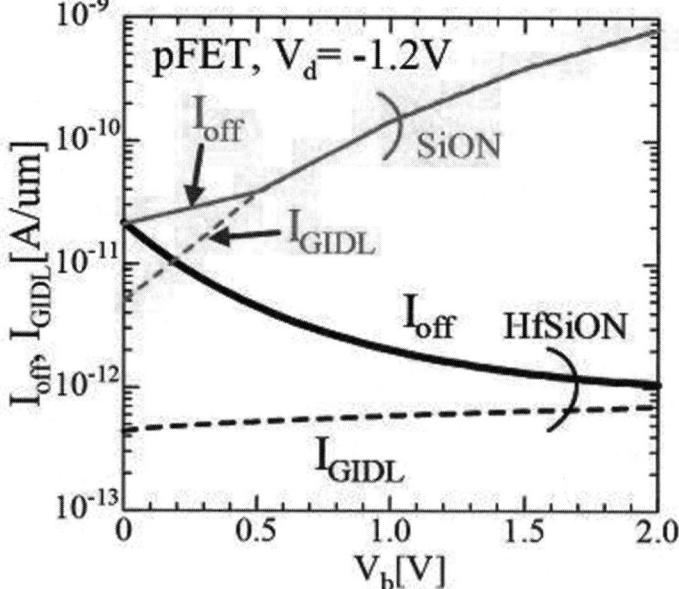

Figure 1.20 A comparison of the GIDL current against the total off current for two different gate oxide materials [20]. (© 2005 IEEE)

1.5 Conclusion

Variability and leakage are major technology challenges for both present and future integrated circuits, and the adoption of adaptive techniques are actually important tools to overcome the technology challenges of variability and leakage. Correct implementation of adaptive techniques in integrated circuits and systems is a significant challenge in itself. In our example, we have shown that while the active leakage reduction techniques reported to date have been successful, they also tend to significantly reduce the designers' error margin.

New transistor design techniques are bright spots on the horizon for designers. Recent innovations in transistor design give designers significant leverage in being able to compensate their circuits for threshold voltage variation, supply voltage variation, and temperature. These innovations serve as the key attack points for making adaptive techniques effective in the design of integrated circuits and systems.

References

[1]   L. Clark et al., "An Embedded 32b Microprocessor Core for Low-Power and High-Performance Applications," IEEE Journal of Solid-State Circuits, Vol. 36, pp. 1599–1608, November 2001.

[2]   M. Meijer, F. Pessolano, and J. P. de Gyvez, "Limits to Performance Spread Tuning Using Adaptive Voltage and Body Biasing," International Symposium on Circuits and Systems, pp. 5–8, May 2005.

[3]   J. T. Kao, M. Miyazaki, and A. P. Chandrakasan, "A 175-mV Multiply-Accumulate Unit Using an Adaptive Supply Voltage and Body Bias Architecture," IEEE Journal of Solid-State Circuits, Vol. 37, No. 11, pp. 1545–1554, November 2002.

[4]   T. McPherson, R. Averill, D. Balazich, K. Barkley, S. Carey, Y. Chan, Y. H. Chan, R. Crea, A. Dansky, R. Dwyer, A. Haen, D. Hoffman, A. Jatkowski, M. Mayo, D. Merrill, T. McNamara, G. Northrop, J. Rawlins, L. Sigal, T. Slegel, D. Webber, P. Williams, and F. Yee, "760MHz G6 S/390 Microprocessor Exploiting Multiple Vt and Copper Interconnects," IEEE International Solid-State Circuits Conference, Vol. XLIII, pp. 96–97, February 2000.

[5]   S. Rusu, S. Tam, H. Muljono, D. Ayers, J. Chang, "A Dual-Core Multi-Threaded Xeon® Processor with 16MB L3 Cache," IEEE International Solid-State Circuits Conference, pp. 102–103, February 2006.

[6]   D. Scott, "Technology Challenges of Adaptive Techniques", Microprocessor Forum, IEEE International Solid-State Circuits Conference, February 2007.

[7]   H. Su, F. Liu, A. Devgan, E. Acar, S. Nassif, "Full Chip Leakage Estimation Considering Power Supply and Temperature Variations," International

Symposium on Low Power Electronics and Design, pp. 78–83, August 25–27, 2003, Seoul, Korea.

[8] K. Shakeri and J. D. Meindl, "Compact Physical IR-Drop Models for Chip/Package Co-Design of Gigascale Integration (GSI)," IEEE Transactions on Electron Devices, Vol. 52, No. 6, pp. 1087–1096.

[9] T.-C. Ong, M. Levi, P.-K. Ko, C. Hu, "Recovery of Threshold Voltage After Hot-Carrier Stressing," IEEE Transactions on Electron Devices, Vol. 35, No. 7, pp. 978–984, July 1988.

[10] A. T. Krishnan, V. Reddy,S. Chakravarthi, J. Rodriguez, S. John, S. Krishnan, "NBTI Impact on Transistor and Circuit: Models, Mechanisms and Scaling Effects [MOSFETs]," IEEE IEDM Technical Digest, pp. 349–352, December 2003.

[11] H. Mizuno, K. Ishibashi, T. Shimura, T. Hattori, S. Narita, K. Shiozawa, S. Ikeda, and K. Uchiyama, "An 18- A Standby Current 1.8-V 200-MHz Microprocessor with Self-Substrate-Biased Data-Retention Mode," IEEE Journal of Solid-State Circuits, Vol. 34, No. 11, pp. 1492–1500, November 1999.

[12] Y. Yasuda, N. Kimizuka, Y. Akiyama, Y. Yamagata, Y. Goto, and K. Imai "System LSI Multi-Vth Transistors Design Methodology for Maximizing Efficiency of Body-Biasing Control to Reduce Vth Variation and Power Consumption," IEDM Technical Digest, pp. 66–71, December 2005.

[13] C. C. Hobbs et al., "Fermi Level Pinning at the Polysilicon/Metal Oxide Interface-Part 1," IEEE Transactions on Electron Devices, Vol. 51, No. 6, pp. 971–977, June 2004.

[14] C.-H. Jan et al., "A 65nm Ultra Low Power Logic Platform Technology Using Uni-Axial Strained Silicon Transistors," IEEE IEDM Technical Digest, pp. 60–63, December 2005.

[15] T. Chen and S. Naffziger, "Comparison of Adaptive Body Bias (ABB) and Adaptive Supply Voltage (ASV) for Improving Delay and Leakage Under the Presence of Process Variation," IEEE Transactions on VLSI Systems, Vol. 11, No. 5, pp. 888–899, October 2003.

[16] K. Ishibashi, "Substrate Bias Techniques for SH4," Short Course on Physical Design for Low Power, High Performance Microprocessor Circuits, 2001 Symposium on VLSI Circuits, 2001.

[17] D. Scott, S. Tang, S. Zhao, and M. Nandakumar, "Device Physics Impact on Low Leakage, High Speed DSP Design Techniques," Proceedings. International Symposium on Quality Electronic Design, pp. 349–354.

[18] R. R. Troutman, "VLSI Limitations from Drain-Induced Barrier Lowering," IEEE Transactions on Electron Devices, Vol. 26, No. 4, pp. 461–469, April 1979.

[19] S. Zhao, S. Tang, M. Nandakumar, D. B. Scott, S. Sridhar, A. Chatterjee, Y. Kim, S.-H. Yang, S.-C. Ai, and S. P. Ashburn, "GIDL Simulation and Optimization for 0.13um, 1.5 V Low Power CMOS Transistor Design," International Conference on Simulation of Semiconductor Processes and Devices, pp. 43–46, 2002.

[20] N. Kimizuka, Y. Yasuda, T. Iwamoto*, I. Yamamoto, K. Takano, Y. Aki-yama, and K. Imai, "Ultra-Low Standby Power (U-LSTP) 65-nm Node CMOS Technology Utilizing HfSiON Dielectric and Body-Biasing Scheme," Symposium on VLSI Technology, Digest of Tech. Papers, pp. 218–219, June 2005.

Chapter 2 Technological Boundaries of Voltage and Frequency Scaling for Power Performance Tuning

Maurice Meijer[1], José Pineda de Gyvez[1,2]

[1] NXP Semiconductors, [2] Eindhoven University of Technology

In this chapter, we concentrate on technological quantitative pointers for adaptive voltage scaling (AVS) and adaptive body biasing (ABB) in modern CMOS digital designs. In particular, we will present the power savings that can be expected, the power-delay trade-offs that can be made, and the implications of these techniques on present semiconductor technologies. Furthermore, we will show to which extent process-dependent performance compensation can be used. Our presentation is a result of extensive analyses based on test-circuits fabricated in the state-of-the-art CMOS processes. Experimental results have been obtained for both 90nm and 65nm CMOS technology nodes.

2.1 Adaptive Power Performance Tuning of ICs

The integration density of Integrated Circuits is doubling every 18 months. Soon, advanced process generations will integrate 1 billion transistors on a single chip. Such chips are the heart of a new generation of devices that are changing our daily life fundamentally. Power consumption of conventional electronic devices is a major concern because the dense devices produce a significant amount of heat imposing constraints on circuit performance and IC packaging. The case for portable devices is obvious, e.g. the goal is to maximize battery time. Designing ICs for low power will be a key practical and competitive advantage in the coming decade.

From a technological standpoint, power consumption can be reduced by downscaling transistor dimensions. CMOS transistor scaling consists of

A. Wang, S. Naffziger (eds.), *Adaptive Techniques for Dynamic Processor Optimization*, DOI: 10.1007/978-0-387-76472-6_2, © Springer Science+Business Media, LLC 2008

reducing all dimensions by a factor k (≈ 1.4), enabling higher integration density [1]. In the constant-field scaling scenario, the circuit speed increases, theoretically, with the amount of scaling k. Constant-field scaling has known benefits such as lower power per circuit, constant power density, and power-delay product that increases by k^3. However, for CMOS technology, over the last 10 years, it has been impossible to scale power supply voltage (V_{DD}) while maintaining speed because of the constraints on the threshold voltage (V_{th}) [2]. Due to increasing leakage current in scaled devices, V_{th} is not lowered to avoid significant static power consumption. Therefore, the electrical field is rising in proportion to k resulting now in almost constant circuit power despite scaling, increased power density by k^2, and power-delay product improvement by a factor of k only. In essence, the limits of a scaling process are caused by physical effects that do not scale properly, among them are quantum-mechanical tunneling, discrete carrier doping, and other voltage-related effects such as the subthreshold swing, and built-in voltage and minimum voltage swings.

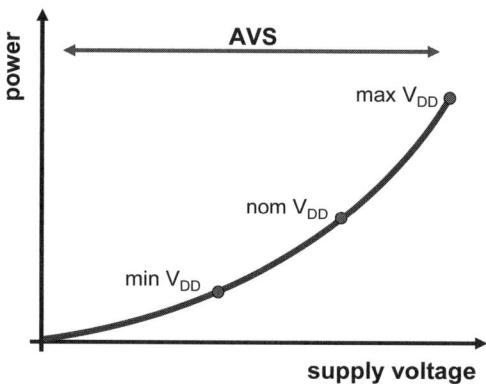

Figure 2.1 Power trends as a function of the supply voltage.

Besides technology scaling, one of the most effective ways to reduce active power consumption is by lowering V_{DD}. Ideally, quadratic power savings are observed as displayed in Figure 2.1. V_{DD} reduction can be applied to a complete chip, but it is most effective when it is applied to local voltage domains with own performance requirements. A common approach is to perform dynamic supply scaling, which exploits the temporal domain to optimize V_{DD} at run-time. This technique dynamically varies both operating frequency and supply voltage in response to workload demands. In this way, a processing unit always operates at the desired performance level while consuming the minimal amount of power. Two basic flavors exist, namely dynamic voltage scaling (DVS) and adaptive voltage scaling (AVS). DVS is an open-loop approach, and it is based on the selection of operating points from a predefined {f,V} table. Alternatively, AVS is a closed-loop approach, and its operating points are based only on the frequency. Software decides on the performance required for the existing workload and selects a target frequency. The voltage is then automatically adjusted to support this frequency. AVS is considered as the most effective technique for achieving power savings through V_{DD} scaling.

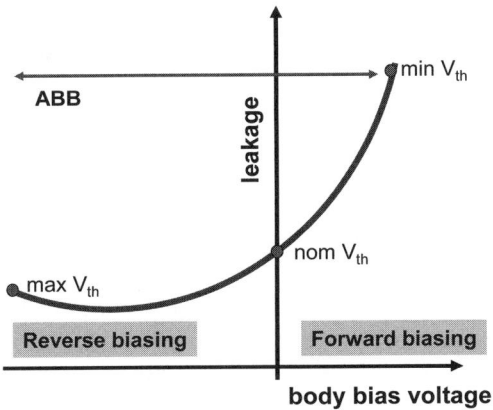

Figure 2.2 Leakage trends as a function of body biasing.

Yet another, but complementary, approach is to adapt to the threshold voltage of MOS devices using transistor body biasing. For NMOS, the V_{th} is increased when its body–source voltage is biased to be negative. This is referred to as reverse body biasing (RBB). Alternatively, the V_{th} is reduced when the body–source voltage is biased to be positive. This is referred to as forward body biasing (FBB). Figure 2.2 illustrates the behavior of leakage as a function of body biasing in modern nanometer technologies. Body biasing can effectively reduce the leakage power of the design, by improving its run-time performance. It is most effective when it is used in conjunction with V_{DD} scaling. Typically, body biasing is done in open-loop to calibrate circuit frequency or leakage for setting a desired mode of operation. Adaptive body biasing (ABB) refers to closed-loop control in which circuit parameters, e.g. speed, are monitored, compared, and controlled against desired values.

Not surprisingly, in recent years, the application of adaptive circuit techniques to control either or both V_{DD} and V_{th} has gained increased attention. This stems from the fact that modern electronics are hampered by the variation of fundamental process and performance parameters such as threshold voltage and power consumption. Design technologies such as

AMD's PowerNow! [3], Transmeta's LongRun [4], Intel's Enhanced SpeedStep [5], are vivid examples of commercial ICs that use power management based on V_{DD} scaling. In addition to these commercial accomplishments, chip demonstrators with V_{DD} and V_{th} scaling capabilities have also been reported in the literature archival [6–8]. Other reported uses of V_{DD} and V_{th} scaling, besides power management in processors, are in testing [9], product binning [10], and yield tuning [11].

2.2 AVS- and ABB-Scaling Operations

As the benefits of V_{DD} and V_{th} scaling are known, we concentrate on quantitative pointers for using such know-how in deep submicron technologies. For this purpose, we have evaluated various process technologies to determine technological boundaries for AVS and ABB when applied to digital logic circuits. Our evaluation is based on an extensive analysis of test-circuits fabricated in 90nm general-purpose (GP), 90nm low-power (LP), and 65nm low-power (LP) triple-well CMOS processes.

For all three CMOS processes, we have designed a clock generator unit (CGU) that consists of multiple independent ring-oscillators and corresponding selection circuitry. We use these CGU designs to determine power-performance trade-offs and leakage reduction factors with AVS and ABB. Each ring-oscillator uses minimum-sized standard-cell inverters as delay elements and a nand-2 gate for enabling control. The power supply of the clock generator can be controlled externally. Body biasing is enabled for N-well and P-well independently through triple-well isolation. The exact same clock generator was laid out in 90nm GP and LP-CMOS using a commercial place-and-route tool with constrained area-routing features. The 65nm LP-CMOS clock generator was designed full-custom using digital standard cells. Our second test-chip is a circular shift-register, which has only been laid out in 90nm LP-CMOS. The design contains 8K flip-flops and 50K logic gates. The logic gates are connected as delay lines between two consecutive flip-flop stages, which have an average logic depth of six cells. One can emulate the activity of any digital core with this circular shift register by shifting in a sequence of zeros and ones. Like the CGU, it has independent bias control over supply voltage, N-well and P-well biasing. The CGU provides the clock to the shift-register. The shift-register is used to perform correlated measurements against the CGU for validation purposes. All measurements have been performed using a Verigy 93K SoC test system in a controlled temperature environment. The temperature is controlled by a Temptronic Thermostream.

Devices in 90nm GP-CMOS operate at a nominal V_{DD} of 1V; their counterparts in LP-CMOS operate at 1.2V. GP-CMOS devices exhibit a lower V_{th} than LP-CMOS devices. On average, the nominal V_{th} is about 0.27V, 0.37V, and 0.43V for 90nm GP, 90nm LP, and 65nm LP-CMOS, respectively. Since ABB enables adaptation of these nominal V_{th} values, we will show the range over which V_{th} can be tuned for one of the considered process technologies. Figure 2.3 puts into perspective V_{th} versus body biasing for 65nm LP-CMOS devices as obtained from circuit simulations. Observe that the actual value of V_{th} and its sensitivity to body bias strongly depend on the process corner: fast, typical, or slow. For the typical NMOS device, body biasing from 0.4V (FBB) down to −1.2V (RBB) spans over a V_{th} range of about 135mV. This range is somewhat larger for PMOS devices (~180mV). Since RBB has a direct impact on leakage reduction, it will become evident that this technique is not very effective because the sensitivity of V_{th} to V_{BS} is small. In the next sections, we quantify the impact of these V_{th} ranges on circuit power-performance tuning.

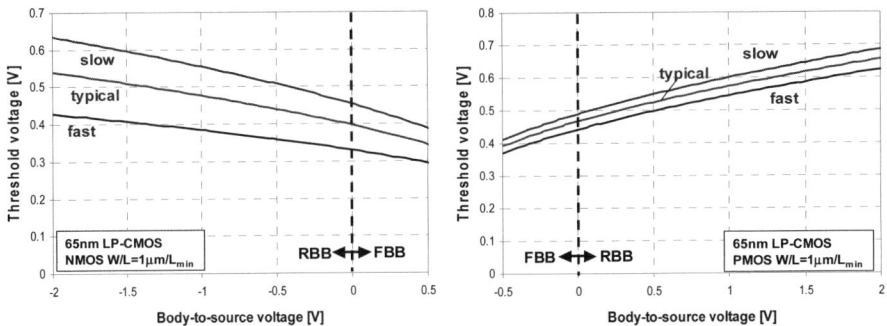

Figure 2.3 V_{th} adaptation through body biasing in 65nm LP-CMOS.

Let us now briefly introduce the conventions used for the AVS and ABB schemes. Figure 2.4 shows a graph of frequency versus power as a function of either or both AVS and ABB. The thick line shows the nominal trend when the supply voltage is varied from its maximum to its minimum value. The AVS operation consists of sweeping the supply voltage while maintaining a nominal constant body bias. The ABB is essentially the contrary approach: the supply voltage is kept constant and the body bias is swept. Here, it holds that frequency and power have an almost linear negative dependence on the threshold voltage. The result is a "cloud" of frequency–power points for a given supply voltage. Finally, AVS+ABB corresponds to the case when both supply voltage and body biasing are swept.

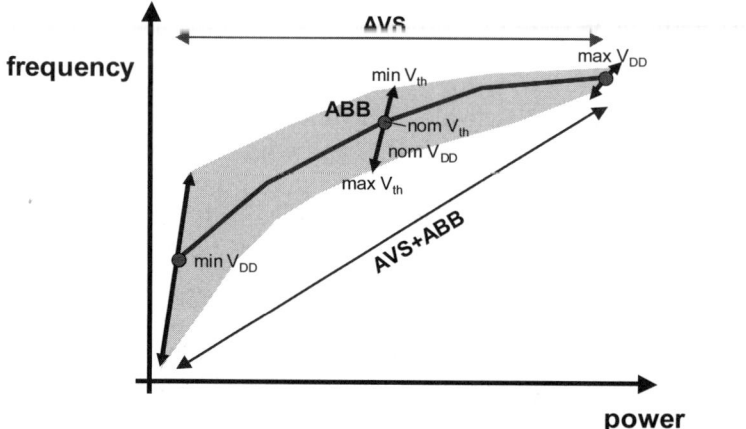

Figure 2.4 AVS and ABB operations.

Table 2.1 presents the voltage ranges that we employed during our measurements. Observe that the wells were forward biased for at most 0.4V and reverse biased by 1V (GP) or 1.2V (LP). Forward biasing is constrained by the turn-on voltage of the transistors' body–source junction diode. Essentially, reverse biasing is unconstrained, but high reverse biasing voltages result in increased gate-induced drain leakage.

Table 2.1 Voltage conventions for scaling operations.

		90nm GP	90nm/65nm LP
AVS	V_{DD}	[0.5,1.0]V	[0.6,1.2]V
ABB	V_{nwell}	[V_{DD}–0.4,V_{DD}+1.0]V	[V_{DD}–0.4,V_{DD}+1.2]V
	V_{pwell}	[–1.0,0.4]V	[–1.2,0.4]V
AVS+ABB	V_{DD}	[0.5,1.0]V	[0.6,1.2]V
	V_{nwell}	[V_{DD}–0.4,V_{DD}+1.0]V	[V_{DD}–0.4,V_{DD}+1.2]V
	V_{pwell}	[–1.0,0.4]V	[–1.2,0.4]V

In the next sections, we will illustrate how these techniques can be used to alter the power performance of integrated circuits. Please note that in the next sections, we will use the term ringo to refer to the ring oscillators in the CGU.

2.3 Frequency Scaling and Tuning

In most applications, there is not always a need for peak performance. In those cases, AVS can be used to lower the supply voltage and to slow down the core's computing power. In fact, operating frequency and supply voltage for a circuit design are coupled. This relationship can be expressed by Sakurai's alpha-power model [12]:

$$f \approx K \cdot \frac{\left(V_{DD} - V_{th}\right)^{\alpha}}{V_{DD}} \tag{2.1}$$

where f is the operating frequency, K is a proportionality factor, and α is a process-dependent parameter that models velocity saturation. In the case of velocity-saturated devices, α is close to 1 and the frequency scales almost linearly with V_{DD}.

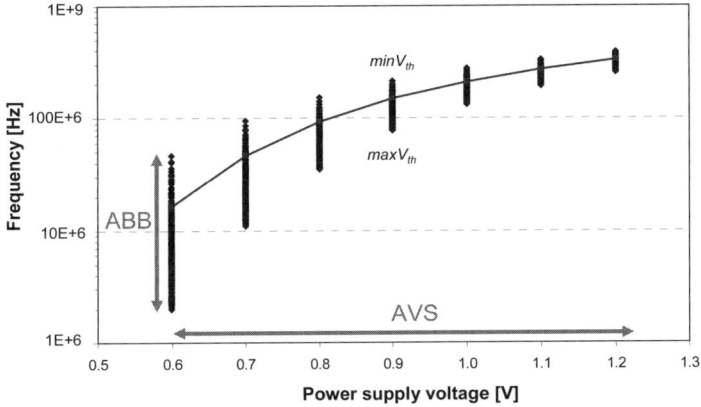

Figure 2.5 Frequency scaling and tuning for the 65nm LP-CMOS ringo.

Let us now investigate the frequency-scaling and tuning ranges offered by AVS and ABB in 65nm LP-CMOS. For this purpose, we determined the dynamic range of a 101-stage ringo that is part of the CGU test-chip. Figure 2.5 shows the ringo frequency as a function of power supply. Each cloud of dots is associated to a unique supply voltage. Each dot in a cloud corresponds to a unique N-well and P-well bias combination, and the line joining the clouds indicates the nominal trend. The ringo frequency at nominal supply (V_{DD}=1.2V) is 327MHz, and 16.2MHz at minimum supply (V_{DD}=0.6V). This results in an AVS tuning range of about 310MHz. Recall that the V_{th} is about 0.43V on average for this technology at nominal V_{DD}. When operating at reduced V_{DD}, the V_{th} increases due to of drain-induced barrier lowering (DIBL). At $V_{DD}=0.6V$, the V_{th} increases by about 100mV. The large frequency reduction with AVS is because the supply voltage becomes close to the V_{th}. For those low V_{DD}s, the transistors are no longer velocity saturated ($\alpha=2$). For the applied range, AVS renders an approximate 20× frequency reduction. If the lower bound of AVS would be set to 0.7V, the frequency reduces by about 7×.

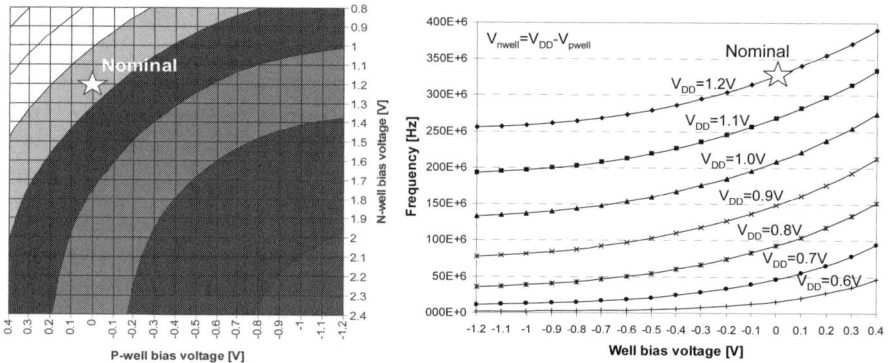

Figure 2.6 Frequency dependence on body-bias voltages; (a) Independent well biasing and $V_{DD}=1.2V$, (b) Symmetrical well biasing and various V_{DD} voltages.

We can now analyze the impact of ABB as a frequency-tuning mechanism at each V_{DD} point. Notice that the relative-tuning range is not the same for all V_{DD} values. In particular, we measured frequency spans of approximately –87% to +188% at $V_{DD}=0.6V$ and approximately ±20% at $V_{DD}=1.2V$ with respect to their nominal frequencies. The larger tuning range of ABB at reduced supply voltages can be explained by the fact that the threshold voltage is a larger portion of the gate drive of the transistors. At such low gate drive, the frequency becomes very sensitive to changes in V_{th}. Notice that a tuning range of –87% at $V_{DD}=0.6V$ implies an 8.1× lower frequency for RBB. In fact, at $V_{DD}=0.6V$, the circuit operates in the subthreshold region for strong reverse body-biasing conditions. In this case, the current is exponentially related to the gate drive voltage, and the frequency is much lower than in case of nominal body biasing. For the measured silicon, ABB gives an absolute tuning range of 135MHz for the chosen N-well and P-well voltages when operating at $V_{DD}=1.2V$. At $V_{DD}=0.6V$, this tuning range is around 45MHz. Figure 2.6a shows a contour plot of the ABB-scaling operation at $V_{DD}=1.2V$. The contours are at 20MHz intervals, and the nominal frequency is at 327MHz. Notice that it is possible to change the V_{th} of the PMOS and NMOS transistors independently and still attain the same frequency. Obviously, the choice of V_{th} has a significant impact on leakage power consumption as we will show later in this chapter. Figure 2.6b shows the frequency tuning for the ABB-scaling operation as function of a symmetrical well bias ($V_{nwell}=V_{DD}-V_{pwell}$) and various supply voltages. Notice that the frequency saturates for strong, reverse body biasing due to its limited V_{th} control range.

The same analysis has been performed for ringos in 90nm CMOS. A summary of the measured frequency-scaling and tuning ranges is given in Table 2.2. Notice the large frequency-scaling range for 65nm LP-CMOS as well as the large frequency-tuning range at reduced V_{DD}. For severe reverse body biasing, the threshold voltage saturates yielding as a result an asymptotic limit on the lowest possible operating frequency. Observe that GP-CMOS shows a lower dependence on V_{DD} and V_{th} as compared to LP-CMOS primarily because the threshold voltage of the former technology is lower.

Table 2.2 Frequency-scaling and tuning ranges for 90nm/65nm CMOS.

		90nm GP	90nm LP	65nm LP
AVS		3.4×	5.9×	20.1×
ABB	$V_{DD}/2$	[−29,24]%	[−81,76]%	[−87,188]%
	V_{DD}	[−8,6]%	[−27,15]%	[−22,19]%
AVS+ABB		5.1×	34.9×	194.1×

2.4 Power and Frequency Tuning

The ultimate use of the AVS and ABB schemes is for performance tuning with performance being the optimal combination of frequency and power, i.e. the lowest power for a given frequency. To investigate the available power–frequency-tuning range offered by AVS and ABB in 65nm LP-CMOS, we consider the same ring oscillator as before. Figure 2.7 presents a plot of the ringo frequency as function of the total power of the CGU, e.g. both CGU-static and dynamic power consumption of the ringo. In our experiments, static power takes into account all sources of leakage, e.g. subthreshold leakage, gate-oxide leakage, etc.

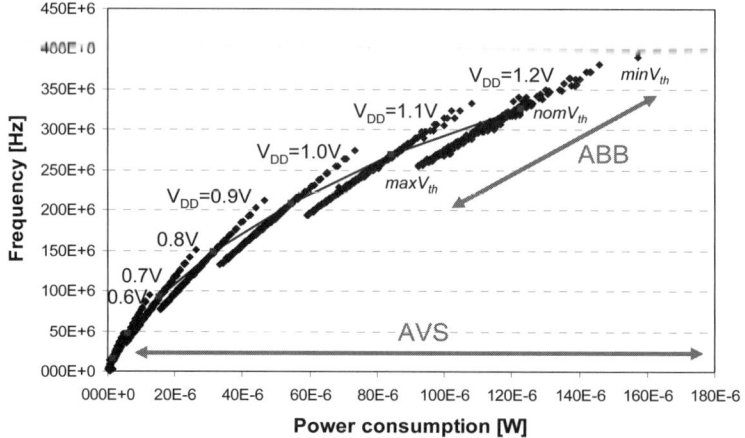

Figure 2.7 Frequency versus total power.

The plot of Figure 2.7 allows us to evaluate power savings and tuning-range control of AVS and ABB. Measurement results indicate 82× power savings by 20.1× frequency downscaling, using AVS when downscaling V_{DD} from 1.2V to 0.6V. The use of ABB at V_{DD} = 1.2V results in ±22% power and ±20% frequency tuning with respect to the nominal operating point. At V_{DD} = 0.6V, we observe a power-tuning range that spans from 78% to +217% and a frequency-tuning range from –87% to +188% with respect to no ABB. The combination of AVS and ABB yields ~790× power savings with ~194× frequency scaling from the highest possible frequency (minimum V_{th}) to the lowest one (maximum V_{th}). These results show the strength of the combined use of AVS and ABB.

Let us now explore possible power-performance tradeoffs by using AVS and ABB. Figure 2.8a shows a zoom-in of Figure 2.7 at V_{DD} =1.2V. If AVS and ABB are applied such that the nominal V_{DD} becomes 1.1V

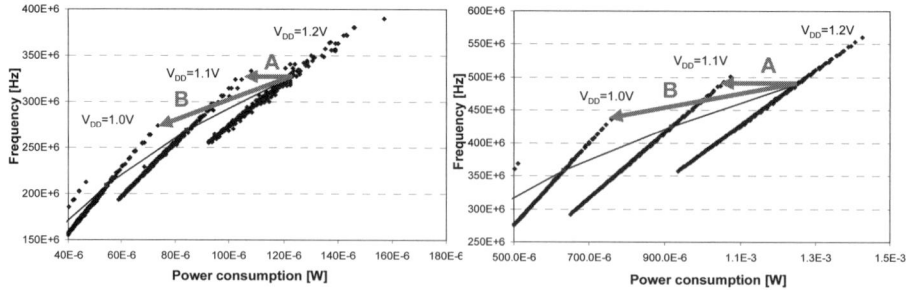

Figure 2.8 Frequency versus total power trade-off; (a) 65nm LP-CMOS, (b) 90nm LP-CMOS.

instead of 1.2V, and the V_{th}s are pulled to a smaller value as indicated by arrow A in Figure 2.8a, we see that it is possible to achieve ~14% power savings with no frequency penalty. A more aggressive V_{DD} downscaling to 1.0V, while pulling the V_{th}s to their minimum value, results in 40% power savings at about 16% frequency penalty as indicated by arrow B. Similar results have been found for 90nm LP-CMOS as shown in Figure 2.8b. In this case, the index factors are 16% power savings with no frequency penalty at V_{DD}=1.1V and 39% power savings with 11% frequency penalty at V_{DD}=1.0V. The benefits of combined AVS+ABB are not found to be technology-node dependent for the considered LP-CMOS process technologies. For 90nm GP-CMOS, however, a slightly larger voltage dependency of performance was observed. Downscaling from its nominal V_{DD} of 1.0V–0.9V, and lowering the V_{th}s a minimum, results in ~23% power savings with ~6% frequency penalty. At V_{DD}=0.8V and minimum V_{th}s, ~48% power savings are achieved with ~18% frequency penalty only. This indicates that there exists a lower frequency-tuning range with ABB for GP-CMOS.

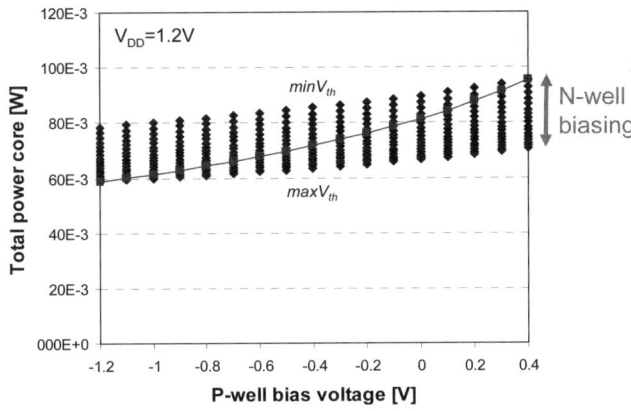

Figure 2.9 Power of 90nm LP-CMOS core as a function of well biasing.

Next we will investigate the properties of ABB in 90nm LP-CMOS on the shift register. Figure 2.9 shows the core's total power for a given circuit activity and V_{DD}=1.2V. Each dot in the clouds is associated to an N-well biasing condition. The line joining the clouds indicates the case when symmetric well biasing is applied. Observe that the well biasing allows a total power-tuning range of about 36mW; this represents about 40% of the nominal power consumption.

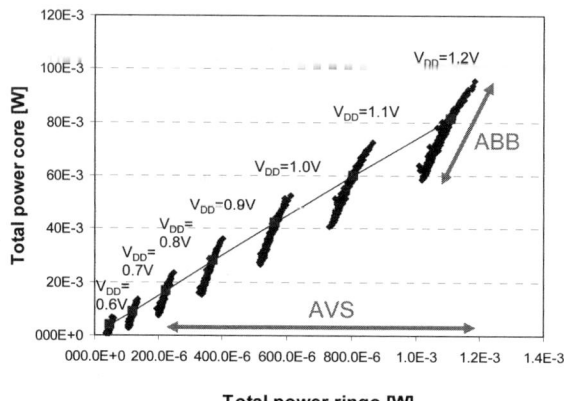

Figure 2.10 Total power correlation for the shift register and the ringo for different V_{DD} values.

Figure 2.10 shows the power consumption correlation between the shift register and the ringo for different V_{DD} values. In this plot, we have used the same conventions as before, i.e. each cloud is associated to a unique V_{DD} value and each point in the cloud corresponds to a unique N-well and P-well bias combination. The shift register operates at the same V_{DD} as the CGU, while its operating frequency is provided by the CGU. The circuit activity of the shift register is kept constant. The dynamic power dominates the total power in both circuit blocks, and therefore, their total power can be estimated by $P \approx aC \cdot V_{DD}^2 \cdot f$, where aC represents the switching circuit capacitance. Since both circuit blocks operate at the same supply voltage and frequency, their power consumption is linearly related by a ratio determined by the switching circuit capacitance. This can be observed in Figure 2.10, where the power consumption of the circuit blocks remains linearly correlated while applying AVS and/or ABB.

Table 2.3 puts into perspective the power–frequency ranges for the ringos in the considered process technologies. Notice that there exist large power–frequency ranges for each process technology. For the cases of AVS only, or AVS+ABB, the ratio of power and frequency shows a factor of 4× energy savings when scaling for the nominal V_{DD} to half of its value. This indicates that the total ringo power is dominated by dynamic power consumption. Furthermore, observe that LP-CMOS offers a larger power- and frequency-tuning range than GP-CMOS when utilizing ABB alone. The frequency-tuning range of GP-CMOS is about 3× lower.

Table 2.3 Power–frequency-tuning ranges for 90nm and 65nm CMOS.

			90nm GP	**90nm LP**	**65nm LP**
AVS		Power savings +	13.7×	23.6×	82.0×
		frequency penalty	3.4×	5.9×	20.1×
ABB	$V_{DD}/2$	Power tuning	[−29,29]%	[−77,65]%	[−78,217]%
		Frequency tuning	[−29,24]%	[−81,76]%	[−87,188]%
	V_{DD}	Power tuning	[−9,10]%	[−25,14]%	[−25,28]%
		Frequency tuning	[−8,6]%	[−27,15]%	[−22,19]%
AVS+ABB		Power savings +	21.2×	117.1×	790.5×
		frequency penalty	5.1×	34.9×	194.1×

2.5 Leakage Power Control

Leakage power is one of the main concerns in deep submicron technologies. In fact, AVS and ABB are often used for leakage reduction purposes. For older process technologies, leakage current is dominated by subthreshold conduction. Subthreshold leakage for a given device strongly depends on threshold voltage choice, process condition, supply voltage, and temperature. For sub-100nm CMOS, other leakage components have become increasingly important [13]. The most prominent ones are direct tunneling currents through the thin gate-oxide and gate-induced drain leakage (GIDL). Both leakage components are strongly V_{DD} dependent. Figure 2.11 puts into perspective leakage current as a function of power supply and temperature for a high-V_{th} NMOS device in 65nm LP-CMOS technology. These results are obtained through circuit simulations for a typical process condition. Observe in Figure 2.11a that subthreshold leakage, gate-oxide tunneling, and GIDL currents are of the same order of magnitude at nominal process–voltage–temperature conditions. Both Figure 2.11a,b show that the dominant leakage component in the total leakage depends on the operating condition.

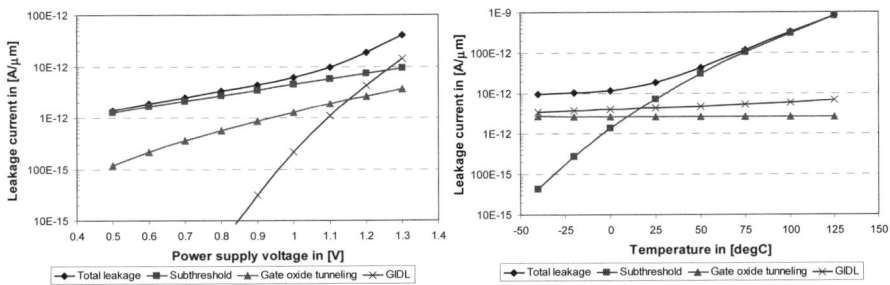

Figure 2.11 Leakage current trends for a 65nm LP-CMOS high-V_{th} NMOS device; (a) V_{DD} dependency at 25°C, (b) temperature dependency at V_{DD}=1.2V.

Figure 2.12 shows the impact of AVS and ABB on the leakage current for our CGU in 65nm LP-CMOS at 25°C. The plot shows measured leakage current versus body bias for three distinct values of power supply. Body biasing is applied symmetrically for N-well and P-well, respectively. The forward and reverse body-biasing ranges are indicated. Clearly, it is shown in Figure 2.12 that the leakage current grows exponentially when applying forward body biasing; this is because of the increased subthreshold leakage when lowering the V_{th}s. In reverse body-biasing operation, the leakage current achieves a minimum value around 500mV RBB. For stronger reverse body biasing, GIDL dominates the leakage current eliminating the ability of ABB to reduce leakage. Observe in Figure 2.12 that applying RBB of 300mV at V_{DD}=1.2V is as effective as lowering V_{DD} by that same amount. For larger RBB at V_{DD}=1.2V, AVS becomes more effective to reduce leakage. This is because GIDL and gate-oxide leakage are strongly reduced for lower V_{DD} operation.

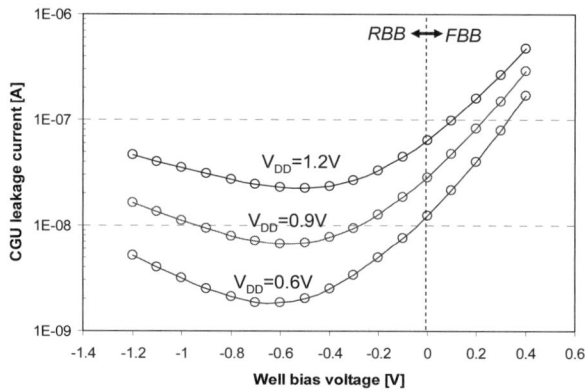

Figure 2.12 Leakage reduction in 65nm LP-CMOS using AVS and ABB.

For the measured die sample, leakage reduces by 5.1× when V_{DD} is scaled down from 1.2V to 0.6V. When using ABB alone at V_{DD} = 1.2V, leakage decreases only by 2.9×. This low impact of ABB is because of a high level of GIDL as explained before. When using ABB alone at V_{DD}=0.6V, leakage decreases by 6.8×. The combination of AVS with ABB renders a leakage reduction of 34.6×. Forward body biasing by 0.4V at V_{DD}=1.2V, 0.9V, or 0.6V increases the leakage current by 7.4×, 10.2×, or 13.7×, respectively.

The actual leakage savings utilizing AVS and ABB are impacted by temperature. At elevated temperatures, the V_{th}s become lower causing subthreshold leakage to become a bigger part of the total leakage current.

GIDL depends only weakly on temperature, and gate-oxide leakage is not temperature dependent. We have also measured temperature dependence of leakage current for various die samples to quantify its impact on the potential of AVS and ABB, to reduce leakage. Figure 2.13 shows experimental results for leakage reduction versus temperature for the same die sample as before. Observe that AVS becomes less effective to reduce leakage with increasing temperature, since the related leakage increase is supply voltage independent. However, the leakage increase is threshold voltage dependent, and therefore, ABB can reduce leakage slightly more effectively when temperature increases. At very high temperatures, i.e. the case of 100°C, the V_{th} is lowered so much that ABB cannot further reduce leakage because of the constrained ABB range we used in our experiments. The trend of AVS+ABB shows the collective effect of reducing leakage by AVS and ABB. In this case, leakage savings are about constant for temperatures up to 75°C.

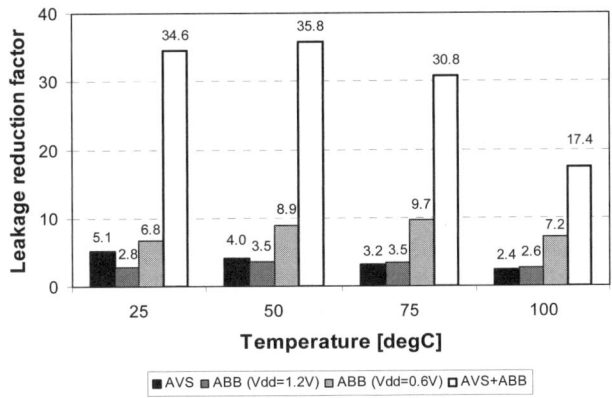

Figure 2.13 Temperature-dependent leakage reduction in 65nm LP-CMOS.

The actual leakage savings achieved by AVS and ABB are also impacted by process parameter variations. Subthreshold leakage strongly depends on process state, while gate-oxide leakage and GIDL are only weakly dependent. Leakage current of the CGU has been measured for 40 die samples from the same silicon wafer at 25°C. We have observed a leakage current ranging from 17.3nA to 322.6nA, depending on the die sample. This corresponds to leakage current variations of about 18.7×.

Table 2.4 shows the average leakage current savings for 65nm LP-CMOS obtained for the measured 40 die samples. The reduction factors for 90nm GP- and LP-CMOS technologies are also shown in this Table. The product of leakage savings with AVS ($V_{DD}/2$) and ABB yields substantial benefits as indicated in row AVS+ABB.

Table 2.4 Leakage current reduction for 90nm and 65nm CMOS at 25°C operation.

		90nm GP	90nm LP	65nm LP
AVS		5.3×	3.3×	5.6×
ABB	$V_{DD}/2$	4.1×	6.6×	4.5×
	V_{DD}	1.2×	3.5×	2.5×
AVS+ABB		21.6×	21.5×	24.8×

2.6 Performance Compensation

Understanding the trade-offs in performance and power is not sufficient to ensure a successful outcome of the IC. The basic problem is that failure of deep submicron process technologies to continue with constant process tolerances opens avenues for new challenging low-power process options and emerging design technologies. Basically, the assimilation of distinct high-performance, low operating power, and low standby power devices requires circuits and systems that concurrently exploit many degrees of freedom in both fabrication and design technologies.

Figure 2.14 shows the impact of process variability on performance spread of a single inverter for various technology nodes. A proportional inverter sizing was done across technology nodes for comparison

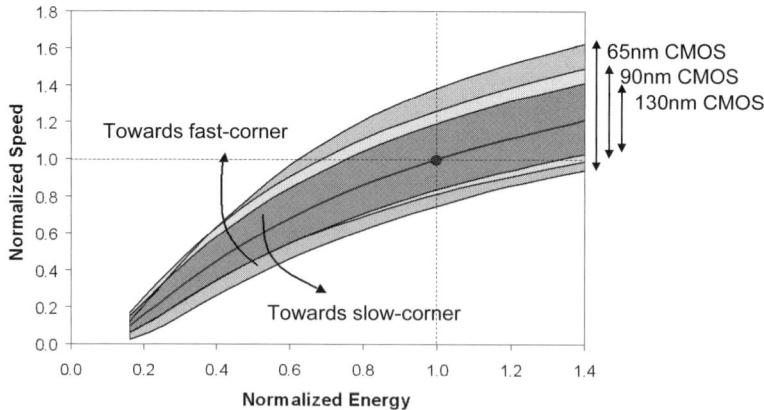

Figure 2.14 Energy spread across various technology nodes.

purposes. The inverter has further a fan-out of four gates. The vertical axis basically shows the spread of speed over three process corners, e.g. typical–slow–fast. The horizontal axis shows the normalized energy per operation. Notice that the performance window spread for 130nm, 90nm, and 65nm CMOS is about 40%, 50%, and 70%, with respect to the nominal operating conditions, respectively. What this graph also shows is that for a constant throughput, the wider the performance spread, the better the opportunities for energy savings are if voltage scaling is applied. For instance, in 65nm CMOS, the normalized speed of "1" can be achieved at an energy of "0.6" instead of at an energy of "1" if the power supply is scaled down. Today's design practices advocate a worst-case design style to ensure a target speed. This brings as implications overhead in area and power as shown in Figure 2.14. Basically, a worst-case design requires stronger cells, which are bigger in area and are also bigger power consumers, to meet timing closure of designs that fall beyond the 3σ due to process variability.

Figure 2.15 shows the impact of process variability on leakage power of the same inverter. One can see that leakage power spread at nominal supply voltage can span over 7×, 9×, and 11× for 130nm, 90nm, and 65nm CMOS, respectively. This spread can be detrimental in ultra low-power designs.

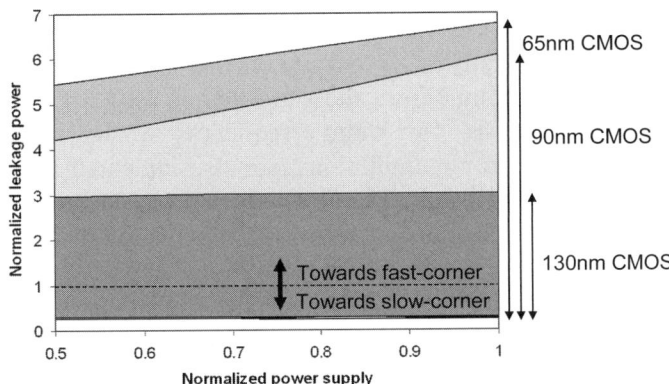

Figure 2.15 Leakage spread across various technology nodes.

As the variation of fundamental parameters such as channel length, threshold voltage, thin oxide thickness, and interconnect dimensions goes well beyond acceptable limits, "on-the-fly" performance compensation is becoming necessary. The influence of process parameter spread on circuit behavior becomes higher and higher. For instance, in older technologies greater than 0.18μm, a V_{th} spread of say 50mV on a nominal V_{th} of 450mV was not that crucial, in nanometer technologies with a nominal V_{th} of 250mV, this variation can make circuit operation quite difficult.

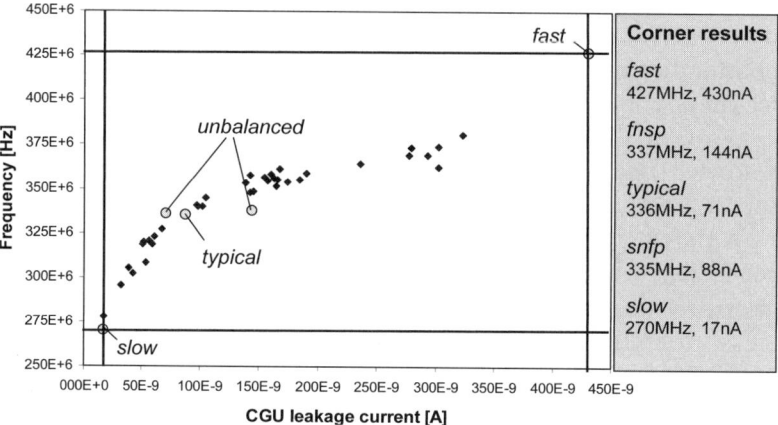

Figure 2.16 Frequency and leakage spread for 40 die samples of the same 65nm LP-CMOS wafer.

Figure 2.16 shows an example of frequency and leakage spread in which ringo frequency versus CGU leakage current is plotted at nominal V_{DD} for 40 die samples coming from the same 65nm LP-CMOS wafer. The five corner specifications for ringo frequency versus CGU leakage, as determined from circuit simulations, are also indicated in Figure 2.16. The total frequency and the leakage spread of the measured die samples are about 100MHz and 305nA, respectively. This translates into a relative frequency spread of ~36% and a relative leakage spread of ~18.7×. Note that we consider the samples with frequencies below "typical" as yield losses, while samples above "typical" are consuming unnecessary extra power. Moreover, the leakage current for a "fast" corner sample is about ~6.1× higher as compared to the "typical" reference, while the leakage current for a "slow" corner sample is about ~4.2× lower.

Next, we will discuss three strategies for compensating the undesired process-dependent frequency and leakage spread by means of post-silicon tuning. A first strategy is to perform post-silicon tuning with ABB only. From experiments, we have determined the tuning ranges for "fast" and "slow" samples. Figure 2.17 shows the potential of ABB to compensate performance for the same die samples as shown before. A 21% frequency increment from the slow corner renders a target frequency of 327MHz, and likewise, a 14% adjustment from the fast corner results in a target frequency of 366MHz. At the same time, the leakage current increases by ~9.8× (from 17nA to 170nA) for a "slow" corner sample, and reduces by ~2.5× (from 430nA to 177nA) for a "fast" corner sample. Observe that in both cases, that is, from slow to typical and from fast to typical, the leakage current of the tuned device is approximately 2.4× higher than the "typical" reference. For the available die sample set, we showed that the application of ABB gives basically a 100% parametric yield improvement. In addition, the leakage spread can be reduced to a factor of ~3.8× as indicated in Figure 2.17 by the dotted line at a typical frequency of 336MHz.

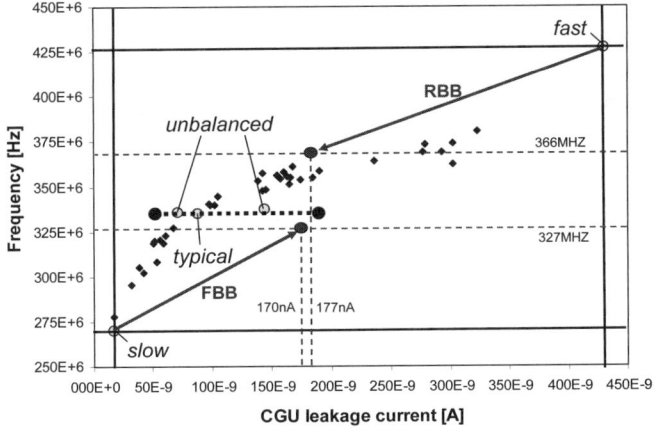

Figure 2.17 Process-dependent performance compensation with ABB.

A second strategy for compensating frequency and leakage spread is based on using ABB and AVS independently. ABB is used to increase the performance of "slow" samples as explained before. AVS is not used in this case because it would require a higher supply voltage than nominal, which may lead to reliability issues for the silicon. Therefore, AVS is only used to reduce the frequency and total power for "fast" samples. This approach is more power-efficient than when using ABB alone because now both dynamic and leakage power are reduced. For a "fast" corner sample, AVS can lower V_{DD} by about 124mV which reduces its switching energy by ~19.6% while still being able to meet the typical frequency specifications. Leakage current reduces less than when using ABB alone; the leakage reduces by ~1.1× (from 430nA to 386nA) for a "fast" corner sample. Consequently, the leakage current of the tuned device is about ~5.44× higher as compared to the "typical" reference.

A third and last strategy consists of setting AVS+ABB jointly. Again, ABB alone is used to increase the performance of "slow" samples. "Fast" samples are biased using AVS+ABB to meet typical frequency specifications while saving power. ABB is used to reduce V_{th} (FBB) such that AVS can reduce V_{DD} more than the case with no FBB, thereby, enabling further overall power savings. Combined AVS+ABB for a "fast" corner sample can lower V_{DD} by about 219mV, which reduces switching energy by about 33.3%. However, this comes at a penalty of increased leakage current. For a "fast" corner sample with 0.4V FBB, the leakage increases by about 3.7× (it becomes 1600nA) as compared to the "fast" corner with no FBB. When comparing against the "typical" reference, the leakage current is about 22.54× higher.

Figure 2.18 puts into perspective the previous results for compensating process-dependent frequency and leakage spread. The values for frequency, power supply voltage, and leakage current are plotted for reference and tuned process corners. The indicated numbers are normalized to the "typical" corner reference. Notice that ABB can effectively reduce frequency and leakage spread, while AVS can trade off higher operating frequency for improved power efficiency. Further total power savings can be achieved with AVS+ABB at the expense of increased leakage.

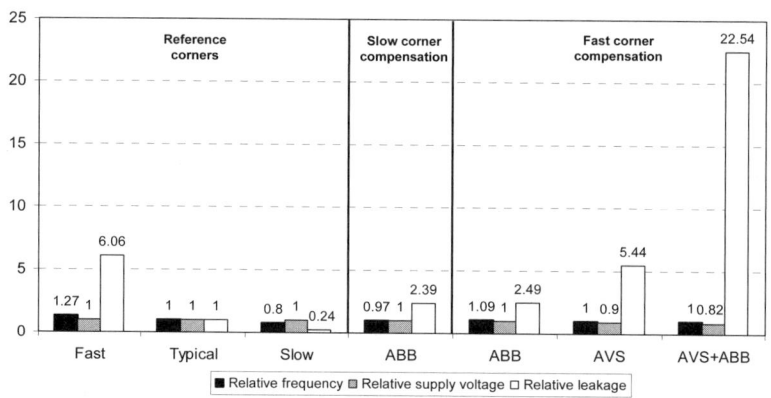

Figure 2.18 Performance compensation in 65nm LP-CMOS.

2.7 Conclusion

The race for low-power devices and the impediments of attaining low power through technology scaling only have opened avenues for design techniques based on voltage and frequency scaling. We presented measurement results that show the extent to which adaptive voltage scaling and adaptive body bias are useful for power and delay tuning in the state-of-the-art CMOS technologies. We observe the benefits of AVS primarily for low power and of ABB for performance tuning. For instance, for a 65nm LP-CMOS, the state-of-the-art technology power savings are in the order of 82× through 20× frequency downscaling. Contrary to the belief that high V_{th} has a considerable impact on leakage power reduction, we observed that reverse-bias ABB alone reduces leakage only by 2.5× at V_{DD}=1.2V. At lower supply voltage (V_{DD}=0.6V), we observed a larger leakage reduction of 6.8×. However, combined AVS and ABB yield ~25× leakage reduction.

With the increased impact of process variability on circuit design, ABB turns out to be a good design technology to keep parametric yield under control. In particular, we observe the means to tune devices with characteristics in the slow or fast process corners to performance specifications of a typical process corner. While at V_{DD}=1.2V, a ±20% frequency and a ±22% power-tuning range of ABB may look limited, the frequency-tuning range proves to be effective for process-dependent performance compensation. In fact, we observed a continuous frequency tuning despite the wide frequency spread. These tuning indices show that the combined use of AVS and ABB offers significant performance control. Of course, this tuning comes at the price of increased static power consumption. In our results, this static power increase is in the order of 2.4× to meet the required specs.

AVS and ABB design technologies have been reported in the technical literature archival as point solutions, usually through custom-based designs. However, the main impact on circuits-and-systems design will show off only when these techniques are methodologically applied. Along with AVS/ABB design techniques come challenges such as the design of supply and well grids, signal integrity at low voltages, voltage-domain crossing, etc. Fortunately, the electronic design automation (EDA) industry is picking up these concepts. Major EDA companies already offer tools for voltage-domain partitioning, multiple static voltage choices, power gating, and leakage control. Yet the dynamic voltage and frequency-scaling techniques have not been totally automated, partly because these techniques are also application dependent. The use of body biasing is slowly making its way into modern designs, yet automation is lacking behind. It is not unusual to see a wrong perception that ABB is used for leakage control only. We also showed in this chapter that in an era where poor V_{th} to V_{SB} sensitivity is evident, the best benefits of ABB design techniques are on parametric yield, i.e. on performance compensation.

References

[1]   W. Haensch, et al., "Silicon CMOS devices beyond Scaling", IBM Journal of Research and Development, July/September 2006, Vol. 50, No. 4/5, pp. 339–361

[2]   D.J. Frank, "Power constrained CMOS scaling limits", IBM Journal of Research and Development, March/May 2002, Vol. 46, No. 23, pp. 235–244

[3]   AMD PowerNOW! Technology, AMD white paper, November 2000, http://www.amd.com

[4]   M. Fleishman, "Longrun power management; Dynamic power management for crusoe processor", Transmeta white paper, January 2001, http://www.transmeta.com

[5]   S. Gochman, et al., "The Intel Pentium M processors: Microarchitecture and performance", Intel Technology Journal, May 2003, Vol. 7, No. 2, pp. 22–36

[6]   T. Kuroda, et al., "Variable supply-voltage scheme for low-power high-speed CMOS digital design", IEEE Journal of Solid-State Circuits, March 1998, Vol. 33, No. 3, pp. 454–462

[7]   K. Nowka, et al., "A 32-bit PowerPC system-on-a-chip with support for dynamic voltage scaling and dynamic frequency scaling", IEEE Journal of Solid-State Circuits, November 2002, Vol. 37, No. 11, pp. 1441–1447

[8]   V. Gutnik and A. Chandrakasan, "Embedded power supply for low-power DSP", IEEE Transactions on Very Large Scale Integration (VLSI) Systems, December 1997, Vol. 5, No. 4, pp.425–435

[9]   T. Miyake, et al., "Design methodology of high performance microprocessor using ultra-low threshold voltage CMOS", Proceedings of IEEE Custom Integrated Circuits Conference, 2001, pp. 275–278

[10] J. Tschanz, J. Kao, S. Narendra, R. Nair, D. Antoniadis, A. Chandrakasan, and Vivek De, "Adaptive body bias for reducing impacts of die-to-die and within-die parameter variations on microprocessor frequency and leakage", IEEE Solid-State Circuits Conference, February 2002, Vol. 1, pp. 422–478

[11] T. Chen and S. Naffziger, "Comparison of Adaptive Body Bias (ABB) and Adaptive Supply Voltage (ASV) for improving delay and leakage under the presence of process variation", IEEE Transactions on VLSI Systems, October 2003, Vol. 11, No. 5, pp. 888–899

[12] T. Sakurai and R. Newton, "Alpha-power law MOSFET model and its applications to CMOS inverter delay and other formulas", IEEE Journal of Solid-State Circuits, April 1990, Vol. 25, No. 2, pp. 584–593

[13] K.Roy, S. Mukhopadhyay, and H. Mahmoodi-Meimand, "Leakage current mechanisms and leakage reduction techniques in deep-submicrometer CMOS circuits ", Proceedings of the IEEE, February 2003, Vol. 91, No. 2 pp. 305–327

[14] M. Meijer, F. Pessolano, and J. Pineda de Gyvez, "Technology exploration for adaptive power and frequency scaling in 90nm CMOS", Proceedings of International Symposium on Low Power Electronic Design, August 2004, pp.14–19

[15] M. Meijer, F. Pessolano, and J. Pineda de Gyvez, "Limits to performance spread tuning using adaptive voltage and body biasing", Proceedings of International Symposium on Circuits and Systems, May 2005, pp.23–26

Chapter 3 Adaptive Circuit Technique for Managing Power Consumption

Tadahiro Kuroda,[1] Takayasu Sakurai[2]

[1]Keio University, [2]University of Tokyo

3.1 Introduction

Adaptive circuit techniques for minimizing power consumption are classified in terms of what is monitored, how it is monitored, what is controlled, how, and in what granularity it is controlled (Figure 3.1).

As for "what is monitored", there are two objects; one is regarding IC operation such as speed, voltage, leakage current, and temperature. The other object is a request to an LSI chip such as workload, quality of service, and error rate. A replica circuit of a critical path, such as a ring oscillator, is often used for monitoring the speed of an LSI chip. In monitoring temperature of a chip, on the other hand, a temperature sensor is placed by an actual circuit.

What is controlled? Clock frequency (f), power supply voltage (V_{DD}), and threshold voltage of a transistor (V_{TH}) are most common targets. The way to control is extending from an analog approach to a digital one and a software-assisted approach. In the digital approach, monitored information can be stored in a register. Since software can use upper system information, more sophisticated control is possible for further power reduction.

A. Wang, S. Naffziger (eds.), *Adaptive Techniques for Dynamic Processor Optimization*,
DOI: 10.1007/978-0-387-76472-6_3, © Springer Science+Business Media, LLC 2008

- **What to monitor**
- **How to monitor**
- **What to control**
- **How to control**
- **Granularity of control**

Figure 3.1 Adaptive control classification.

Granularity of the control is another aspect. The finer the granularity in terms of time and space, the further the power reduction, but at a cost of increase in layout area and other associated penalties. Since power consumption is becoming a serious problem, the granularity tends to be finer. The granularity has changed timewise from a millisecond order to a microsecond order and spatially from a chip level to a block level.

In this chapter, circuit techniques for the adaptive control are presented. They are reviewed from perspectives of what to monitor, how to monitor, what to control, how to control, and the granularity of the control. Adaptive V_{DD} and V_{TH} controls and cooperative control with software and operating system will be discussed in detail.

3.2 Adaptive V_{DD} Control

3.2.1 Dynamic Voltage Scaling

Dynamic voltage scaling (DVS) [1] is one of the most popular approaches in power reduction. V_{DD} is dynamically lowered to an extent where required performance of the target system is ensured. Significant power reduction is possible with DVS, since dynamic power of CMOS circuits is proportional to the square of V_{DD}.

Power consumption due to leakage current is also reduced effectively by DVS in scaled devices [2], as shown in Figure 3.2. Since the subthreshold leakage current is caused by a drain-induced barrier lowering (DIBL) effect, the lower V_{DD} results in the higher V_{TH}, and the smaller subthreshold leakage current. Gate leakage current is also reduced as well.

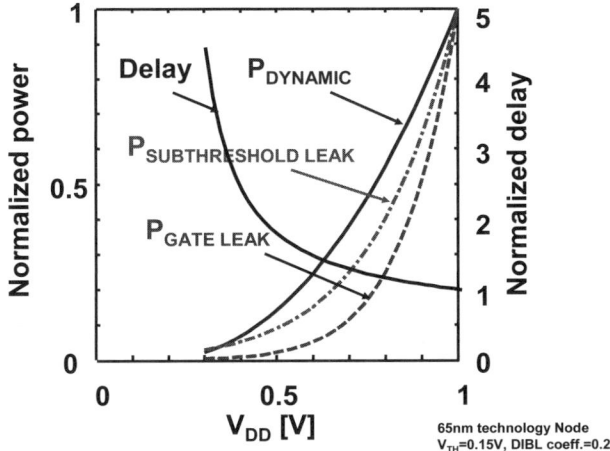

Figure 3.2 Power dissipation dependence on V_{DD}. Lowering V_{DD} is effective in reducing not only active power but also leakage power.

3.2.2 Frequency and Voltage Hopping

Cooperative control of both clock frequency (f) and supply voltage (V_{DD}) generates a multiplier effect in power reduction. Power consumption (P) dependence on clock frequency in a frequency–voltage cooperative power control (FVC) [3] differs from design to design. Figure 3.3 shows a typical P–f curve. The P–f curve is generally expressed as [4]

$$P = k'f \quad \text{when } f \leq f_m,$$

$$P = kf^{\gamma} \quad \text{when } f \geq f_m, \tag{3.1}$$

where f_m is clock frequency at the lowest power supply voltage, V_{min}, and k, k', and γ are constants determined by design parameters. γ is larger than 1 and typically smaller than 2.5. The P–f curve is composed of two parts: a linear region when $f < f_m$, and a γ-power region when $f > f_m$. In the linear region, P is directly proportional to f, since V_{DD} is constant. In the γ-power region, P is proportional to the γth power of f. We know through our experience that Equation (3.1) gives a good approximation in real designs.

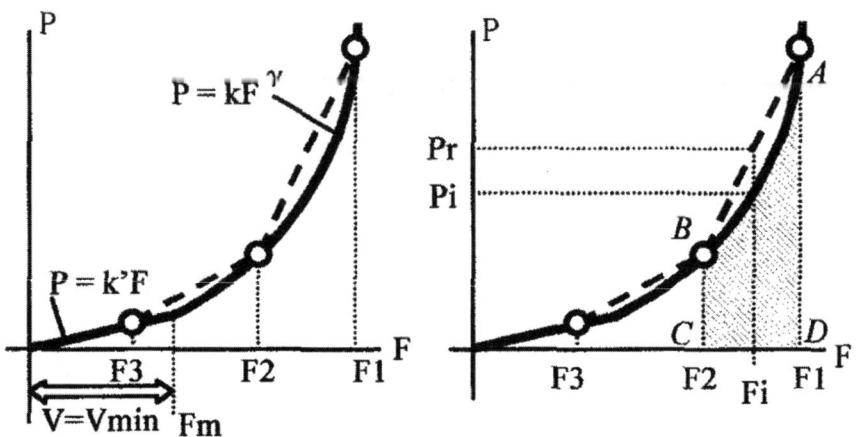

Figure 3.3 Power-frequency relation; (a) P–f curve in continuous DVS (solid line) and piecewise linear relation in frequency–voltage hopping (dashed line); (b) power waste by introducing frequency–voltage hopping.

In practical design, f and V take discrete values, since otherwise circuit design and testing become so complicated that large associated penalties need to be paid. Let us assume that f changes in a discrete fashion, such as f_1, f_2, f_3, and so on. Let us call this frequency change as a frequency–voltage hopping. The P–f curve is represented by piecewise linear function, as shown by the dashed line in Figure 3.3. Figure 3.3b depicts a waste of power dissipation, P_r–P_i, in the frequency–voltage hopping, compared to the case where the clock frequency changes in a continuous fashion.

Relative value of the waste, P_r/P_i, for the region of $f > f_m$ is given by

$$\frac{P_r}{P_i} = \frac{(\alpha-1)\beta^\gamma + K(\beta-\alpha)}{(\beta-1)\alpha^\gamma}, \qquad (3.2)$$

where $\alpha = \dfrac{f_i}{f_2}$, $\beta = \dfrac{f_1}{f_2}$, and $K = \left(\dfrac{f_m}{f_2}\right)^{\gamma-1}$.

By differentiating Equation (3.2) in terms of α and setting the result to zero, it is found that the waste becomes the largest at

$$\alpha_0 = \frac{(\gamma-1)(\beta^\gamma - K)}{\gamma\beta(\beta^{\gamma-1} - K)} \qquad (3.3)$$

The maximum of P_r/P_i is then given by substituting α_0 for α in Equation (3.2).

If f_i takes values uniformly from f_2 to f_1, average of the waste, which is given by $\dfrac{\sum\limits_{n} P_r\big(f_i(n)\big)}{\sum\limits_{n} P_i\big(f_i(n)\big)}$, can be approximately calculated as a ratio of area under the dashed line as defined by trapezoid *ABCD* in Figure 3.3b over area under the solid curve as depicted by hatched area. The average waste is calculated by

$$\frac{\sum\limits_{n} P_r\big(f_i(n)\big)}{\sum\limits_{n} P_i\big(f_i(n)\big)} \approx \frac{(\beta-1)(\gamma+1)\big(\eta^{\gamma-1}+\beta\big)}{(\gamma+1)\eta^{\gamma-1}\big(\beta^2\eta^2-1\big)-2\beta^2\big(\eta^{\gamma+1}-1\big)}, \tag{3.4}$$

where $\eta = f_1/f_m$.

From Equations (3.2)–(3.4), we can calculate the waste of power in introducing the frequency–voltage hopping compared to the case where we employ the continuous DVC. Table 3.1 shows the calculation results. Suppose a case where $f_m = f_2$, in other words, V_{DD} changes from its maximum to minimum values accordingly as f changes from f_1 to f_2. If f_2 is chosen larger than half of f_1, the average waste of power is smaller than 13%. Remember that γ is typically smaller than 2.5. Let us next suppose a case where $f_m = (f_1 + f_2)/2$; in other words, V_{DD} changes from its maximum to minimum values, and V_{DD} stays at V_{min} after f is lowered beyond f_m. The average waste of power is bigger than the previous case, but still it is smaller than 20%.

From these discussions, it is concluded that in the frequency–voltage cooperative power control, hopping in two levels of the clock frequency (f_1 and f_2) with the corresponding changes in V_{DD} yields almost as good effect (with over 80% efficiency) in power reduction as the continuous control. You can remember it, as a rule of thumb, that f_2 should be chosen as half of f_1.

The frequency and voltage hopping scheme is employed for MPEG-4 decoding in the Hitachi SH-4 CPU [4]. Table 3.2 summarizes the measured performance. From the measurement of the P–f characteristics, γ is 1.6. Since f_1 is 200MHz, f_2 is chosen to be 100MHz by applying the rule of thumb. Since V_{DD} reaches V_{min} (=1.2V) before f reaches f_2, no more f_i is needed. Therefore, there are three operational modes: a high-speed mode at 200MHz, a low-speed mode at 100MHz, and a sleep mode. The average of the power dissipation is reduced to 22.6% by introducing the low-power mode and sleep mode.

Table 3.1 Waste of power in frequency and voltage hopping, compared to the continuous DVC; (a) when $f_m = f_2$ (i.e., V_{DD} changes from its maximum to minimum values accordingly as f changes from f_1 to f_2); (b) when $f_m = (f_2 + f_1)/2$ (i.e., V_{DD} changes from its maximum to minimum values, and V_{DD} stays at V_{min} after f is lowered beyond f_m). Upper and lower numbers in each column of the table denote the average waste and the maximum waste, respectively.

(a) $f_m = f_2$

γ f1/f2	1.5	2.0	2.5	3.0
1.5	1.01	1.03	1.05	1.08
	1.02	1.04	1.08	1.13
2.0	1.03	1.07	1.13	1.20
	1.05	1.13	1.24	1.41
3.0	1.06	1.15	1.27	1.40
	1.12	1.33	1.69	2.26

(b) $f_m = (f_1 + f_2)/2$

γ f1/f2	1.5	2.0	2.5	3.0
1.5	1.03	1.06	1.09	1.13
	1.06	1.12	1.19	1.26
2.0	1.05	1.11	1.17	1.24
	1.10	1.22	1.36	1.52
3.0	1.09	1.18	1.28	1.39
	1.17	1.38	1.63	1.94

Table 3.2 Experimental results of frequency and voltage hopping for MPEG-4 decoding in the Hitachi SH-4 CPU. Average power dissipation was reduced to 22.6%.

Operation mode	High speed	Low speed	Sleep
Voltage (V)	2.0	1.2	1.2
Frequency (MHz)	200	100	0
Power (mW)	600	200	20
Execution time (%)	3.3	53.5	43.2
Average power	135.6 (22.6% of the power in HS mode)		

3.3 Adaptive V_{TH} Control

Delay variation (ΔT_{pd}) due to V_{TH} variation (ΔV_{TH}) is substantially increased at low V_{DD}'s. The increased variation of the gate propagation delay degrades the chip performance. In order to keep the delay variation percentage constant in low V_{DD}'s, ΔV_{TH} should be reduced approximately by [5]

$$\frac{\Delta V'_{TH}}{\Delta V_{TH}} = \left(\frac{T'_{pd}}{T_{pd}} \cdot \frac{V'_{DD}}{V_{DD}} \right)^{\frac{1}{\alpha}} ,$$

(3.5)

where α represents the velocity saturation effect and typically is 1.3 [6], and T_{pd} is CMOS gate propagation delay. For example, when V_{DD} is lowered from 1.5V to 1.0V and V_{TH} is lowered to maintain circuit speed (i.e., $T_{pd}=T_{pd}'$), ΔV_{TH} should be reduced by 27%. It is very difficult, however, to lower ΔV_{TH} by this much by means of process and device refinement. In this section, circuit techniques for adapting V_{TH} control are discussed.

3.3.1 Reverse Body Bias (VTCMOS)

A variable threshold voltage CMOS technology (VTCMOS) [5, 7–11] controls V_{TH} by means of substrate bias control. In this technique, devices are fabricated for lower V_{TH} than a design target, and V_{TH} is set to the target by adjusting reverse body bias (RBB), V_{BB}. Since subthreshold leakage current depends very strongly on V_{TH}, V_{TH} can be compensated for variations by feedback control of V_{BB} such that monitored leakage current is set to a target value.

3.3.1.1 Self-Adjusting Threshold Voltage (SAT) Scheme

A self-adjusting threshold voltage (SAT) scheme, depicted in Figure 3.4, compensates for the V_{TH} variation [6, 7]. The subthreshold leakage current is monitored by a leakage current monitor (LCM). The substrate bias is generated by a self-substrate bias circuit (SSB). LCM activates SSB when a monitored leakage current in LCM, $I_{leak.LCM}$, is larger than a target preset value, I_{ref}. SSB lowers V_{BB} by pumping out current from the substrate [12]. Accordingly, V_{TH} is raised and $I_{leak.LCM}$ is reduced.

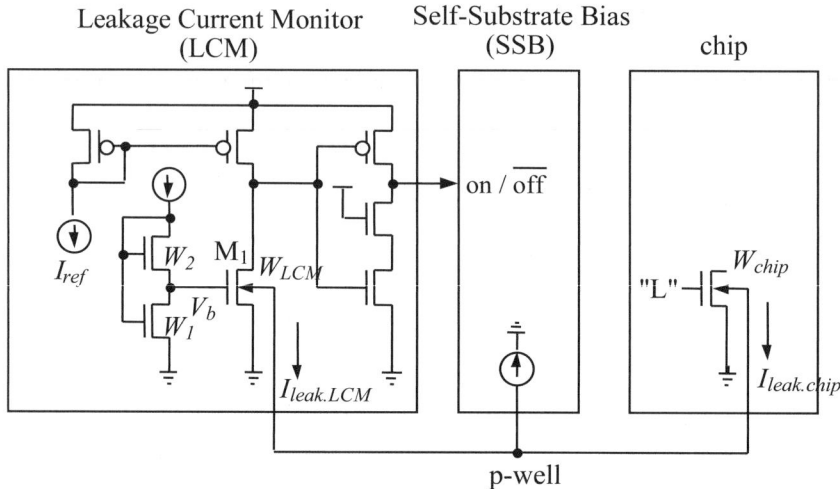

Figure 3.4 Self-adjusting threshold voltage (SAT) scheme.

When $I_{leak.LCM}$ becomes smaller than I_{ref}, LCM stops SSB. However, the substrate current due to the impact ionization and the junction leakage raises V_{BB} gradually again. Accordingly, V_{TH} is lowered gradually and $I_{leak.LCM}$ increases. When $I_{leak.LCM}$ becomes larger than I_{ref}, LCM activates SSB again. By activating SSB intermittently in this way, V_{TH} can be set to the target value, and consequently, its process-induced variation can be compensated to be smaller.

3.3.1.2 Leakage Current Monitor

In Figure 3.4, the ratio of $I_{leak.LCM}$ to the total leakage current in a chip, $I_{leak.chip}$, is given by

$$X_{LCM} \equiv \frac{I_{leak.LCM}}{I_{leak.chip}} = \frac{W_{LCM}10^{(V_b-V_{TH})/S}}{W_{chip}\ 10^{-V_{TH}/S}} = \frac{W_{LCM}}{W_{chip}} \cdot 10^{\frac{V_v}{S}}, \qquad (3.6)$$

where W_{chip} is effective total channel width corresponding to the total leakage current in the chip, W_{LCM} is channel width of a monitor transistor in LCM, S is the subthreshold slope, and V_b is its gate potential. Since $I_{leak.LCM}$ leads to a power penalty of LCM, it should be as small as possible. Too small $I_{leak.LCM}$, however, slows LCM response speed, which enlarges fluctuation of V_{BB} caused by the on–off control of SSB, resulting in larger dynamic error of V_{TH}. When $I_{leak.LCM}$ is 1μA for the chip leakage current of 1mA, the leakage current detection ratio, X_{LCM}, is 0.1%. Given $V_b = 2S$, which is approximately 0.2V, the size of the monitor transistor can be designed as small as approximately 0.001% of the effective total transistors in the chip.

A bias circuit for V_b is depicted in Figure 3.4. A current source is designed such that the two transistors are operated in the subthreshold region. As the drain currents of the two transistors are equal,

$$W_2 \cdot 10^{(V_1 - V_b - V_{TH})/S} = W_1 \cdot 10^{(V_1 - V_{TH})/S},$$

$$\therefore V_b = s \cdot \log \frac{W_2}{W_1}. \tag{3.7}$$

Substituting Equation (3.7) into Equation (3.6),

$$X_{LCM} = \frac{W_{LCM}}{W_{chip}} \cdot \frac{W_2}{W_1}. \tag{3.8}$$

X_{LCM} can be determined only by transistor size ratio and independent of V_{DD}, temperature, and process variation. If V_b is generated by dividing voltages between V_{DD} and V_{SS} by resistors ($V_b = \lambda V_{DD}$), and consequently, X_{LCM} is a function of V_{DD} and S. Since S is a function of temperature, X_{LCM} depends on V_{DD} and temperature, which is not desirable. Variation in X_{LCM}, analyzed by SPICE simulation, is within 15%, which results in less than 1% error in V_{TH} controllability.

3.3.1.3 V_{TH} Controllability

An MPEG-4 video codec chip [13] is fabricated in two runs. The target of V_{TH} in one run is 0.05V and that for the other is 0.15V by changing conditions of ion implantation. About 40 chips are measured for each V_{TH} condition in the following three ways: (1) V_{TH} as processed without body biasing, (2) V_{TH} controlled by VTCMOS in the active mode, and (3) V_{TH} controlled by VTCMOS in the standby mode. In (2), the MPEG-4 chip is operated with test vector inputs so that the measurements include dynamic errors, such as those due to substrate noise influence. The measured results at 27°C and 70°C are plotted in Figure 3.5a–d. Statistics of the distribution such as the average (x) and the standard deviation (σ) are presented in Tables 3.3a and b. The VTCMOS technology reduces V_{TH} variation from ±0.1V to ±0.05V in both the active and the standby modes and raises V_{TH} by 0.25V in the standby mode.

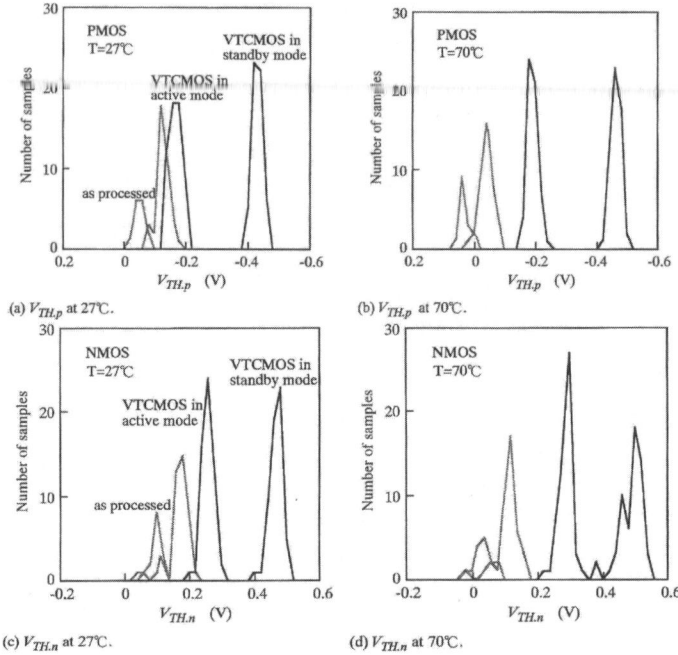

(a) $V_{TH.p}$ at 27°C.   (b) $V_{TH.p}$ at 70°C.

(c) $V_{TH.n}$ at 27°C.   (d) $V_{TH.n}$ at 70°C.

$\bar{x}$: average, σ: standard deviation.

Figure 3.5 Measured V_{TH}: (a) $V_{TH.p}$ at 27°C, (b) $V_{TH.p}$ at 70°C, (c) $V_{TH.n}$ at 27°C, and (d) $V_{TH.n}$ at 70°C.

Table 3.3a Measured V_{TH} as processed.

	$V_{TH.p}$ (V)				$V_{TH.n}$ (V)			
	27°C		70°C		27°C		70°C	
Target V_{TH}	$\bar{x}$	σ	$\bar{x}$	σ	$\bar{x}$	σ	$\bar{x}$	σ
0.05	−0.06	0.014	0.03	0.016	0.09	0.022	0.03	0.028
0.15	−0.13	0.022	−0.05	0.021	0.16	0.029	0.11	0.031

$\bar{x}$: average, σ: standard deviation.

Table 3.3b Measured V_{TH} controlled by VTCMOS technology.

	$V_{TH.p}$ (V)				$V_{TH.n}$ (V)			
	27°C		70°C		27°C		70°C	
VTCMOS	$\bar{x}$	σ	$\bar{x}$	σ	$\bar{x}$	σ	$\bar{x}$	σ
Active mode	−0.17	0.018	−0.20	0.016	0.25	0.019	0.28	0.019
Standby mode	−0.44	0.015	−0.47	0.016	0.46	0.019	0.48	0.036

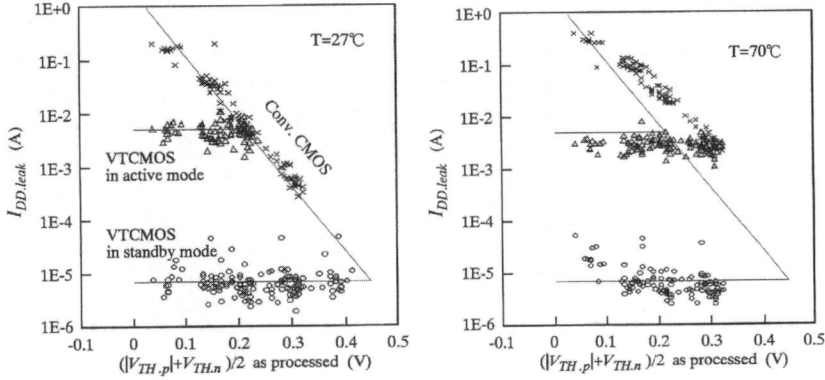

Figure 3.6 Measured chip leakage current.

Measured temperature dependence of V_{TH} is 0.7mV/°C for an NMOS and −0.7mV/°C for a PMOS under the VTCMOS control, whereas the values in the conventional CMOS device are −1.3mV/°C and 2.0mV/°C, respectively. When V_{DD} is around 0.5V, the drain current shows positive temperature dependence, since the increase in the drain current by V_{TH} decrease surmounts the mobility degradation [14]. This may cause thermal runaway if the subthreshold leakage becomes the dominant component in power dissipation at low V_{TH}. In a scaled device with low V_{DD} and low V_{TH}, temperature dependence control becomes indispensable. The temperature dependence of V_{TH} in VTCMOS can be controlled by controlling the temperature dependence of I_{ref} in LCM.

Chip leakage current is measured at 27°C and 70°C, and the results are plotted in Figure 3.6. The horizontal axes is the average of $|V_{TH.p}|+V_{TH.n}$. The VTCMOS technology sets the leakage current below 10mA in the active mode and below 10µA in the standby mode, independently from processed V_{TH} and temperature.

3.3.1.4 Device Perspective

In applying RBB, the drain-substrate depletion layer extends, which worsens the short-channel effect (SCE) and the V_{TH} variations across a die. Furthermore, the body effect coefficient, γ, is reduced more in a shorter channel transistor, since channel potential is more influenced by drain than by substrate due to the DIBL effect. Coupled with SCE, the V_{TH} variation across a die is increased by the substrate bias. Measurement in 0.18µm single-V_{TH} and 0.13µm dual-V_{TH} logic technologies for high-performance microprocessors shows that [15] (1) RBB becomes less effective for leakage reduction at shorter channel lengths and lowers V_{TH} at both high and room temperatures when leakage currents are large and (2) RBB effectiveness also diminishes with technology scaling primarily because of worsening SCE, especially when the target V_{TH} value is low.

The simplified scaling theory predicts that it will eventually be difficult to cause a large-enough change in V_{TH} through RBB. In practice, however, RBB is still effective in the 65nm technology generation by careful channel engineering and V_{DD} control [16].

3.3.2 Forward Body Bias

From the observations on device scaling in the previous section, the range of substrate biasing is extended from RBB to forward body bias (FBB) [17]–[19]. FBB is applied to a transistor with high V_{TH} to bring V_{TH} down to the target value.

Since FBB improves the device short-channel effects, it reduces sensitivity of V_{TH} to variation in gate length, oxide thickness, and channel doping. As a result, it is reported in [19] that die-to-die V_{TH} variation is 36% smaller in a PMOS and 48% smaller in an NMOS when FBB is used, even with ±20% variation in the body bias value.

Even though FBB lowers V_{TH} and improves circuit performance, FBB increases leakage current due to parasitic bipolar current and forward source–body junction current. This determines an optimum FBB value. The optimum FBB value, between 400 and 500mV at 110°C, provides maximum frequency improvement (13%). The total switched capacitance and switching energy are 10% higher because of larger junction capacitance, larger average gate capacitance at lower V_{TH}, and increased short-circuit current. Although active leakage power, including subthreshold leakage, parasitic bipolar current, and forward source–body junction current, increases by 10–100×, it remains sufficiently small compared to switching power. For bias values larger than this optimum, junction capacitance, body effect, and source–body junction forward current increase rapidly and fully negate any delay improvements induced by further V_{TH} reduction. Active leakage power also becomes an unacceptably large fraction of the total power. For designs operating at a maximum junction temperature of 110°C, the desired FBB value is 450mV with ±50mV tolerance.

3.3.3 Control Method and Granularity

As one of the early examples where the VTCMOS technology was employed, Figure 3.7 shows a microphotograph of an MPEG-4 video codec chip that was presented in 1998 [13]. The chip was fabricated in a 0.3μm CMOS n-well/p-sub technology. Three million transistors are integrated on the chip, including a 52-kB SRAM. The chip size is 9mm by 9mm. Leakage current is monitored by using a replica circuit in Figure 3.4. RBB is applied by an analog control in granularity of a chip level and a millisecond order.

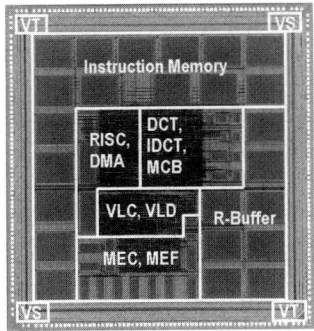

Figure 3.7 MPEG-4 video codec chip with VTCMOS technology. Leakage current is monitored by a replica circuit. RBB is applied by analog control in granularity of a chip level and a millisecond order.

The monitor objects have been extended from leakage current to speed, the voltage ranges of substrate biasing from RBB to forward body bias (FBB), and the control method from analog to digital.

Figure 3.8 shows a microphotograph of a microprocessor with a speed-adaptive threshold voltage (SA-Vt) CMOS scheme [20]. The chip was fabricated in a 0.2μm CMOS triple-well technology. The body bias is continuously controlled from −1.5V (RBB) to +0.5V (FBB) by digital control to compensate for fluctuations in fabrication and changes in V_{DD} and operating temperature.

Since circuit speed depends on both a PMOS V_{TH} and an NMOS V_{TH}, they cannot be determined uniquely by monitoring only speed. As shown in Figure 3.9, logical threshold voltage of a CMOS gate is also monitored to keep it for a prefixed value. Both V_{TH}'s of PMOS and NMOS can be uniquely determined [21].

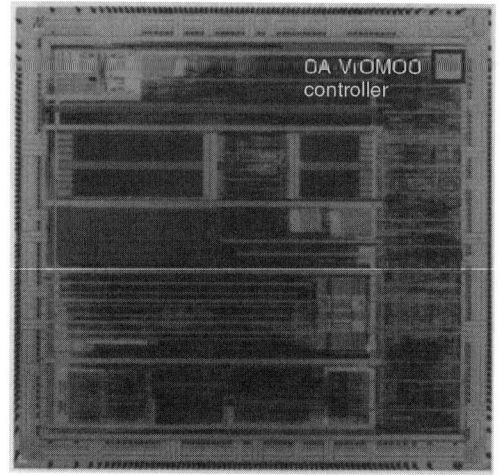

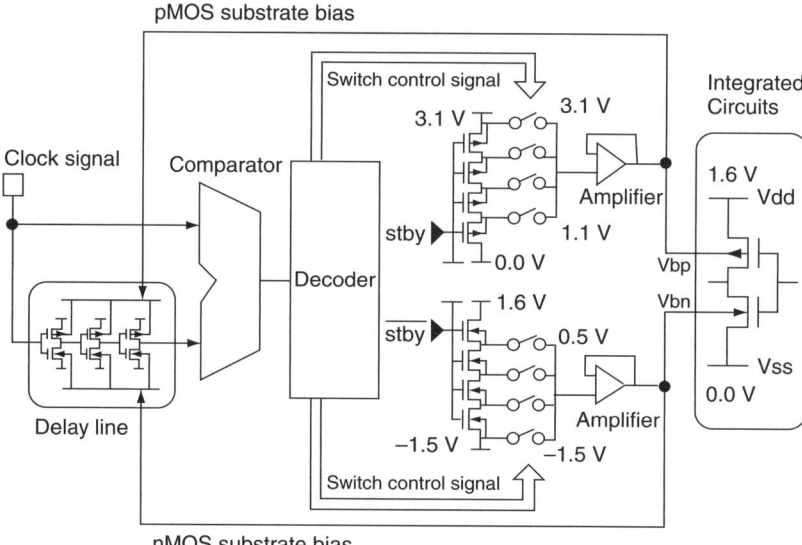

Figure 3.8 Microprocessor chip and speed-adaptive threshold voltage (SA-Vt) CMOS scheme. Speed is monitored by a replica circuit. Body bias is extended from RBB to FBB and controlled by digital in granularity of a chip level and a millisecond order.

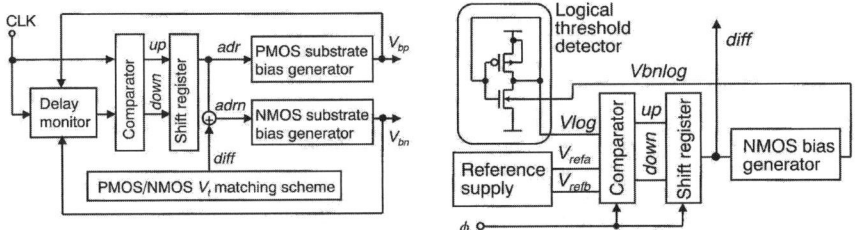

Figure 3.9 Logical threshold monitor.

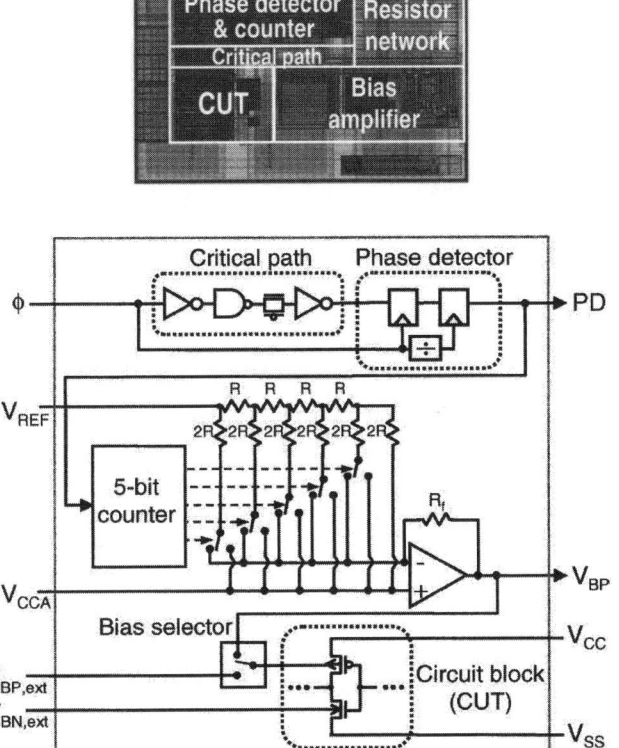

Figure 3.10 Microphotograph of sub-site and block diagram of adaptive body bias circuit.

Granularity of control in terms of space and time is becoming finer; from chip to block levels [22], and from microsecond to nanosecond ranges [23]. For instance, in Figure 3.10, body of a chip is biased separately

in 21 sub-sites that are distributed over 4.5mm by 6.7mm [22]. The chip
was fabricated in a 0.15μm CMOS technology. N-well for PMOS is continuously controlled from −0.5V (RBB) to +0.5V (FBB) by digital control.
Each sub-site has a replica of the critical path whose delay is compared
against an externally applied target clock frequency.

A self-adjusted forward body bias (SAFBB) scheme [23] in Figure 3.11
is employed for gated body. The total current for generating FBB is limited
by a current source in a controller such that the DC current does not dominate the total current dissipation, independent of the number of transistors
in a block under the FBB control. The chip was fabricated in a 0.13μm
CMOS p-substrate twin-well technology. FBB is applied by analog control
in granularity of a block level. The body bias for PMOS changes within
1μs. Such a short changing time is possible because of two reasons; the
current source continues to charge the body until body voltage reaches its
final value for FBB, and the sub-site is as small as a block.

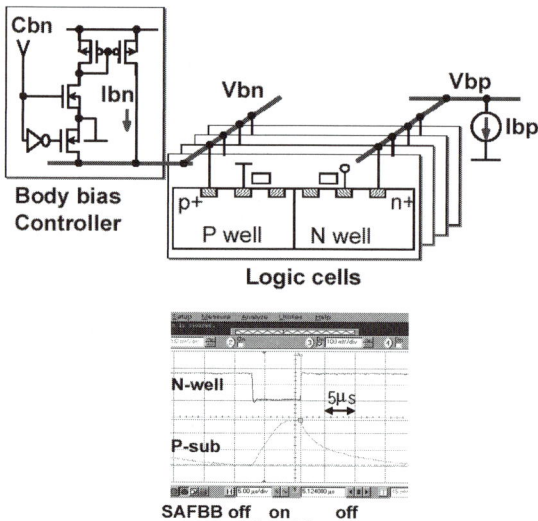

Figure 3.11 Self-adjusted forward body bias (SAFBB) scheme and
body waveforms.

3.3.4 V_{TH} Control Under Variations

Although the spatial granularity of the body biasing will be finer, it shall
be very difficult to control each V_{TH} transistor by transistor. Still the adaptive V_{TH} control shall keep its effectiveness with the following reason.

Suppose a circuit with many transistors whose V_{TH}'s receive random variation, and the variation is expressed by the normal distribution as shown in Figure 3.12, with average value, V_{TH0}, and standard deviation, σ_{VTH}.

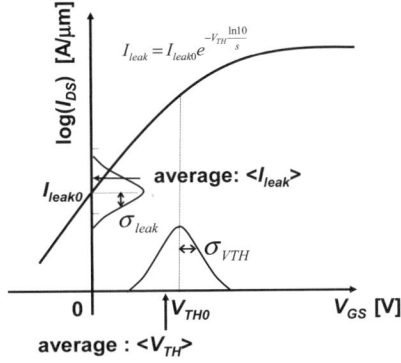

Figure 3.12 Random variation of V_{TH} and I_{leak}.

Average of subthreshold leakage current is given by [24]

$$\left\langle I_{leak} \right\rangle = \int_{-\infty}^{\infty} I_{leak}\left(V_{TH}\right) f\left(V_{TH}\right) dV_{TH}$$

$$= \int_{-\infty}^{\infty} I_{leak0} e^{-V_{TH}\frac{\ln 10}{S}} \frac{1}{\sqrt{2\pi}\sigma_{VTH}} e^{-\frac{\left(V_{TH}-V_{TH0}\right)^2}{2\sigma_{VTH}^2}} dV_{TH}$$

$$= I_{leak0} e^{\frac{\left(\sigma_{VTH}\frac{\ln 10}{S}\right)^2}{2}} = I_{leak0} 10^{\sigma_{VTH}^2 \frac{\ln 10}{2S}}, \qquad (3.9)$$

where I_{leak0} is leakage current at V_{TH0}. Corresponding average V_{TH} that yields $\langle I_{leak}\rangle$ is given by

$$\left\langle V_{TH} \right\rangle = V_{TH0} - \sigma_{VTH}^2 \frac{\ln 10}{2S} \qquad (3.10)$$

The relation in Equation (3.10) is plotted in Figure 3.13. The figure shows that even if V_{TH} fluctuates randomly by σ_{VTH} of 30mV, average of the total leakage of a circuit increases only by the equivalent amount when V_{TH} is lowered only by 10mV. In other words, random fluctuation of V_{TH} in each transistor does not bring a significant impact in leakage current of the circuit. This sounds quite natural if you notice a fact that a transistor with V_{TH} lowered by $3\sigma_{VTH}$ has around 10 times larger leakage current, but since such a transistor exists only at a rate of 1.5 per 1000 transistors, it brings small impact to the total leakage current. Random fluctuation in V_{TH} brings less impact on leakage current of a circuit than inter-chip V_{TH} fluctuations that can be compensated effectively by the adaptive V_{TH} control. Adaptive control to compensate for fluctuations in transistor level is not needed.

The same effect of statistical distribution can be found in a path delay. For instance, delay time of a path that is composed of n-stage gates is given as sum total of delay time of the n gates. Suppose delay time of each gate receives random variation of the Gaussian distribution, relative variation of the path delay is reduced to $1/\sqrt{n}$ of that of the gate delay, according to the central limit theorem. Speed variation caused by random V_{TH} variation becomes smaller as n increases.

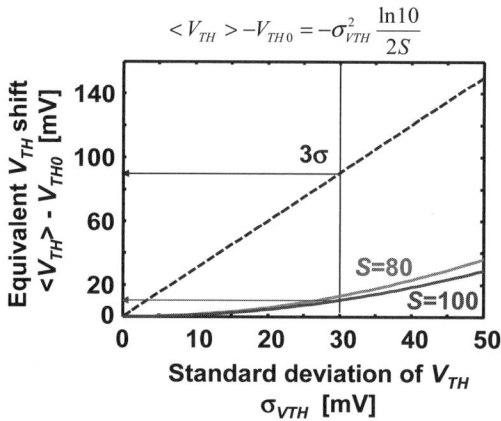

Figure 3.13 Average leakage dependence on V_{TH} variation.

On the other hand, if V_{TH} varies in clusters, lot by lot, chip by chip, or block by block, it brings large impact on circuit speed and leakage current because all V_{TH}'s are shifted to the same direction. These systematic variations can be reduced effectively by the adaptive V_{TH} control.

3.3.5 V_{TH} Control vs. V_{DD} Control

Variations in path delay can be compensated by the adaptive control of V_{TH} and/or V_{DD}. Which control is more efficient?

Power dissipation of a CMOS circuit is given by

$$P_{total} = P_{dynamic} + P_{leak} = \alpha f C V_{DD}^2 + V_{DD} I_0 W 10^{-\frac{V_{TH}}{S}} \qquad (3.11)$$

Let us suppose V_{TH} and V_{DD} are changed, while other parameters are constant. The power dissipation becomes the largest ($P_{total.max}$) under the maximum V_{DD} and minimum V_{TH}. A ratio of P_{total} over $P_{total.max}$ is given by

$$\frac{P_{total}}{P_{total.max}} = (1-\eta_L)\left(\frac{V_{DD}}{V_{DD.max}}\right)^2 + \eta_L 10^{\frac{V_{TH.min}-V_{TH}}{S}} \frac{V_{DD}}{V_{DD.max}}, \quad (3.12)$$

where η_L is a ratio of leakage power to the total power dissipation.

$$\eta_L = \frac{P_{leak.max}}{P_{total.max}} \quad (3.13)$$

It is known that P_{total} becomes minimum at around η_L=0.3 when V_{TH} and V_{DD} are lowered such that circuit speed is unchanged [25].

The same kind of equation for circuit speed is similarly derived and given by

$$\frac{Speed}{Speed_{max}} = \frac{1}{\left(\dfrac{V_{DD}}{V_{DD.max}}\right)\left(\dfrac{V_{DD.max}-V_{TH.min}}{V_{DD}-V_{TH}}\right)^\alpha}, \quad (3.14)$$

where α represents the velocity saturation effect [6].

Now let us suppose a case where V_{TH} is lower by 0.1V than a target value due to process fluctuation. Circuit speed becomes 20% faster, while

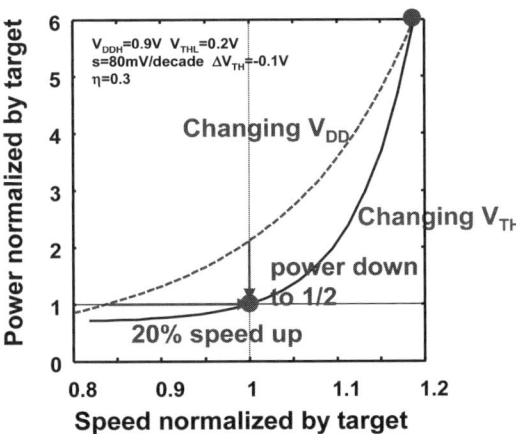

Figure 3.14 Comparison between V_{TH} control and V_{DD} control. The V_{TH} control, compared to the V_{DD} control, lowers power dissipation to half for the same circuit speed or increases circuit speed by 20% for the same power dissipation.

power dissipation becomes six times larger. Let us next apply the adaptive V_{TH} control and the adaptive V_{DD} control. The calculation results by using the above equations are plotted in Figure 3.14. When V_{TH} is raised by the adaptive V_{TH} control, power dissipation is lowered to half compared to the case where V_{DD} is lowered by the V_{DD} control. When V_{TH} is lowered, circuit speed is increased by 20% compared to the case where V_{DD} is raised. The adaptive V_{TH} scheme works more effectively to compensate for variations in power and speed that are caused by fluctuations in V_{TH}.

3.4 Hardware and Software Cooperative Control

The control method is extended from analog to digital and from hardware to software. In this section, hardware–software cooperative control is presented.

3.4.1 Cooperation Between Hardware and Application Software

In real-time systems, utilization of a processor is frequently less than one, even if all tasks run at their worst-case execution time (*WCET*). There is always some slack time (worst-case slack time). Moreover, workload of each task may vary from time to time, which results in another kind of slack time (workload-variation slack time).

A run-time voltage hopping (RVH) scheme [26] exploits both the worst-case slack time and the workload-variation slack time. Clock frequency (f_{CLK}) and hence supply voltage (V_{DD}) are scheduled as depicted in Figure 3.15 with the following steps.

(1) A task is divided into N timeslots. Following parameters are obtained through static analysis or direct measurement; *WCET* of whole task (T_{WC}), ith timeslot (T_{WCi}), and *WCET* from ($i+1$)th to Nth timeslots (T_{Ri}).

(2) For each timeslot, target execution time (T_{TAR}) is calculated as $T_{TAR} = T_{WC} - T_{WCi} - T_{ACC} - T_{TD}$, where T_{ACC} is accumulated execution time from 1st to ($i-1$)th timeslots, and T_{TD} is transition delay to change f_{CLK} and V_{DD}.

(3) For each candidate clock frequency, $f_j = f_{CLK}/j$ (j=1, 2, 3...), estimated maximum execution time Tj is calculated as $T_j = T_{Wi} * j$. If f_j is not equal to clock frequency of ($i-1$)th timeslot, $T_j = T_j + T_{TD}$.

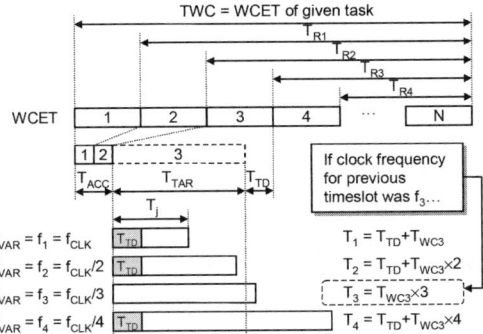

Figure 3.15 f_{CLK} and V_{DD} scheduling in RVH scheme.

(4) Clock frequency f_{VAR} is determined as minimum clock frequency f_j whose estimated maximum execution time T_j does not exceed target time T_{TAR}, as shown in Figure 3.15.

(5) Supply voltage V_{VAR} is determined from the lookup table.

Steps (1) and (2) are performed at compile, while steps (3)–(5) are carried out at run time.

Figure 3.16 shows measured power dissipation reduction ratio when the scheme is employed to an MPEG-4 SP@L1 video encoding application. It is seen that power dissipation is reduced to 6%. Only two discrete levels of clock frequency (f, $f/2$) are sufficient, meaning that the scheme is very simple in both hardware and software designs.

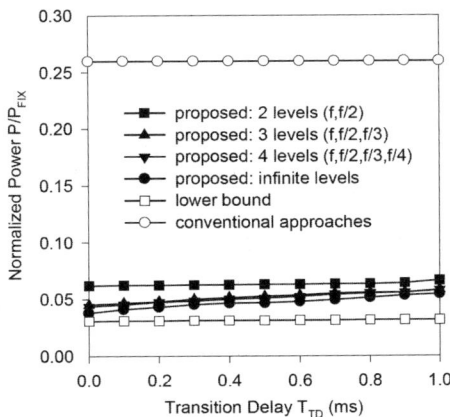

Figure 3.16 Power reduction of MPEF-4 encoding by RVH scheme.

3.4.2 Cooperation Between Hardware and Operating System

The RVH scheme is limited to a single application. A cooperative power optimization method among operation system (OS), applications, and hardware platform is essential [27, 28]. Cooperation is needed because OS only knows global timing information among tasks, while each application has knowledge about its own structure and behavior.

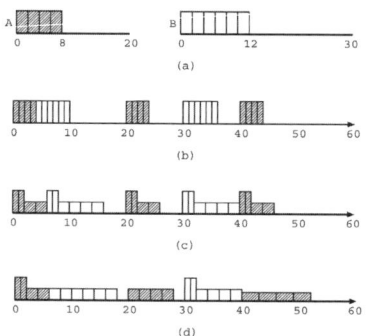

Figure 3.17 Scheduling; (a) task set, (b) conventional rate-monotonic scheduling, (c) slice-level control of speed without interaction with OS, (d) cooperative scheduling.

OS controls the execution flow of tasks with off-the-shelf microprocessor and custom chips that provide power-down mode and discrete levels of speed (i.e., f and V_{DD}). The main function of OS consists of (1) providing virtual deadline to each task in such a way that deadlines of all tasks are always guaranteed and (2) predicting the exact time interval during which there is no activity on the processor and bringing the processor into power down. This is done based on status of queues (ready queue and dominant queue).

An example is shown in Figure 3.17 [27]. Consider the two tasks shown in Figure 3.17a. Suppose that they consist of four and six slices, respectively, with each slice requesting 2 time units for its *WCET*. If we assume that period is equal to deadline, rate monotonic priority assignment is a natural choice meaning that A gets higher priority. A typical schedule, when each slice runs at half of its *WCET*, is shown in Figure 3.17b. Suppose that there are three speed levels; 1, 1/2, and 1/3. The cooperative scheduling is shown in Figure 3.17d. At time 0, A is forced to complete its execution within its *WCET* at 8 because B is in RUN state. This is similar to having virtual deadline at 8. At time 6, A goes to DORMANT state. Thus, the virtual deadline of B is set to 20, which is the minimum of its deadline at 30 and the next arrival time of A at 20. The remaining schedule can be verified similarly. For comparison, Figure 3.17c shows a schedule when the method in [26] is applied to a multitasking environment if proper support from OS is possible.

Experimental results with a prototype system in [28] show that 74% power saving is possible in multitask multimedia environment compared to the conventional real-time OS (μITRON) when workload is 38%.

3.5 Conclusion

Adaptive circuit techniques for reducing power consumption are presented from perspectives of what to monitor, how to monitor, what to control, how to control, and the granularity of the control.

The monitor object is extended from leakage current to speed, voltage, and temperature. Replica circuits such as a leakage current monitor, a ring oscillator, and a logical threshold monitor are used.

The control objects are clock frequency, V_{DD}, and V_{TH}. In the frequency–voltage cooperative control, hopping in two levels of the clock frequency (f_1 and f_2) with corresponding changes in V_{DD} yields almost as good effect in power reduction as their continuous control. f_2 should be set at half of f_1.

V_{TH} can be controlled by body bias (VTCMOS). V_{TH} variations can be compensated by feedback control of the body bias such that monitored leakage current is set to a target value. The range of the body biasing is extended from reverse body bias to forward body bias. The adaptive V_{TH} control continues to work effectively under random variation of V_{TH} in scaled devices.

The control method is extended from analog to digital and from hardware to software. The granularity of the control in terms of space and time is becoming finer, from chip to block levels and from microsecond to nanosecond ranges.

References

[1] T. Kuroda, K. Suzuki, S. Mita, T. Fujita, F. Yamane, F. Sano, A. Chiba, Y. Watanabe, K. Matsuda, T. Maeda, T. Sakurai, and T. Furuyama, "Variable supply-voltage scheme for low-power high-speed CMOS digital design," IEEE J. Solid-State Circuits, vol. 33, no. 3, pp. 454–462, Mar. 1998.

[2] T. Sakurai, "Low power digital circuit design (keynote)," ESSCIRC'04, pp. 11–18, Sept. 2004. T. Sakurai, "Perspectives of low-power VLSI's," IEICE Transactions on Electronics, vol. E87-C, no. 4, pp. 429–437, Apr. 2004.

[3]   A. Chandrakasan, V. Gutnik, and T. Xanthopoulos, "Data driven signal processing: an approach for energy efficient computing," Proc. ISLPED'96, pp 347–352, Aug. 1996.

[4]   K. Aisaka, T. Aritsuka, S. Misaka, K. Toyama, K. Uchiyam, K. Ishibashi, H. Kawaguchi, and T. Sakurai, "Design rule for frequency-voltage cooperative power control and its application to an MPEG-4 decoder," Symp. on VLSI Circuits Digest of Technical Papers, pp. 216–217, Jun. 2002.

[5]   T. Kuroda, T. Fujita, S. Mita, T. Nagamatu, S. Yoshioka, K. Suzuki, F. Sano, M. Norishima, M. Murota, M. Kako, M. Kinugawa, M. Kakumu, and T. Sakurai, "A 0.9V 150MHz 10mW 4mm^2 2-D discrete cosine transform core processor with variable-threshold-voltage scheme," IEEE J. Solid-State Circuits, vol. 31, no. 11, pp. 1770–1779, Nov. 1996.

[6]   T. Sakurai and A. R. Newton, "Alpha-power law MOSFET model and its applications to CMOS inverter delay and other formulas," IEEE J. Solid-State Circuits, vol. 25, no. 2, pp. 584–594, Apr. 1990.

[7]   T. Kobayashi and T. Sakurai, "Self-adjusting threshold-voltage scheme (SATS) for low-voltage high-speed operation," Proc. CICC'94, pp. 271–274, May 1994.

[8]   K. Seta, H. Hara, T. Kuroda, M. Kakumu, and T. Sakurai, "50% active-power saving without speed degradation using standby power reduction (SPR) circuit," ISSCC Dig. Tech. Papers, pp. 318–319, Feb. 1995.

[9]   T. Kuroda, T. Fujita, T. Nagamatu, S. Yoshioka, T. Sei, K. Matsuo, Y. Hamura, T. Mori, M. Murota, M. Kakumu, and T. Sakurai, "A high-speed low-power 0.3μm CMOS gate array with variable threshold voltage (VT) scheme," Proc. CICC'96, pp. 53–56, May 1996.

[10] T. Kuroda, T. Fujita, S. Mita, T. Mori, K. Matsuo, M. Kakumu, and T. Sakurai, "Substrate noise influence on circuit performance in variable threshold-voltage scheme," Proc. ISLPED'96, pp. 309–312, Aug. 1996.

[11] T. Kuroda and T. Sakurai, "Threshold-voltage control schemes through substrate-bias for low-power high-speed CMOS LSI design," J. VLSI Signal Processing Systems, Kluwer Academic Publishers, vol. 13, no. 2/3, pp. 191–201, Aug./Sep. 1996.

[12] R. D. Pashley and G. A. McCormick, "A 70-ns 1K MOS RAM," ISSCC Dig. Tech. Papers, pp. 138–139, Feb. 1976.

[13] M. Takahashi, M. Hamada, T. Nishikawa, H. Arakida, Y. Tsuboi, T. Fujita, F. Hatori, S. Mita, K. Suzuki, A. Chiba, T. Terasawa, F. Sano, Y. Watanabe, H. Momose, K. Usami, M. Igarashi, T. Ishikawa, M. Kanazawa, T. Kuroda, and T. Furuyama, "A 60mW MPEG4 video codec using clustered voltage scaling with variable supply-voltage scheme," ISSCC Dig. Tech. Papers, pp. 34–35, Feb. 1998.

[14] K. Kanda, K. Nose, H. Kawaguchi, and T. Sakurai, "Design impact of positive temperature dependence of drain current in sub 1V CMOS VLSI's," Proc. CICC'99, pp. 563–566, May 1999.

[15] A. Keshavarzi, S. Ma, S. Narendra, B. Bloechel, K. Mistry, T. Ghani, S. Borkar, and V. De, "Effectiveness of reverse body bias for leakage control in scaled dual Vt CMOS ICs," Proc. LPED'01, pp. 207–212, Aug. 2001.

[16] M. Togo, T. Fukai, Y. Nakahara, S. Koyama, M. Makabe, E. Hasegawa, M. Nagase, T. Matsuda, K. Sakamoto, S. Fujiwara, Y. Goto, T. Yamamoto, T. Mogami, M. Ikeda, Y. Yamagata, and K. Imai, "Power-aware 65nm node CMOS technology using variable V_{DD} and back-bias control with reliability consideration for back-bias mode," Symp. on VLSI Technology Dig. Tech. Papers, pp. 88–89, June 2004.

[17] S. Narendra, M. Haycock, V. Govindarajulu, V. Erraguntla, H. Wilson, S. Vangal, A. Pangal, E. Seligman, R. Nair, A. Keshavarzi, B. Bloechel, G. Dermer, R. Mooney, N. Borkar, S. Borkar, and V. De, "1.1 V 1 GHz communications router with on-chip body bias in 150 nm CMOS," ISSCC Dig. Tech. Papers, pp. 270–271, Feb. 2002.

[18] S. Vangal, M. A. Anders, N. Borkar, E. Seligman, V. Govindarajulu, V. Erraguntla, H. Wilson, A. Pangal, V. Veeramachaneni, J. Tschanz, Y. Ye, D. Somasekhar, B. Bloechel, G. Dermer, R. K. Krishnamurthy, K. Soumyanath, S. Mathew, S. Narendra, M. Stan, S. Thompson, V. De, and S. Borkar, "5-GHz 32-bit integer execution core in 130-nm dual-V/sub T/ CMOS," IEEE J. Solid-State Circuits, vol. 37, no. 11, pp. 1421–1432, Nov. 2002.

[19] S. Narendra, A. Keshavarzi, B. A. Bloechel, S. Borkar, and V. De, "Forward body bias for microprocessors in 130-nm technology generation and beyond," IEEE J. Solid-State Circuits, vol. 38, no. 5, pp. 696–701, May 2003.

[20] M. Miyazaki, G. Ono, T. Hattori, K. Shiozawa, K. Uchiyama, and K. Ishibashi, "A 1000-MIPS/W microprocessor using speed-adaptive threshold-voltage CMOS with forward bias," ISSCC Dig. Tech. Papers, pp. 420–421, Feb. 2000.

[21] G. Ono and M. Miyazaki, "Threshold-voltage balance for minimum supply operation," Symp. VLSI Circuits Dig. 16, pp. 206–209, June 2002.

[22] J. Tschanz, J. Kao, S. Narendra, R. Nair, D. Antonladls, A. Chandrakasan, and V. De, "Adaptive body bias for reducing impacts of doe-to-deiand within-die parameter variations on microprocessor frequency and leakage," IEEE J. Solid-State Circuits, vol. 37, no. 11, pp. 1396–1402, Nov. 2002.

[23] K. Ishibashi, T. Yamashita, Y. Arima, I. Minematsu, and T. Fujimoto, "A 9µW 50MHz 32b adder using a self-adjusted forward body bias in SoCs," ISSCC Dig. Tech. Papers, pp. 116–117, Feb. 2003.

[24] Q. Liu, T. Sakurai, and T. Hiramoto, "Optimum device consideration for standby power reduction scheme using drain-induced barrier lowering," Jpn. J. Apply. Phys. vol. 42, no. 4B, pp. 2171–2175, Apr. 2003.

[25] T. Kuroda, "Optimization and control of VDD and VTH for low-power, high-speed CMOS design (invited)," ICCAD'02 Dig. Tech. Papers, pp. 28–34, Nov. 2002.

[26] S. Lee and T. Sakurai, "Run-time voltage hopping for low-power real-time systems," Proc. DAC'00, pp. 806–809, June 2000.

[27] Y. Shin, H. Kawaguchi, and T. Sakurai, "Cooperative Voltage Scaling (CVS) between OS and applications for low-power real-time systems," Proc. CICC'01, pp. 553–556, May 2001.

[28] H. Kawaguchi, Y. Shin, and T. Sakurai, " μ ITRON-LP: power-conscious real-time OS based on cooperative voltage scaling for multimedia applications," IEEE Transaction on Multimedia, vol. 7, no. 1, pp. 67–74, Feb. 2005.

Chapter 4 Dynamic Adaptation Using Body Bias, Supply Voltage, and Frequency

James Tschanz

Intel Corporation

4.1 Introduction

Continued technology scaling, while providing ever-increasing transistor density and reduced cost per transistor, has the unwanted side effects of increasing variations. Process variations can be due to many non-idealities that occur during the manufacturing process; however, chief among these is the difficulty of patterning line dimensions which are much smaller than the wavelength of light used during lithography. The resulting variation in channel length across the die (and across the wafer, from lot to lot, etc.) is one of the dominant causes of delay and leakage variation in high-performance microprocessors [1]. Other effects such as line-edge roughness and random dopant fluctuation also contribute to the variations, especially in circuits with small transistors, or circuits in which matching of devices is important. Die-to-die variations can be considered to impact all devices on the same die equally and cause differences among dies on the same wafer, as well as from wafer to wafer and lot to lot. These variations can be mitigated in some products by binning – that is, selling the microprocessors at multiple price/performance points. Within-die variations, on the other hand, result in differing transistor characteristics within the same die. These cannot be reduced by binning or by any other die-level technique, and are typically guardbanded. Because within-die variations are becoming more prominent as technology scales, and because design margins are

A. Wang, S. Naffziger (eds.), *Adaptive Techniques for Dynamic Processor Optimization*,
DOI: 10.1007/978-0-387-76472-6_4, © Springer Science+Business Media, LLC 2008

continually shrinking, it is necessary to develop intelligent techniques for tolerating or compensating within-die variations.

On top of the static process variations which occur, however, microprocessors experience a wide range of dynamic variations (Table 4.1). These dynamic variations are a result of the environment in which the processor is used, as well as the applications and workload which are run. Dynamic variations include temperature changes, voltage droops, noise events, as well as transistor degradation and aging. While these variations can be mitigated as much as possible through careful design, this is often done at considerable cost (for example, overly conservative design rules, additional power consumption, or expensive package decoupling capacitors). Those effects that cannot be handled through design must be guardbanded, resulting in a power overhead or performance penalty. Because both performance and power are more important now than ever before, guardbanding these variations is expensive and undesirable. Dynamic techniques for sensing and responding to these variations can therefore be used to significantly improve the efficiency of the design as compared to a worst-case design methodology.

Table 4.1 Examples of dynamic variations.

Parameter	Time Scale	Impact
Supply voltage	Nanoseconds to microseconds	Droop: impacts Fmax Overshoot: impacts reliability
Temperature	Microseconds	Fmax and reliability
Transistor degradation	Hours to days	Fmax degradation SRAM stability

4.2 Static Compensation with Body Bias and Supply Voltage

Variations that are static in nature (for example, process variations) can be compensated using static techniques which are calibrated once after fabrication and then remain constant throughout the lifetime of the part. An example of a static compensation technique is clock skew compensation [2], in which clock delay buffers are tuned post-fabrication to optimize clock skew and improve clock timing. The settings for these adaptive techniques may be saved in nonvolatile fuse memory, loaded from the system as part of the boot-up routine, or determined on each power-up through the use of self-test circuitry. In this section, we describe two common knobs for tuning system performance after fabrication: body bias and supply voltage.

4.2.1 Adaptive Body Bias

Body bias refers to a nonzero voltage which is applied between the source and body (substrate or n-well) of a MOS transistor. Because typically the substrate of the die is connected to ground, and the n-wells are connected to the supply voltage, transistors are either zero biased or reverse biased (if, for example, the transistor is part of a stack). This voltage difference between the source and body of a transistor impacts the width of the depletion region around the source, drain, and gate of the device, and therefore modulates the threshold voltage. If the body–source junction is reverse biased ($V_{body}<0$ for NMOS, $V_{body}>V_{CC}$ for PMOS), the magnitude of the threshold voltage increases. If the body–source junction is forward biased ($V_{body}>0$ for NMOS, $V_{body}<V_{CC}$ for PMOS), the magnitude of the threshold voltage reduces. Therefore, body bias can be viewed as a "knob" for tuning the threshold voltage of MOS devices.

The sensitivity of MOS devices to body bias and the range of bias voltages that can be applied are a function of the process technology and device design. In the reverse direction, applying larger and larger amounts of reverse body bias (RBB) continually causes the threshold voltage to increase. This increase in V_T reduces the subthreshold component of leakage power (Figure 4.1). However, as the reverse bias increases, reverse junction current increases as well. Therefore, if the goal is to minimize the leakage current of a circuit, the optimum reverse bias voltage is the point at which the increase in reverse junction current balances out the reduction in subthreshold leakage. Previous studies have shown that this optimum can range from –0.5V to –1.5V and below, depending on the process technology and device channel length [3, 4].

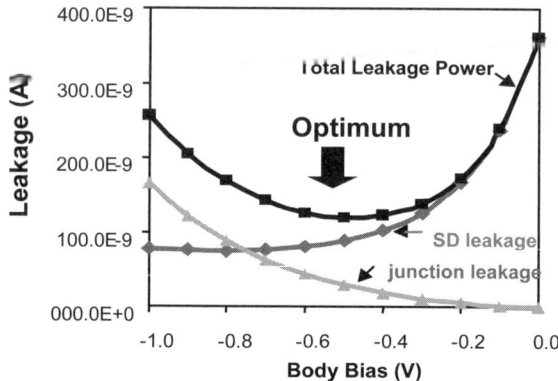

Figure 4.1 Leakage change with reverse body bias [3]. (© 1999 IEEE)

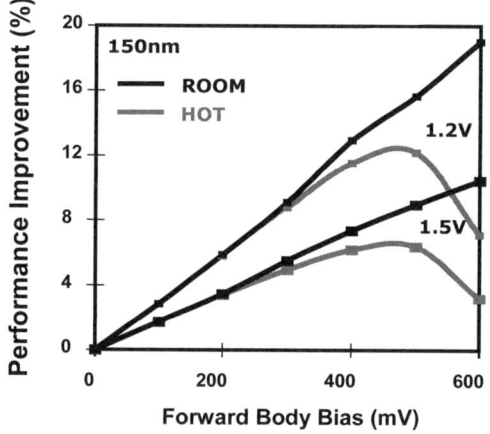

Figure 4.2 Performance improvement with forward body bias [5]. (© 2003 IEEE)

In the forward direction, there is a similar trade-off. As the forward body bias (FBB) voltage increases, the threshold voltage reduces, resulting in reduced switching delay for the circuit. At the same time, the forward junction current across the body–source diode increases as well. If this current becomes too large, it can result in non-full-rail switching for the circuit and be subtracted from the switching current. Again, this optimum voltage depends strongly on temperature, and the test-chip measurements (Figure 4.2) have shown that, at high temperature, the optimum forward body bias for maximum frequency is in the range of 400–500mV [5, 6].

Because body bias provides a way of changing the threshold voltage of fabricated transistors, it can be used to compensate the effects of static process variations. Bidirectional adaptive body bias uses both forward and reverse body biases to bring the fabricated dies to their desired threshold voltage – high-V_T dies receive forward body bias while low-V_T dies are reverse biased. This approach is shown in Figure 4.3. If only die-to-die variations are considered, an optimal body bias can be found for each die to completely compensate the process variations, resulting in a population of dies with identical threshold voltages (assuming sufficient body bias range). Because in reality within-die variations in threshold voltage exist as well, the compensated dies will still show a distribution of threshold voltages, however this distribution will be significantly tightened from the original case.

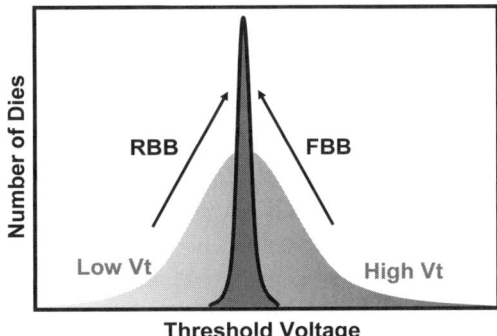

Figure 4.3 Variation compensation using adaptive body bias.

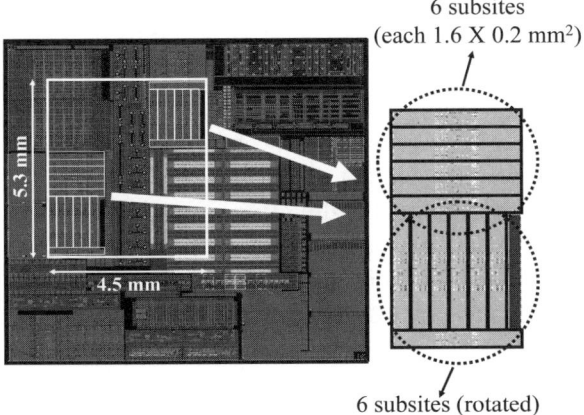

Figure 4.4 Adaptive body bias test-chip [7]. (© 2002 IEEE)

Figure 4.4 shows an adaptive body bias test-chip implemented in the 150nm CMOS technology generation [7]. Each test-chip die contains 21 "subsites" distributed over a 4.5×5.3mm^2 area in two orthogonal

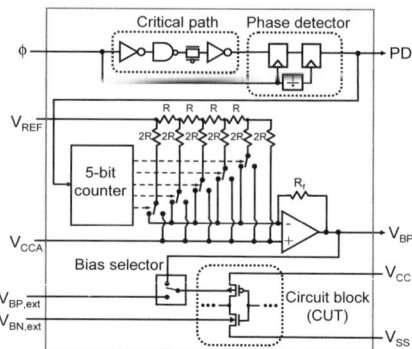

Figure 4.5 Key circuit elements of one subsite of adaptive body bias test-chip [7].
(© 2002 IEEE)

orientations. Each of these subsites (Figure 4.5) represents a circuit block of a microprocessor design and contains a complete adaptive body bias (ABB) generator and control circuit in addition to critical path blocks. One critical path from this circuit block is replicated and a target clock frequency ϕ is applied externally. This represents the desired frequency of operation for the circuit block. The delay of the critical path replica is compared to the incoming clock period through the use of a phase detector circuit, and the output from this phase detector drives a counter and D/A converter to generate the body bias voltage. This forms a feedback circuit which automatically adjusts the body bias until the delay of the critical path matches the incoming clock period. To find the optimum body bias voltage for each die, different target frequencies can be applied to the input clock, and after the body bias adapts to meet the target frequency, the leakage of the die is measured. The body bias voltage which gives the maximum performance subject to the power constraint is the optimum voltage which is chosen for that die sample.

Measurement results for adaptive body bias (ABB) as compared to no body bias (NBB) are shown in Figure 4.6. All fabricated dies must meet a minimum performance specification, as shown by the vertical dashed line

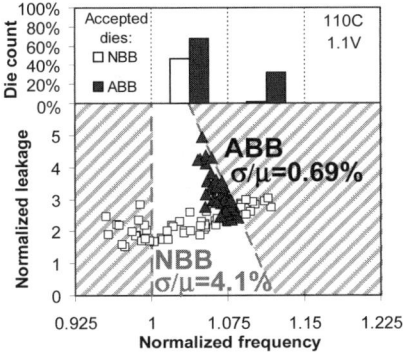

Figure 4.6 Measurement results: comparison of no body bias (NBB) and adaptive body bias (ABB) [7]. (© 2002 IEEE)

at a frequency of 1, as well as a maximum leakage specification dictated by the platform total power requirements. This maximum leakage line is slanted reflecting that the fast dies run at higher frequency which results in higher dynamic power consumption – therefore their allowed leakage power is low. Application of ABB reduces die-to-die frequency variations (σ/μ) by an order of magnitude, and 100% of the dies become acceptable as compared with only 50% accepted dies for NBB. In addition, 30% of the dies are now in the highest frequency bin allowed by the power density limit.

The above procedure is very effective for compensating the die-to-die parameter variations; however, since only one bias voltage is used per die, it is not possible to compensate any variations across the die. In order to reduce the impacts of within-die variations as well, multiple bias voltages can be employed and individually tuned. The number of body bias regions used across the die depends on the correlation distance of the within-die variation components as well as the area overhead and testing complexity involved in generating multiple bias voltages on the die. Figure 4.7 demonstrates the gains possible by using multiple bias voltages – in this example, each of the 21 subsites on the test-chip receives its own unique body bias voltage. In this case, frequency variation is reduced by another 4× as compared to the die-to-die ABB, and 99% of the dies are now in the highest-revenue bin.

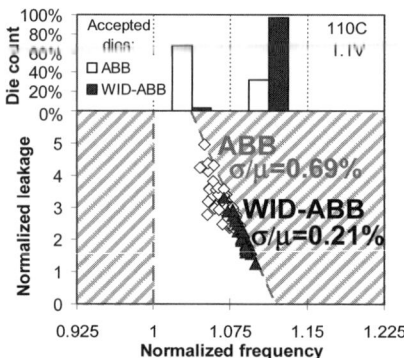

Figure 4.7 Measurement results: comparison of ABB and within-die ABB [7].
(© 2002 IEEE)

4.2.2 Adaptive Supply Voltage

Supply voltage can be used in the same way as body bias to counteract the effects of process variations. While frequency binning is the simplest way to compensate die-to-die variations and recover dies which exceed the power requirement, adaptive V_{CC} provides two significant benefits over simple frequency binning. First, dies that violate the power constraint will have V_{CC} reduced in tandem with their natural operating frequencies, which provides better power savings than frequency reduction alone. In contrast to simple frequency reduction, lowering V_{CC} reduces standby leakage power as well, while switching power is reduced in a cubic manner. Therefore, lowering V_{CC} and frequency together allows dies to be accepted in a higher frequency bin than with simple frequency binning alone. Second, dies which are too slow can be recovered by increasing their V_{CC} to increase their natural operating frequency and move them to the highest frequency bin allowed by the active power limit. Gate-oxide reliability considerations limit the maximum allowed V_{CC}; however, this constraint is not usually a problem for mobile processors with V_{CC} lower than the maximum allowed by the process.

Evaluation of the impact of adaptive supply voltage has been performed on the same 150nm CMOS test-chip as was described above in the body bias section [8]. For these measurements, it is assumed that the processor is a low-power product which is running at a V_{CC} below the V_{MAX} limit. Therefore, slow dies can be sped up by increasing the V_{CC}, while leaky dies can be recovered by reducing the V_{CC}. As shown in Figure 4.8, applying adaptive V_{CC} improves the mean die frequency as well as the number of parts in the highest frequency bin. However, effectiveness of adaptive V_{CC} depends critically on the voltage resolution provided by the voltage regulator module. Using 50mV resolution instead of 20mV renders the technique ineffective.

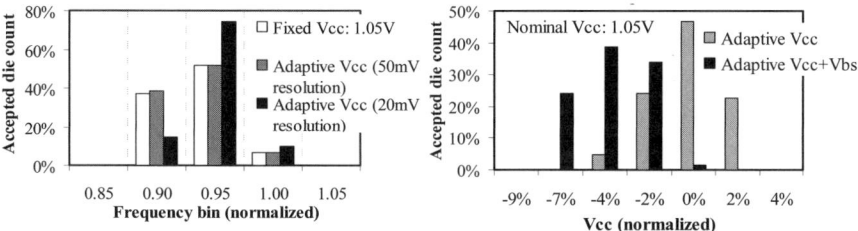

Figure 4.8 (a) Comparison of fixed V_{CC} and adaptive V_{CC}, (b) Comparison of adaptive V_{CC} and adaptive $V_{CC}+V_{BS}$ [8]. (© 2003 IEEE)

Using adaptive V_{CC} in conjunction with adaptive body bias (adaptive V_{BS}) is more effective than using either of them individually (Figure 4.8b). In this combined scheme (adaptive $V_{CC}+V_{BS}$), a single V_{CC} and NMOS/PMOS V_{BS} combination is used per die to move it to the highest frequency bin subject to the active power limit. Adaptive V_{BS} uses FBB to speed up dies that are too slow, and RBB to reduce frequency and leakage power of dies that are too fast and leaky. Adaptive $V_{CC}+V_{BS}$, on the other hand, recovers these dies above the active power limit by (1) first lowering V_{CC} and natural operating frequency together to bring the sum total of their switching and leakage powers well below the active power limit and (2) then applying FBB to speed them up and move them to the highest frequency bin allowed by the active power limit. As a result, more dies use lower V_{CC} values than adaptive V_{CC}. In addition, more dies use FBB, instead of RBB, compared to adaptive V_{BS} (Figure 4.9). Since the effectiveness of RBB for leakage power reduction diminishes with technology scaling [4], adaptive $V_{CC}+V_{BS}$ will be more effective in future technology generations than adaptive V_{BS} alone. Bias voltages for NMOS and PMOS transistors are typically generated using on-die circuitry and routed to transistor wells using a separate bias grid, incurring an area overhead of 2–4%.

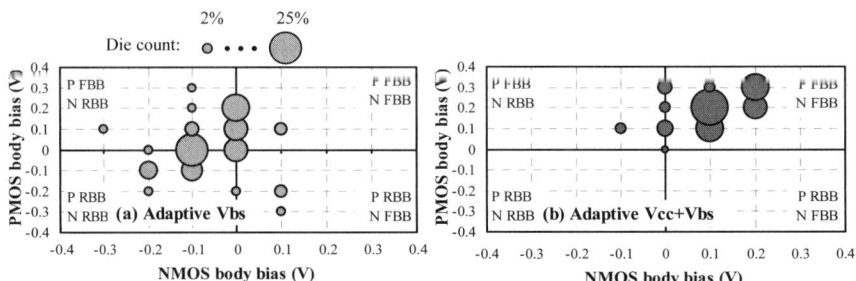

Figure 4.9 Optimal body bias voltages chosen for (a) adaptive V_{BS}, (b) adaptive $V_{CC}+V_{BS}$ [8]. (© 2003 IEEE)

4.3 Dynamic Variation Compensation

4.3.1 Dynamic Body Bias

Body bias can also be used in a dynamic sense as part of a power management scheme or to compensate dynamic variations. Due to advanced power control features, microprocessors can experience a very wide range of activity factors during normal operation – ranging from very high activity for tasks which are heavily computationally intensive to very low activity when the processor is in standby mode. Therefore it is impossible to find the device threshold voltage, supply voltage, and frequency which is energy optimal across all usage conditions. Body bias provides a way to adjust the threshold voltage dynamically to improve performance during active mode while saving power in standby mode.

When the processor is actively running computations, the activity factor is high, and typically dynamic power dominates over the leakage power. In this case, forward body bias can be applied to lower the threshold voltage and improve performance. Alternately, the device threshold voltage can be increased in the process so that when FBB is applied, it is lowered to the original target value. Applying FBB in this manner also has the advantage of improving the short-channel effects of the devices compared to lowering the V_T through process only. When the processor goes into an idle or standby mode, the power is dominated by transistor leakage. Zero or reverse body bias can then be applied to raise the threshold voltage and reduce the leakage. In this manner, the processor operates much more efficiently in both active and standby modes.

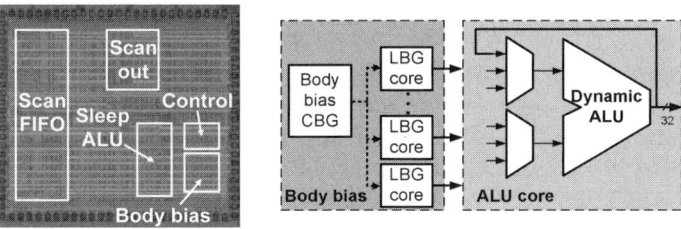

Figure 4.10 Dynamic ALU test-chip with on-chip PMOS body bias [9].
(© 2003 IEEE)

An implementation of dynamic body bias for power control is shown in Figure 4.10. This test-chip in 130nm CMOS technology [9] includes a 32-bit dynamic ALU with on-chip dynamic body bias for the PMOS transistors. The body bias circuitry consists of two main blocks: a central bias generator (CBG) and many distributed local bias generators (LBGs) (Figure 4.11). The function of the CBG is to generate a process, voltage, and temperature-invariant reference voltage which is then routed to the local bias generators. The CBG uses a scaled bandgap circuit to generate a reference voltage which is 450mV below the bandgap supply V_{CCA} – this represents the amount of forward bias to apply in active mode. This reference voltage is then routed to all of the distributed local bias generators, shielded on both sides by V_{CCA}. The function of the LBG is to translate this voltage, referenced to V_{CCA}, to a body voltage which is referenced to the local block V_{CC}. This ensures that any variations in the local V_{CC} will be tracked by the body voltage, maintaining a constant 450mV of FBB. Translation of the reference is accomplished through the use of a current mirror followed by a voltage buffer to drive the final n-well load. Low-frequency tracking of supply variations is handled by the current mirror while a capacitor provides the high-frequency tracking. In idle mode, the current mirror is disabled and a zero-bias switch transistor connects the body to V_{CC}, applying zero body bias for leakage reduction. A total of 40 distributed LBGs are used to bias the ALU, and the total area overhead for this body bias technique is 6–8%, including the bias generators as well as the additional routing required to separate the body terminals from the supply.

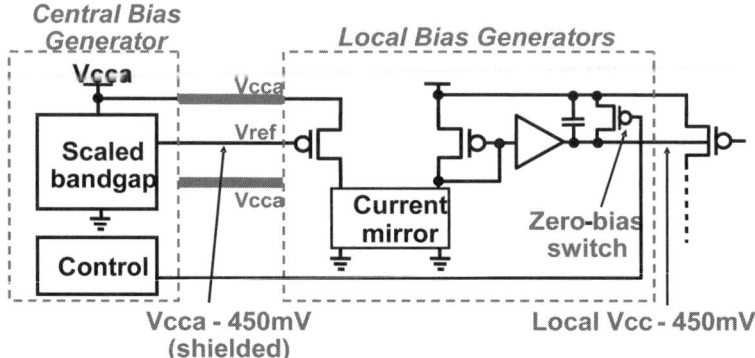

Figure 4.11 Bias generator circuits for dynamic ALU test-chip [9].
(© 2003 IEEE)

The adder operational frequency ranges from 3GHz (1.05V) to 4.2GHz (1.4V) when zero body bias (ZBB) is applied to the PMOS transistors in the core (Figure 4.12a). If the dynamic body bias circuitry is enabled to apply 450mV FBB to the core, the frequency improves by 3–7%. To achieve a target frequency of 4.05GHz, the supply voltage must be set to 1.35V when no body bias is used but can be lowered to 1.28V with FBB. This supply voltage reduction results in lower switching power for the FBB design at the same clock frequency. When the adder is put into standby mode, ZBB is used for the core, and this results in a leakage reduction of 2×. Total power savings for the ALU at a typical activity profile are shown in Figure 4.12b – for this example, the dynamic bias achieves 8% total power reduction. Therefore dynamic body biasing allows the frequency improvement due to FBB coupled with the reduced leakage power of ZBB.

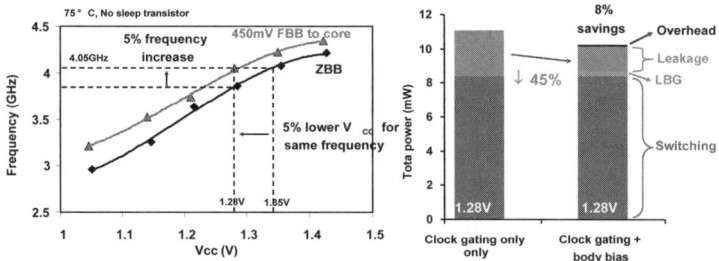

Figure 4.12 (a) Maximum frequency vs. supply voltage for ALU with and without body bias. (b) Typical power savings due to dynamic body bias [9].
(© 2003 IEEE)

4.3.2 Dynamic Supply Voltage, Body Bias, and Frequency

While static techniques such as clock tuning, adaptive body bias, and adaptive supply voltage can effectively compensate process variations, other variations such as temperature, voltage droops, noise, and transistor aging are dynamic and change throughout the lifetime of the processor. These cannot be compensated using a static technique and are typically guardbanded using either reduced frequency or higher supply voltage. This guardbanding is expensive in terms of performance and power and is becoming prohibitive as design margins shrink. To achieve an energy-efficient microprocessor which operates correctly in the presence of these variations, a method of sensing the environment and responding by changing voltage, body bias, or frequency is necessary. In this section, we describe one implementation of a dynamic adaptive processor design.

4.3.2.1 Design Details

The test-chip in 90nm CMOS technology (Figure 4.13) contains a TCP offload accelerator core, a data input buffer, V_{CC} droop sensors, thermal sensors, a dynamic adaptive biasing (DAB) control unit, distributed noise injectors, body bias generators, and a three-PLL dynamic clocking unit [10]. The DAB controller receives inputs from the thermal sensors and droop detectors. Average supply current is sensed by the off-chip voltage regulator module (VRM), and digitally communicated to the DAB controller on chip. The programmable noise injectors are used to generate various supply noises and load currents, in addition to that generated by

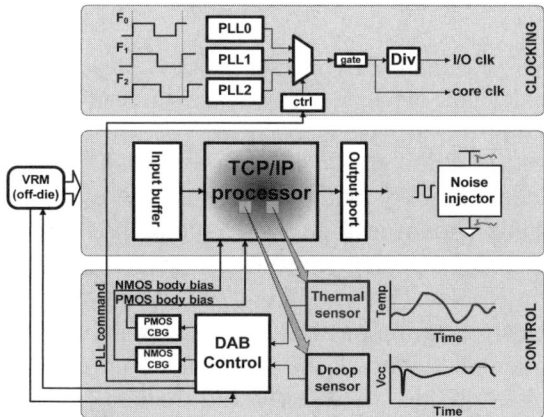

Figure 4.13 Block diagram of the dynamic adaptive TCP/IP processor [10].
(© 2007 IEEE)

the core during normal operation. The DAB controller drives the dynamic frequency unit, body bias generators, and voltage setting of the off-chip VRM to dynamically adapt frequency, body bias, and V_{CC} to achieve optimum settings for the given conditions. This DAB controller (Figure 4.14) is based on a lookup table which is indexed by the output of the thermal, droop, and current sensors and is loaded with pre-characterized data representing the optimum V_{CC}, body bias, and frequency for each of the sensor combinations. The control also includes programmable timers and logic to ensure that transitions in V_{CC}, body bias, and frequency happen in the correct sequence needed for fault-free operation and to eliminate instability around the sensor trip points. The control is designed to be fast enough to respond to 2nd and 3rd droops in voltage as well as changes in temperature and overall chip activity factor.

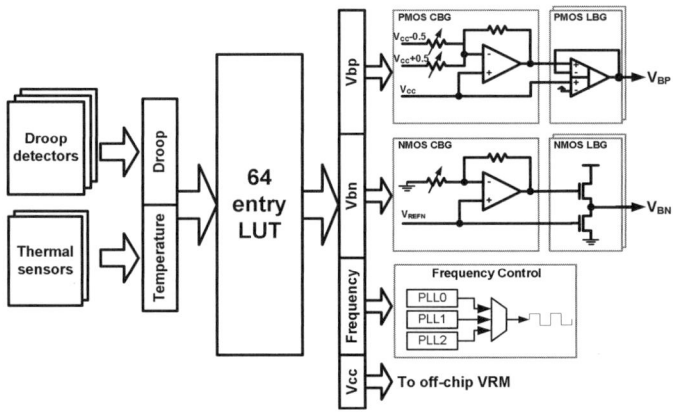

Figure 4.14 Organization of the dynamic adaptive bias controller, and the interface to the dynamic clocking and body bias circuits [10]. (© 2007 IEEE)

Responding to the relatively fast V_{CC} droops also requires a method for changing frequency quickly without waiting for a PLL to relock. The clocking subsystem, shown in Figure 4.15, contains three PLLs running at independent frequencies and a multiplexer to select between them in a single cycle while ensuring that there are no shortened clock cycles. Several algorithms for changing frequency by switching between multiple PLLs are implemented as part of the frequency control, including a simple algorithm which switches between three locked PLLs, to a flexible algorithm which keeps one PLL always locked at a frequency higher and lower than the current frequency. When a frequency change is requested, a switch is made to the slower (or faster) PLL, and then the other two PLLs are relocked and the process repeated. This allows the entire frequency space to be covered in 3% steps. The dynamic frequency algorithms are implemented in the DAB control, and commands are sent to the PLL block to switch between PLLs and update PLL divider values. Clock gating is also implemented to reduce active power consumption of the core when the TCP/IP header has finished processing and the core is idle. Both NMOS and PMOS body bias generators are implemented on the die and each includes a central bias generator (CBG) which is controlled by the DAB control, and many local bias generators (LBGs) distributed throughout the die. The PMOS bias implementation includes a differential difference amplifier (DDA) which allows both reverse and forward bias values to be generated with 32mV resolution. The NMOS bias implementation uses a simpler matched source-follower LBG for forward body bias only. Input header data to the core is supplied from the on-chip input buffer, and all arrays and programmable features are loaded through JTAG scan.

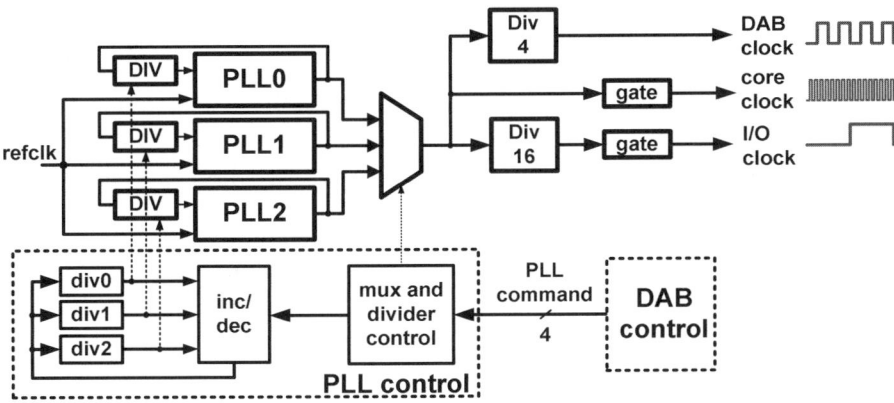

Figure 4.15 Dynamic clocking circuitry using multiple PLLs for fast frequency control [10]. (© 2007 IEEE)

4.3.2.2 Measurement Results

Maximum frequency of the design ranges from 2.2GHz at 1V to 3.4GHz at 1.4V, and total power consumption at 1.2V is 1.3W for a high-activity test. Frequency can be increased by 9–22% through application of NMOS and PMOS forward body bias. F_{MAX} and power measurements are taken across a range of voltages, body biases, and temperatures and the results loaded into the DAB control lookup table. Dynamic response of the chip to

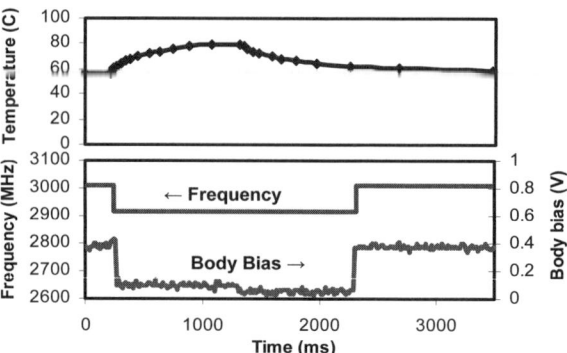

Figure 4.16 Response of frequency and body bias to dynamic temperature change [10]. (© 2007 IEEE)

temperature changes during a high-workload test (Figure 4.16) shows that while the worst-case frequency is set by the highest expected temperature, as the temperature drops, the core frequency can be increased. At the same time, at low temperature, the leakage component of power is reduced, and forward body bias (in this example, NMOS forward body bias) can be applied to further increase the performance. This combination reduces the guardband needed for maximum temperature and, in this example, results in a 1.4% increase in average frequency over the duration of the test.

In a similar way, clock frequency can be adjusted in response to dynamic voltage droops that occur due to step changes in current demand by the processor (Figure 4.17). In this case, a sudden increase in current demand causes a voltage droop to occur, after which the voltage settles to a lower voltage determined by the IR drop of the power delivery network. While a standard design would have to operate at a frequency determined by the worst-case voltage during the droop, the adaptive processor can detect the droop and dynamically respond by lowering frequency. The maximum frequency can then by increased by 32% for this large voltage droop, improving average performance for the workload.

Dynamic frequency and body bias capabilities also allow the design to respond to frequency degradation that results from device-aging mechanisms such as NBTI [11]. The threshold voltage increase in the PMOS devices due to aging can be compensated by applying increasing

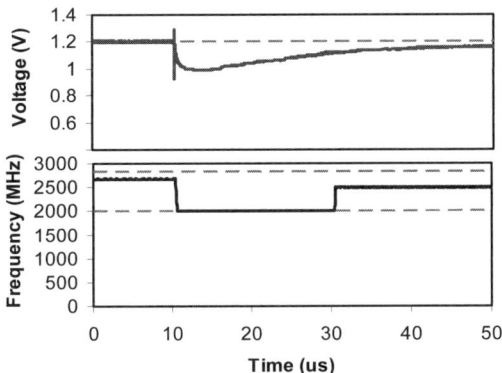

Figure 4.17 Response of clock frequency to dynamic voltage droops [10].
(© 2007 IEEE)

amounts of PMOS forward body bias over the lifetime of the part. Measurements (Figure 4.18) show that the maximum frequency of the part degrades by ~3% over its lifetime, requiring an initial frequency guardband of more than 3% due to process variations. By applying the correct amount of PMOS body bias, the threshold voltage can be reduced back to its initial value, counteracting the effects of aging and allowing the part to remain at a constant frequency over its lifetime. This allows the aging guardband to be removed and the performance of the part to be increased.

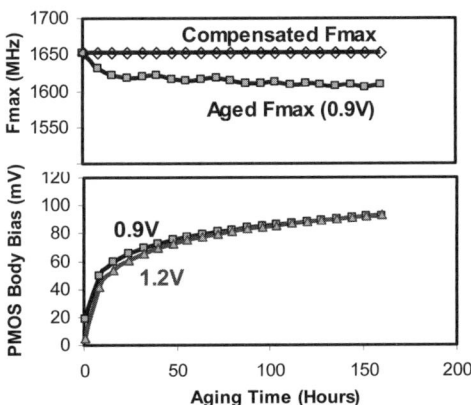

Figure 4.18 Aging compensation using dynamic body bias. The amount of FBB required to completely compensate aging is similar for both 0.9V and 1.2V supply [10]. (© 2007 IEEE)

4.4 Conclusion

Both static variations such as process fluctuation and dynamic variations in voltage, temperature, and aging are increasing with each technology generation. Simply worst-casing these variations during the design phase is no longer viable as this results in a design which is nonoptimal in power and performance. These variations need to be handled using a combination of variation-tolerant circuit techniques, architecture innovations, and system-level dynamic response.

Body bias can be used for both static variation compensation during active mode and leakage reduction for a low-power standby mode. Body bias can also be used as a method of dynamic response – maintaining circuit operation through a voltage droop for compensating transistor degradation due to aging. In much the same way, supply voltage can be statically set to compensate the die-to-die variations, or dynamically changed in response to temperature and power fluctuations. Finally, clock frequency can be modulated in a processor to adapt to the current environmental conditions. These three techniques can be combined to handle both static and dynamic variations in an efficient and low-overhead way.

References

[1]   K. A. Bowman, S. G. Duvall, and J. D. Meindl, "Impact of die-to-die and within-die parameter fluctuations on the maximum clock frequency distribution for gigascale integration", *IEEE J. Solid-State Circuits,* Vol. 37, pp. 183–190, Feb. 2002.

[2]   N. A. Kurd, J. S. Barkatullah, R. O. Dizon, T. D. Fletcher, and P. D. Madland, "A multigigahertz clocking scheme for Pentium® 4 micro-processor", *IEEE J. Solid-State Circuits*, Vol. 36, pp. 1647–1653, Nov. 2001.

[3]   A. Keshavarzi et al., "Technology scaling behavior of optimum reverse body bias for standby leakage power reduction in CMOS IC's", *Proc. ISLPED*, pp. 252–254, Aug. 1999.

[4]   A. Keshavarzi, S. Ma, S. Narendra, B. Bloechel, K. Mistry, T. Ghani, S. Borkar, and V. De, "Effectiveness of reverse body bias for leakage control in scaled dual V_T CMOS ICs", *Proc. ISLPED*, pp. 207–212, Aug. 2001.

[5]   S. Narendra et al., "Forward body bias for microprocessors in 130nm technology generation and beyond", *IEEE J. Solid-State Circuits*, Vol. 38, No. 5, May 2003.

[6]   S. Narendra, M. Haycock, V. Govindarajulu, V. Erraguntla, H. Wilson, S. Vangal, A. Pangal, E. Seligman, R. Nair, A. Keshavarzi, B. Bloechel, G. Dermer, R. Mooney, N. Borkar, S. Borkar, and V. De, "1.1V 1GHz communications router with on-chip body bias in 150nm CMOS", *IEEE ISSCC Dig. Tech. Papers*, pp. 270–271, Feb. 2002.

[7]   J. Tschanz, J. Kao, S. Narendra, R. Nair, D. Antoniadis, A. Chandrakasan, and V. De, "Adaptive body bias for reducing impacts of die-to-die and within-die parameter variations on microprocessor frequency and leakage", *IEEE J. Solid-State Circuits*, Vol. 37, Issue 11, pp. 1396–1402, Nov. 2002.

[8]   J. Tschanz et al., "Effectiveness of adaptive supply voltage and body bias for reducing impact of parameter variations in low-power and high-performance microprocessors", *IEEE J. Solid State Circuits*, Vol. 38, No. 5, May 2003.

[9]   J. Tschanz et al., "Dynamic sleep transistor and body bias for active leakage power control of microprocessors", *IEEE J. Solid State Circuits,* Vol. 38, No. 11, Nov 2003.

[10] J. Tschanz et al., "Adaptive frequency and biasing techniques for tolerance to dynamic temperature-voltage variations and aging", *IEEE ISSCC Dig. Tech. Papers*, Feb. 2007.

[11] D. Schroder et al., *J. Appl. Phys.*, Vol. 94, No. 1, July 2003.

Chapter 5 Adaptive Supply Voltage Delivery for Ultra-dynamic Voltage Scaled Systems

Yogesh K. Ramadass, Joyce Kwong, Naveen Verma, Anantha Chandrakasan

Massachusetts Institute of Technology

Minimizing the power consumption of battery-powered systems is a key focus in integrated circuit design. The increased importance of power is even more notable for a new class of energy-constrained systems. These systems must achieve long system lifetimes from a limited energy source, so the need to reduce energy consumption whenever possible is paramount. Dynamic voltage scaling (DVS) [1] is a popular method to achieve energy efficiency in systems that have widely variant performance demands. As V_{DD} decreases, transistor drive currents decrease, bringing down the speed of operation of a circuit. A DVS system adjusts the supply voltage, operating the circuit at just enough voltage to meet performance, thereby achieving overall savings in total power consumed.

Figure 5.1a plots the required rate of the system versus the normalized energy required to process one generic block of data. The most straightforward method for saving energy when the workload decreases is to operate at the maximum rate until all of the required processing is complete and then to shutdown. This approach only requires a single power supply voltage (corresponding to full rate operation), and it results in linear energy savings. A variable supply voltage with infinite allowable levels provides the optimum curve for reducing energy. The energy savings that can be obtained out of dithering the voltage supplies will be explained in Section 5.3.1.

While DVS is a popular method to minimize power consumption in digital circuits given a performance constraint, certain emerging applications like wireless micro-sensor networks [2, 3] and implantable medical electronics [4] are severely energy-constrained. For applications like implantable medical devices that are battery-operated, though the required speed of operation is low, the battery is expected to last till the lifetime of

A. Wang, S. Naffziger (eds.), *Adaptive Techniques for Dynamic Processor Optimization*,
DOI: 10.1007/978-0-387-76472-6_5, © Springer Science+Business Media, LLC 2008

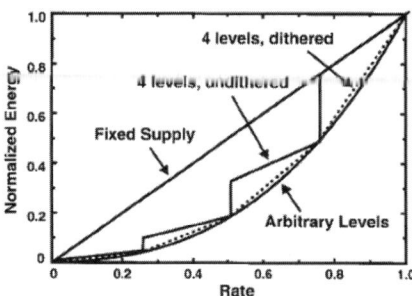

Figure 5.1a Theoretical energy consumption versus rate for different power supply strategies [1]. (© [1997] IEEE)

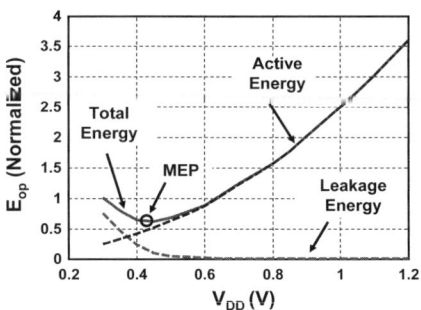

Figure 5.1b Active, leakage, and total energy per operation curves showing the minimum energy point (0.42V) for a 7-tap FIR filter implemented in 65nm CMOS.

the device, without the possibility of a recharge. On the other hand, a key requirement in the design of sensor systems is constraining the power dissipation of the system below 10μW [5] which will allow operation strictly using scavenged energy. So, irrespective of the mode of power delivery, there is a severe constraint on the energy consumed per desired operation of these devices. By introducing the capability of sub-threshold operation, DVS systems can be made to operate at their minimum energy operating voltage [6] in periods of very little activity, leading to further savings in total energy consumed. This way ultra-dynamic voltage scaling (U-DVS) can be achieved. Figure 5.1b shows the minimum energy operating voltage for a 7-tap FIR filter implemented in a 65nm CMOS process. It can be seen that close to 6× savings in energy can be obtained by operating at the minimum energy point (MEP) as opposed to the nominal voltage of 1.2V. Most energy-constrained applications work at their MEP primarily and only jump to higher voltages when high performance is demanded by certain cases.

The minimum energy operating voltage usually falls in the sub-threshold regime of operation of the circuits. While sub-threshold operation helps in decreasing the overall power and energy consumed, there are several challenges involved in designing circuits suitable for sub-threshold operation. First, the circuits are very sensitive to process variations as the delay is exponentially dependent on the operating voltage. Second, robust operation of memory circuits is particularly challenging across process corners. Furthermore, the optimum energy point is sensitive to operating conditions such as temperature, load, and data dependencies, thereby requiring a control circuit to track the MEP as it changes. This chapter talks about a robust design methodology for sub-threshold operation that re-duces energy dissipation of digital circuits, in exchange for slower per-formance, and about designing memory cells that can work at ultra-low voltages. The chapter also talks about a feedback circuit which includes the appropriate power conversion circuitry necessary to operate digital cir-cuits at the minimum energy point.

5.1 Logic Design for U-DVS Systems

In order to adapt to widely varying performance constraints in an energy-efficient manner, logic circuits must be voltage scalable from the above-threshold to the sub-threshold regime. During strong inversion operation, logic circuits can trade off energy consumption to meet performance tar-gets. In sub-threshold, however, circuits display heightened sensitivity to process variation, particularly in the threshold voltage, which can ad-versely affect functionality. Figure 5.2 illustrates the effect of global and local process variation on active currents in a 65nm process, where the relative NMOS and PMOS strengths may be significantly skewed. The spread of the distributions, or the standard deviation normalized by the mean, is an order of magnitude higher in sub-threshold. Furthermore, de-vice "on" currents become comparable in magnitude to the "off" currents such that static CMOS logic structures behave as ratioed circuits [7]. Con-sequently, robustness at the low-voltage corner is the primary design con-sideration for logic circuits in U-DVS systems. This section will discuss statistical techniques for designing logic circuits to function in sub-threshold.

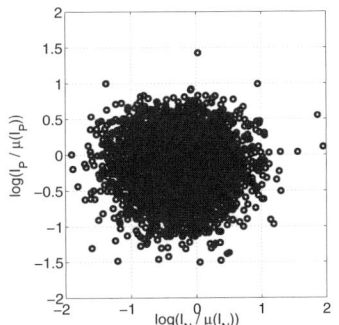

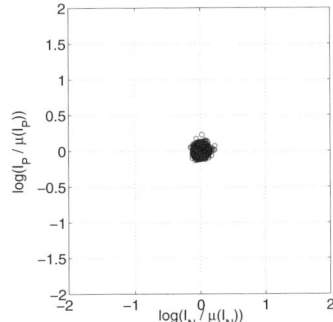

Figure 5.2a Normalized active current distribution at $V_{DD} = 0.3V$.

Figure 5.2b Normalized active current distribution at $V_{DD} = 1.2V$.

5.1.1 Device Sizing

Process variation affects functionality of a logic gate by shifting its voltage transfer characteristic (VTC). In this context, the worst-case variation causes the NMOS to be much weaker than PMOS, or vice versa, thereby degrading output levels of the logic gate. Random local variation can be reduced by increasing the device channel area [8] at the expense of higher energy consumption. To address this trade-off, devices should be upsized only as necessary to achieve the desired functional yield.

The butterfly plot is useful in modeling the effect of variation on proper logic operation [9]. This plot is formed by simulating two logic gates back to back and therefore corresponds to superimposing the VTC of one gate on the inverted VTC of the other. As shown in Figure 5.3a, a plot with two bi-stable points and one meta-stable point implies that the logic structure can support high and low voltage levels. However, V_t variation can be modeled as series noise sources, which in the worst case have opposite polarities. Now, the VTCs in Figure 5.3b have only a mono-stable point, which implies such severe V_t variation that a logic path formed from the two gates, by unrolling the back-to-back structure, cannot support two stable logic levels. The butterfly plot thus indicates whether logic gates under V_t variation provide proper logic levels for correct functionality.

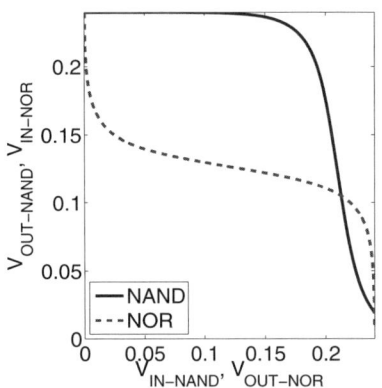
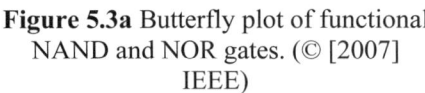
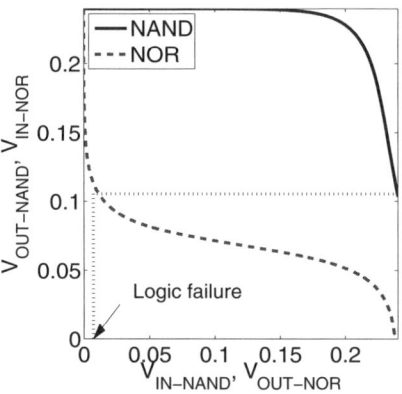

Figure 5.3a Butterfly plot of functional NAND and NOR gates. (© [2007] IEEE)

Figure 5.3b Butterfly plot of gates with failing output levels due to V_t variation. (© [2007] IEEE)

Defining a logic failure as having a mono-stable point in the butterfly plot, logic gates can be designed to achieve a desired functional yield under process variation. Figures 5.4a and 5.4b plot the failure rate of an inverter from Monte Carlo simulations, where global and local process parameters are varied such that the Monte Carlo runs are analogous to sampling inverters across multiple chips. It is important to note that the failure rate decreases exponentially as V_{DD} or device width is increased. The same trends are observed in other logic primitives such as a stack of two NMOS devices [9]. At a given V_{DD}, this analysis provides the minimum device sizing constraints necessary for logic gates to meet the target functional yield.

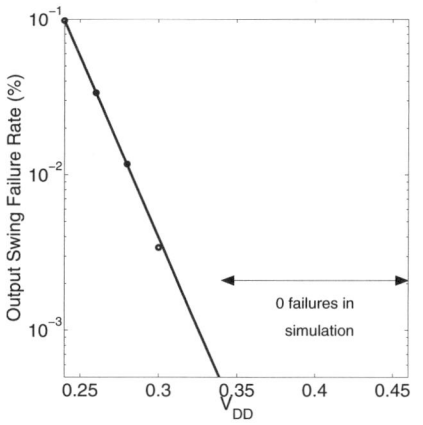

Figure 5.4a Failure rate versus V_{DD} of an inverter under global and local process variations.

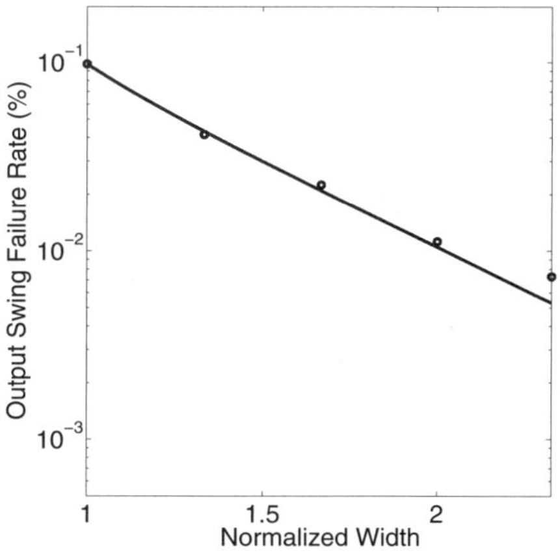

Figure 5.4b Failure rate versus device width of an inverter under global and local process variations.

Register operation in sub-threshold is similarly susceptible to reduced logic levels. This section considers design issues in the classic multiplexer-based transmission gate register. Its ability to retain data is measured by the hold static-noise margin (SNM) of the master and slave latches. The hold SNM is characterized by finding the butterfly plot of the equivalent circuit shown in Figure 5.5, taking into account the voltage drop across TG_2 and worst-case leakage across TG_1. As with logic design, sizing constraints for proper data retention can be found by observing the failure rate due to negative hold SNM in Monte Carlo simulations.

Local variation may also adversely impact the transient behavior of registers, imposing further design considerations. One particular failure mechanism occurs when V_t mismatch in the input data buffer I_1 (Figure 5.5) produces a reduced output swing at node T1 during the low phase of CLK, resulting in degraded signal levels at nodes NT1 and T2. Consequently, the master latch does not settle to the correct state after the rising edge of CLK. Improper functionality is also possible when the local clock buffers do not produce a clock signal of sufficient swing, preventing transmission gates from turning completely off, thus impeding signal propagation. Transient simulations accounting for process variation will reveal the extent to which these effects limit the robustness of a particular register design.

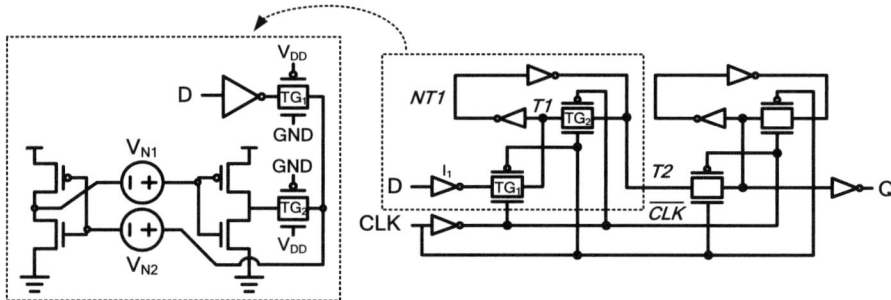

Figure 5.5 Multiplexer-based transmission gate register, with equivalent circuit for verifying hold SNM in sub-threshold shown on the left.

5.1.2 Timing Analysis

With heightened variation in device currents, delay uncertainty correspondingly increases in sub-threshold, which must be considered in circuit timing analysis. Figure 5.6 characterizes delay variation through a uniformly sized NAND-NOR chain, plotting equal σ/μ variability contours as device sizes and logic depth are varied. An increase in either parameter reduces delay variability, which suggests that long timing paths, latch-based designs, and a minimum sizing constraint for clock buffers can improve timing robustness. Importantly, the bottom and left edges of the plot show diminishing returns, implying that a small increase in one parameter can be traded off for a large decrease in the other.

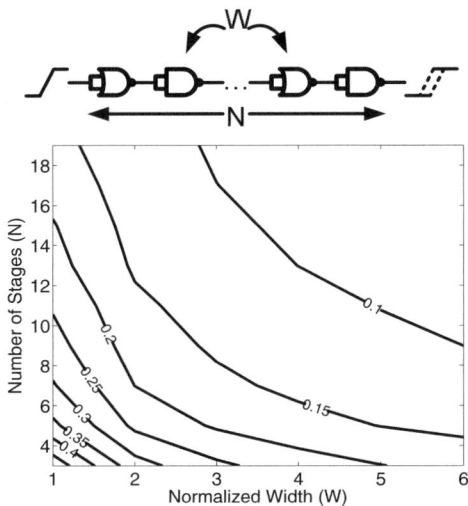

Figure 5.6 Equal σ/μ variability contours of NAND-NOR chain. (© [2007] IEEE)

Given the wide delay distributions in sub-threshold, traditional static timing analysis tools, which only use points at the tails of the distributions to verify timing, will provide unrealistic results. Instead, several block-based, path-based, and parameter space approaches, as discussed in [10], propagate delay distributions through a circuit for better accuracy. Reference [11] focuses specifically on the lognormal distributions seen in sub-threshold, deriving analytical models of their sum and maximum. Similar variation-aware analysis techniques are necessary in designing U-DVS logic circuits with minimal energy overhead.

5.2 SRAM Design for Ultra-scalable Supply Voltages

Although dynamic voltage scaling is extremely valuable for power management, ensuring stable SRAM operation has become so difficult that modern designs often incorporate separate static, full-voltage supplies to bias the memory arrays [12]. In emerging portable applications, however, SRAMs are occupying a dominating portion of the total power and cannot be excused. This is particularly true since, in addition to affording CV_{DD}^{2} savings, voltage scaling alleviates drain-induced barrier lowering and, thus, significantly reduces the total leakage current: an important component of power consumption in SRAMs. It has been shown, for instance, that reducing the supply voltage from 1V to 0.35V in a 65nm CMOS design reduces the total leakage power by over 20× [13]. Of course, the need to achieve

ultra-scalable voltage operation in SRAMs does not preclude the requirement of maximum density. In fact, the increase in the quantity and complexity of features being integrated in portable devices stresses density as much
as energy efficiency. Accordingly, for these designs, minimizing bit-cell
area and maximizing array efficiency, with respect to peripheral circuits,
remain paramount design concerns. Specifically, this implies that constituent
devices in the bit-cell must be kept small, and, where possible, read, write,
and voltage adaptability assists should employ area-efficient peripheral techniques. Finally, to maintain array efficiency, it is desirable to integrate a
maximum number of bit-cells in each column and row.

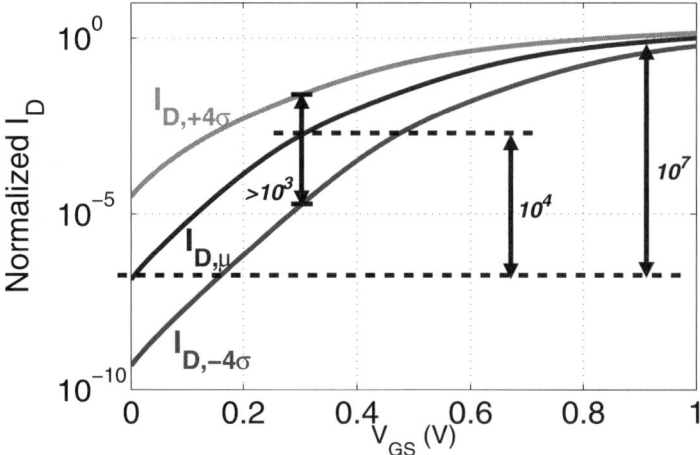

Figure 5.7 I_D versus V_{GS} behavior of a 65nm MOSFET showing increased variation and reduced I_{ON}/I_{OFF} at low voltages. (© [2007] IEEE)

Fundamental device characteristics critical to SRAMs are degraded by
several orders of magnitude at reduced voltages. Accordingly, the primary
challenge of ultra-dynamic voltage scaling in SRAMs is achieving low-
voltage, sub-threshold operation. Figure 5.7 shows the I_D versus V_{GS} characteristic of a MOSFET (in a 65nm technology) and elucidates two critical
effects that oppose cell area scaling and array integration efficiency at low
voltages. First, at 0.3V, threshold voltage variation, commonly observed at
+/–4σ in array sizes of interest, results in over three orders of magnitude
change in I_D. Increasing device sizes reduces the variation, but this level of
severity implies that, generally, device strengths cannot be set reliably, as
has been required in traditional bit-cell design. Second, at 0.3V, the on-to-
off ratio of the current is nominally just 10^4, whereas at higher voltages it

is 10^7. Consequently, both "on" and "off" devices figure prominently in setting the voltage level of shared nodes.

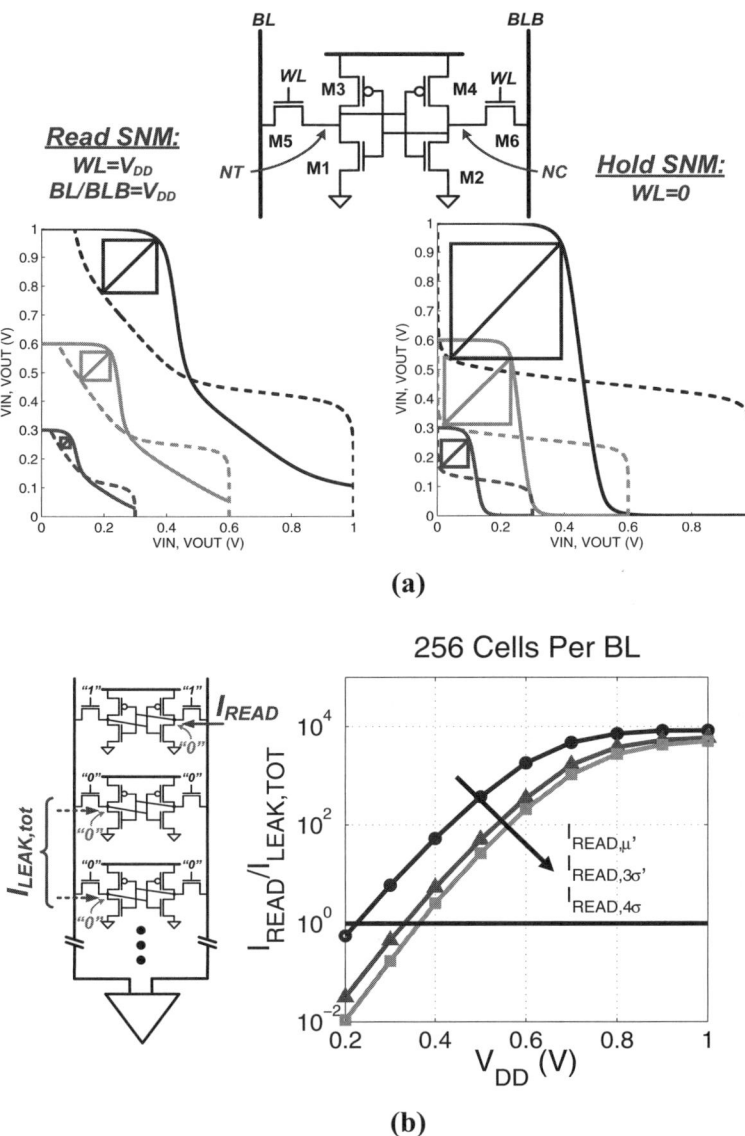

(a)

(b)

Figure 5.8 Conventional SRAM (a) static-noise margin and (b) bit-line leakage with respect to supply voltage. (© [2007] IEEE)

Relating these effects to SRAMs, variation in the 6T cell of Figure 5.8a can skew the relative strength of the pull-down devices, M1/M2, which must be stronger than the access devices, M5/M6, for correct read operation. The transfer curves from NT–NC and NC–NT are shown for various V_{DD}'s; in all cases, they nominally intersect at two stable points near V_{DD} and ground, representing the storable data states, as well as one metastable point at mid-V_{DD}. However, if variation is severe enough to skew both transfer curves by an amount equal to the edge length of the largest embedded square, called the static-noise margin (SNM), one of the required storage states is lost [14]. While the read SNM is precariously degraded at low voltages, Figure 5.8a shows that the hold SNM, which considers the case where the word-line (WL) is low, can be more easily retained. Similarly, the reduced on-to-off ratio of the device currents at low voltages has the problematic effect shown in Figure 5.8b, where the leakage currents from the unaccessed cells sharing the bit-lines can exceed the read-current from the accessed cell. As a result, the droop on the two bit-lines is indistinguishable. The following sections describe circuit techniques to address these limitations.

5.2.1 Low-Voltage Bit-Cell Design

As described above, low-voltage operation requires an improvement in both read SNM, to avoid bit flipping, and read-current, to avoid sensing failures due to bit-line leakage. Unfortunately, however, the 6T bit-cell, shown in Figure 5.8a, imposes an inherent trade-off between these two. This comes about as a result of the access devices, M5/M6, which should be weak for good read SNM but strong for good read-current. Of course, the pull-down devices can be strengthened; however, soft gate-oxide breakdown effects in these devices oppose an improvement in the read SNM [15, 16], and the area increase required to manage variation is overwhelming.

Alternatively, the 8T bit-cell shown in Figure 5.9 uses a read-buffer (M7/M8) to break the trade-off between read SNM and read-current. Of course, the addition of extra devices can result in reduced density; however, the resulting structure can be free of the read SNM limitation, and its minimum operating voltage can be set by the hold SNM, which, as mentioned, is preserved to very low voltages.

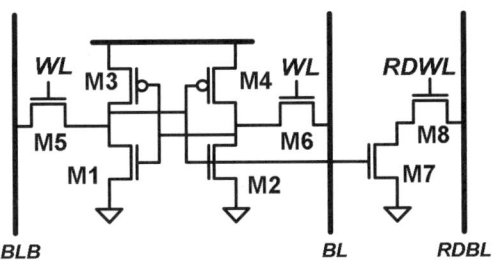

Figure 5.9 8T bit-cell with a 2 transistor read-buffer formed by M7/M8.
(© [2007] IEEE)

Lastly, for an ultra-dynamic voltage scaling design, it is important to note that the trade-off between cell area and read-current/read SNM changes dramatically with operating voltage. Specifically, Figure 5.10 shows the improvement in 4σ read-current at low voltages as a result of read-buffer upsizing. Consequently, as the performance of reduced voltage modes in an application becomes more critical, device upsizing has enhanced appeal.

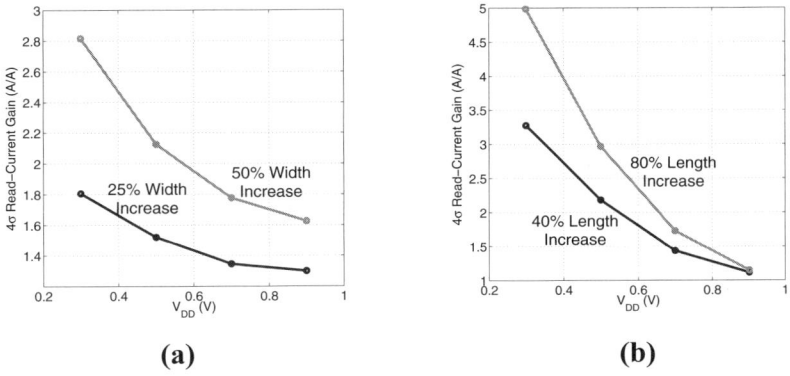

Figure 5.10 4-σ read-current gain due (a) width upsizing and (b) length upsizing of read-buffer devices. (© [2007] IEEE)

5.2.2 Periphery Design

Since the trade-off between read-current and read SNM is built into the 6T cell as a result of the access devices, the bit-cell itself must be modified to simultaneously address those limitations at low operating voltages. Most other limitations, however, can be addressed using peripheral or architectural assists that impose minimal density penalty.

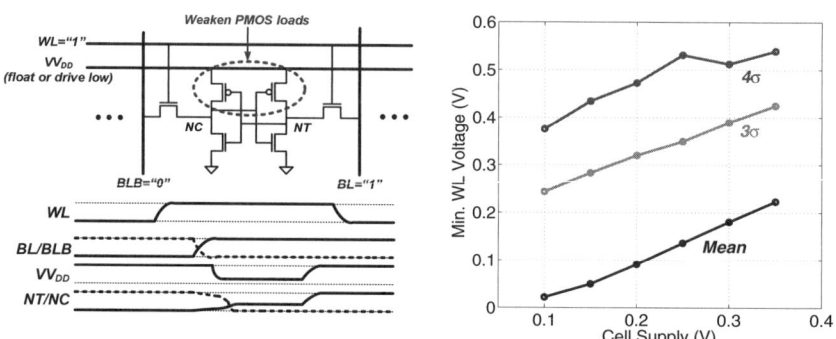

Figure 5.11 Reducing cell supply eases strength requirement of access devices, as reflected by reduction in minimum word-line voltage required for successful write. (© [2007] IEEE)

For instance, enhanced error correction coding (ECC) is required in order to take full advantage of the 8T cell's wider operating margin (i.e., hold SNM instead of read SNM). Soft-errors exhibit spatial locality, so SRAMs conventionally employ column-interleaved layouts to avoid multi-bit errors in logical words. During write operations, some cells are row selected but not column selected (commonly called half-accessed cells), and, consequently, they must be read SNM stable. Alternatively, in non-interleaved layouts [13], only cells from the addressed word need to be selected, and no read SNM limitation exists. However, since bits from a logical word are adjacent, additional ECC complexity is required to tolerate multi-bit soft-errors [17].

An additional difficulty during write operations arises from device variation increasing the strength of the pull-up devices, which must be overcome by the access devices in order to ensure successful write. However, the required relative strengths can be enforced; for example, the word-line voltage can be boosted above V_{DD}, or the appropriate bit-line voltage can be pulled below ground to strengthen the access devices. Unfortunately, both of these strategies involve the complexity of driving a large capacitance beyond one of the rail voltages. Instead, the bit-cell supply voltage can be floated [18] or driven low [13] to weaken the pull-up PMOS load devices. Figure 5.11 shows that as the cell supply, VV_{DD}, is reduced, the strength requirement of the access device during a write operation is reduced, which is represented by a decrease in the minimum word-line voltage that still results in a successful write.

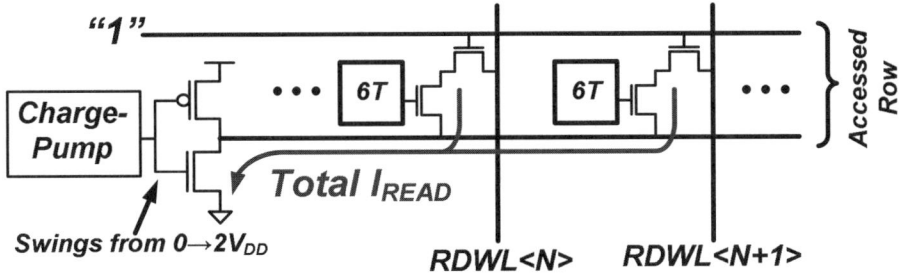

Figure 5.12 Read-buffer foot-driver limitation can be alleviated in sub-V_t designs by driving the peripheral footer with a charge-pump circuit. (© [2007] IEEE)

Finally, the problematic sub-threshold leakage currents from the unaccessed cells that result in excessive bit-line leakage can be eliminated by pulling the foot of the 8T cell read-buffer up to V_{DD}. Of course this imposes a severe current drive requirement on the peripheral foot driver shown in Figure 5.12, since, when accessed, it must sink the read-current from all cells in the row. For sub-threshold supply voltages, the peripheral footer can be driven with a charge-pump circuit, resulting in an exponential increase in its drive strength [13]. This technique, however, does not scale well to higher voltages in a U-DVS system. Nonetheless, despite the overhead, footer upsizing is a practical solution in this case since the cell read-current is dominantly limited by the bit-cells themselves which face up to 5σ degradation. The foot driver can be much larger, thereby suffering much less degradation from variation, and since it is in the periphery, only 2 or 3σ degradation must be attributed.

5.3 Intelligent Power Delivery

5.3.1 Deriving V_{DD} for Given Speed Requirement

To effectively use DVS to reduce power consumption, a system controller that determines the required operating speed of the processor at run-time is needed. The system controller makes use of algorithms, termed *voltage schedulers,* to determine the operating speed of the processor at run-time. For general-purpose processors, these algorithms effectively determine the overall workload of the processor and suggest the required operating speed to handle the user requests. Some of the commonly used algorithms have been described in [19]. For DSP systems like video processors, the speed of the system is typically measured by looking at the buffer length occupied. Once this operating speed has been determined, the operating voltage of the circuit needs to be changed so that it can meet the required speed of operation.

The simplest way to change the rate of the processor is to let it operate at full speed for a fraction of the time and to then shut it down completely. The fixed power supply curve in Figure 5.1a shows the linear energy savings that can be obtained by this process. A variable supply voltage on the other hand can provide with super-linear savings in energy consumed. The curve with infinite allowable levels provides the optimum curve for reducing energy. The change in supply voltage can be achieved through several means. Supply voltage dithering, which uses discrete voltage and frequency pairs, was proposed as a solution to achieve DVS [1]. Local voltage dithering (LVD) [20] improves on existing voltage dithering systems by taking advantage of faster changes in workload and by allowing each block to optimize based on its own workload. While dithering can provide close to the optimal savings in energy consumed, it requires an efficient system controller that can time-share between the different voltage levels adding to the overall complexity of the system. This is of specific concern in ultra-low-power applications. Also, voltage dithered systems that achieve U-DVS require at least two voltage levels different from the battery voltage to achieve the stated power savings. This increases the number of DC–DC converters to supply these voltage levels.

Having a DC–DC converter that can supply scalable voltages as demanded by the system it is catering to can be of great advantage in terms of both simplicity of the overall solution and cost. This requires a DC–DC converter that can firstly deliver variable load voltages. A suitable control strategy is needed to change the load voltage supplied by the DC–DC converter to maintain the operating speed. Reference [21] presents a closed loop architecture to change the output voltage of a voltage scalable DC–DC converter to make the load circuit operate at the desired rate. Reference [1] uses a hybrid approach employing both look-up tables and a phase-locked loop (PLL) to enable fast transitions in load voltage with change in the desired rate. While the look-up table aids in the fast transition, the PLL helps in tracking process variations and operating conditions. Both these approaches use switching regulators with off-chip inductors. The next section talks about some of the commonly used topologies for U-DVS DC–DC converters.

5.3.2 DC–DC Converter Topologies for U-DVS

5.3.2.1 Linear Regulators

Low-dropout (LDO) linear regulators [22] are widely used to supply analog and digital circuits and feature in several standalone or embedded
power management ICs. The main advantage of LDO's is that they can be
completely on-chip, occupy very little area, and offer good transient and
ripple characteristics, together with being a low-cost solution. Using
LDO's for U-DVS, however, is detrimental because of the linear loss of
efficiency in an LDO. A linear regulator essentially controls the resistance
of a transistor in order to regulate the output voltage. As a result, the current delivered to the load flows directly from the battery and hence the
maximum efficiency achievable is limited to the ratio of the output voltage
to the input voltage. Thus, the farther away the load voltage is from the
battery voltage, the lower the efficiency of the LDO. This hampers the potential savings in power consumption that can be achieved by lowering the
voltage through DVS.

5.3.2.2 Inductor-Based DC–DC Converter

The most efficient DC–DC voltage converters are inductor-based switching regulators, which normally generate a reduced DC voltage level by filtering a pulse-width modulated (PWM) signal through a simple LC filter.
A buck-type regulator can generate different DC voltage levels by varying
the duty-cycle of the PWM signal. Given ideal devices and passives, an
inductor-based DC–DC converter can theoretically achieve 100% efficiency independent of the load voltage being delivered. Moreover, in the
context of DVS systems, scaling the output voltage can be done with completely digital control circuitry [21] which consumes very little overhead
power. An implementation of an inductor-based switching regulator for
minimum energy operation is described in Section 5.3.3.1C. While buck
converters [23] can operate at very high efficiencies (>90%), they generally require off-chip filter components. This might limit their usefulness
for integrated power converter applications. Integrating the filter inductor
on-chip requires very high switching frequencies (>100MHz) in order to
minimize area consumed. This increases the switching losses in the converter and together with the increase in conduction losses due to the low
inductor Q-factors achievable on-chip severely affects the efficiency that
can be obtained out of the converter.

5.3.2.3 Switched Capacitor-Based DC–DC Converter

U-DVS systems often require multiple on-chip voltage domains with each domain having specific power requirements. A switched capacitor (SC) DC–DC converter is a good choice for such battery-operated systems because it can minimize the number of off-chip components and does not require any inductors. Previous implementations of SC converters (charge pumps) have commonly used off-chip charge-transfer capacitors [24] to output high load power levels. A SC DC–DC converter which integrates the charge-transfer capacitors was described in [25].

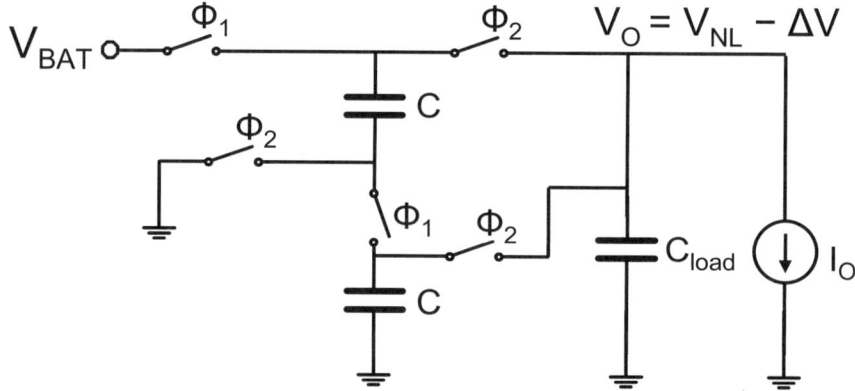

Figure 5.13 A switched capacitor voltage divide-by-2 circuit.

Consider the divide-by-2 circuit shown in Figure 5.13. The charge-transfer (flying) capacitors are equal in value and help in transferring charge from the battery to the load. During phase Φ_1 of the system clock, the charge-transfer capacitors get charged from the battery (V_{BAT}). In the Φ_2 phase of the clock, they dump the charge gained onto the load. At no load, this circuit tries to maintain the output voltage V_O at $V_{BAT}/2$, where V_{BAT} is the battery voltage. The actual value of V_O that the circuit settles down to is dependent on the load current I_O, the switching frequency, and C. Let the circuit deliver a load voltage $V_O = V_{NL} - \Delta V$, where V_{NL} is the no-load voltage for this topology. The SC converter limits the maximum efficiency that can be achieved in this case to $\eta_{lin} = (1 - \Delta V/V_{NL})$. Thus, the farther away V_O is from V_{NL} (i.e., higher ΔV), the smaller the maximum efficiency that can be achieved by this topology. This is a fundamental problem with charge transfer using only capacitors and switches. The linear efficiency loss is similar to linear regulators. However, with SC converters, it is possible to switch in different gain-settings whose no-load output voltage is closer to the load voltage desired. Apart from the linear conduction loss, losses due to bottom-plate parasitics of on-chip capacitors and switching losses limit the efficiency of the SC DC–DC converter [26]. The efficiency achievable in a switched capacitor system is in general smaller than that can be achieved in an inductor-based switching regulator with off-chip passives. Furthermore, multiple gain-settings and associated control circuitry are required in a SC DC–DC converter to maintain efficiency over a wide voltage range. However, for on-chip DC–DC converters, a SC solution might be a better choice, when the trade-offs relating to area and efficiency are considered. Furthermore, the area occupied by the switched capacitor DC–DC converter is scalable with the load power demand, and hence the switched capacitor DC–DC converter is a good solution for low-power on-chip applications.

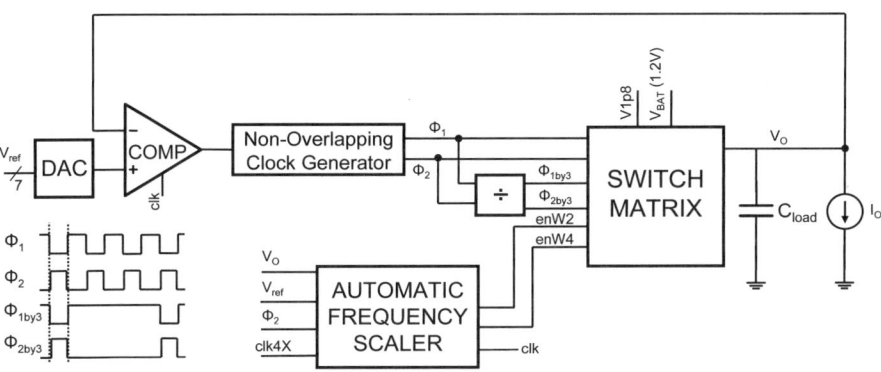

Figure 5.14 Architecture of a switched capacitor DC–DC converter with on-chip charge-transfer capacitors. (© [2007] IEEE)

A SC DC–DC converter that employs five different gain-settings with ratios 1:1, 3:4, 2:3, 1:2, and 1:3, is described in [26]. The switchable gain-settings help the converter to maintain a good efficiency as the load voltage delivered varies from 300mV to 1.1V. Figure 5.14 shows the architecture of the SC DC–DC converter. At the core of the system is the switch matrix which contains the charge-transfer capacitors and the charge-transfer switches. A suitable gain-setting is chosen depending on the reference voltage V_{ref}, which is set digitally. A pulse frequency modulation (PFM) mode control is used to regulate the output voltage to the desired value. Bottom-plate parasitics of the on-chip capacitors significantly affect the efficiency of the converter. A divide-by-3 switching scheme [26] was employed to mitigate the effect due to bottom-plate parasitics and improve efficiency. The switching losses are scaled with change in load power by the help of the automatic frequency scaler block. This block changes the switching frequency as the load power delivered changes, thereby reducing the switching losses at low load.

The efficiency of the SC converter with change in load voltage while delivering 100μW to the load from a 1.2V supply is shown in Figure 5.15. The converter was able to achieve >70% efficiency over a wide range of load voltages. An increase in efficiency of close to 5% can be achieved by using divide-by-3 switching.

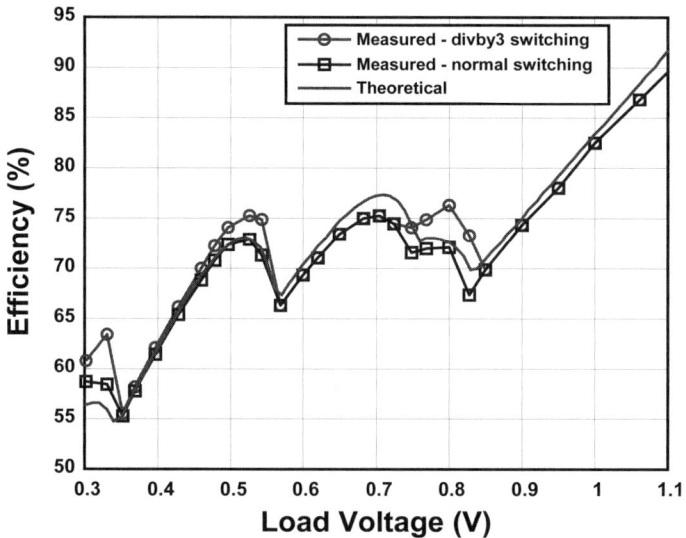

Figure 5.15 Efficiency of the switched capacitor DC–DC converter with change in load voltage. (© [2007] IEEE)

5.3.3 DC–DC Converter Design and Reference Voltage Selection for Highly Energy-Constrained Applications

While dynamic voltage scaling is a popular method to minimize power consumption in digital circuits given a performance constraint, the same circuits are not always constrained to their performance-intensive mode during regular operation. There are long spans of time when the performance requirement is highly relaxed. There are also certain emerging energy-constrained applications where minimizing the energy required to complete operations is the main concern. For both these scenarios, operating at the minimum energy operating voltage of digital circuits has been proposed as a solution to minimize energy. The minimum energy point

(MEP) is defined as the operating voltage at which the total energy consumed per desired operation of a digital circuit is minimized. Switching energy of digital circuits reduces quadratically as V_{DD} is decreased below V_T (i.e., sub-threshold operation), while the leakage energy increases exponentially. These opposing trends result in the minimum energy point. The MEP is not a fixed voltage for a given circuit and can vary widely depending on its workload and environmental conditions (e.g., temperature). Any relative increase in the active energy component of the circuit due to an increase in the workload or activity of the circuit decreases the minimum energy operating voltage. On the other hand, a relative increase of the leakage energy component due to an increase in temperature or the duration of leakage over an operation pushes the minimum energy operating voltage to go up. This makes the circuit go faster, thereby not allowing the circuit to leak for a longer time. By tracking the MEP as it varies, energy savings of 50–100% has been demonstrated [27] and even greater savings can be achieved in circuits dominated by leakage. This motivates the design of a minimum energy tracking loop that can dynamically adjust the operating voltage of arbitrary digital circuits to their MEP.

5.3.3.1 Minimum Energy Tracking Loop

Figure 5.16 shows the architecture of the minimum energy tracking loop. The objective of this loop is to track the minimum energy operating voltage of the load circuit. The load circuit (FIR filter) is powered from an off-chip voltage source through a DC–DC converter and is clocked by a

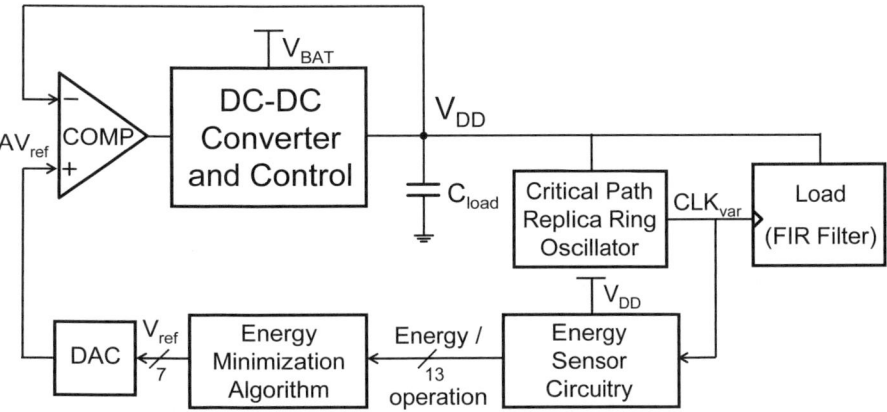

Figure 5.16 Architecture of the minimum energy tracking loop. (© [2007] IEEE)

critical path replica ring oscillator which automatically scales the clock frequency of the FIR filter with change in load voltage. The energy sensor circuitry calculates on chip, the energy consumed per operation of the load circuit at a particular operating voltage. It then passes the estimate of the energy/operation (E_{op}) to the energy minimizing algorithm, which uses the E_{op} to suitably adjust the reference voltage to the DC–DC converter. The DC–DC converter then tries to get V_{DD} close to the new reference voltage, and the cycle repeats till the minimum energy point is achieved. The only off-chip components of this entire loop are the filter passives of the inductor-based switching DC–DC converter.

A. Energy Sensing Technique

The key element in the minimum energy tracking loop is the energy sensor circuit which computes the E_{op} of the load circuit at a given reference voltage. Methods to measure E_{op}, by sensing the current flow through the DC–DC converter's inductor [28], dissipate a significant amount of overhead power. The approach is more complicated at sub-threshold voltages because the current levels are very low. Furthermore, an estimate of the energy consumed per operation is what is required and not just the current which only gives an idea of the load power. The methodology used here, to estimate E_{op}, does not require any high-gain amplifiers or analog circuit blocks.

The DC–DC converter while operating in steady state keeps the output voltage close to the reference voltage. Just before the energy sense cycle begins, the DC–DC converter is disabled. The energy sense cycle consists of N operations of the digital circuit where the value N can be 32 or 64. Assuming that the voltage across the storage capacitor of the DC–DC converter, C_{load}, falls from the reference voltage V_1 to V_2 in the course of N operations of the digital circuit, E_{op} at the voltage V_1 is equal to

$$E_{op} = \frac{C_{load}\left(V_1^2 - V_2^2\right)}{2N} \tag{5.1}$$

To measure E_{op} accurately, V_2 should be close in value (within 20mV) to V_1. Measuring E_{op} by digitizing V_1 and V_2 using conventional ADCs would require at least 11 bits of precision in the ADC. This could prove costly in terms of power consumed. An energy-efficient approach to obtain E_{op} is to observe that, by design, V_1 is very close to V_2. Thus, the following simplification can be applied within an acceptable error:

$$E_{op} = \frac{C_{load}\ (V_1 + V_2)(V_1 - V_2)}{2N} \approx \frac{C_{load}\ V_1(V_1 - V_2)}{N} \qquad (5.2)$$

$$E_{op} \propto V_1(V_1 - V_2) \qquad (5.3)$$

From Equation (5.3), it can be seen that the energy consumed per operation is directly proportional to the product of V_1 and $V_1 - V_2$. Since, the digital representation of V_1, which is the reference voltage to the DC–DC converter, is already known, only the digital value for the voltage difference $(V_1 - V_2)$ is required to estimate E_{op}. This voltage difference is obtained digitally using a fixed frequency clock, a constant current sink, a comparator, and a counter [27]. These blocks help in quantizing voltage into time steps, as in an integrating ADC [29]. The number of fixed frequency clock cycles obtained from the counter is directly proportional to $V_1 - V_2$. This quantity is then digitally multiplied with V_1 which is the reference voltage V_{ref} to the DC–DC converter. The product of these two quantities gives an estimate of the energy consumed per operation by the digital circuit at voltage V_1. The estimate obtained is a normalized representation of the absolute value of the energy consumed per operation. This estimate is passed on to the energy minimization algorithm block.

B. Energy Minimization Algorithm

Once the estimate of the energy per operation is obtained, the minimum energy tracking algorithm uses this to suitably adjust the reference voltage to the DC–DC converter. The minimum energy tracking algorithm is a slope-tracking algorithm which makes use of the single minimum, concave nature of the E_{op} versus V_{DD} curve (see Figure 5.1b). The algorithm starts by setting the reference voltage V_{ref} to some initial value. The energy per operation at this voltage is computed and stored in a minimum energy register ($E_{op,min}$). The tracking loop then automatically increments V_{ref} by one voltage step. Once V_{DD} settles at this newly incremented voltage, E_{op} is computed again and is compared with the value stored in the minimum energy register. At this point, if the newly computed E_{op} is found to be smaller, the loop then just keeps incrementing V_{ref} at fixed voltage steps, while at the same time updating $E_{op,min}$ till the minimum is achieved. The other possibility is that the newly computed energy per operation is higher than that stored in the minimum energy register. In this case, the loop changes direction and begins to decrement V_{ref}. The loop keeps decrementing V_{ref} till the E_{op} calculated is higher than $E_{op,\ min}$ at which time the loop increments V_{ref} by one voltage step to get to the MEP and shuts down. Figure 5.17 shows the minimum energy tracking loop in operation for a 7-tap FIR filter load circuit.

The voltage step used by the tracking algorithm is usually set to 50mV A large voltage step leads to coarse tracking of the MEP, with the possibility of missing the MEP. On the other hand, keeping the voltage step too small might lead to the loop settling at the non-minimum voltage due to errors involved in computing E_{op} [30]. The E_{op} versus V_{DD} curve is shallow near the MEP, and hence a 50mV step leads to a very close approximation of the actual minimum energy consumed per operation. The MEP tracking loop can be enabled by a system controller as needed depending on the application, or periodically by a timer to track temperature variations.

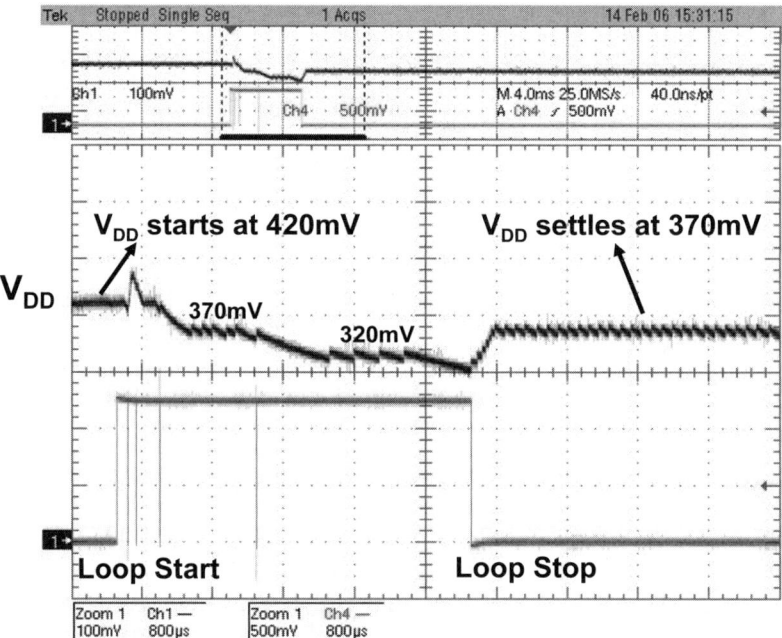

Figure 5.17 Measured waveform showing the minimum energy tracking loop in operation. (© [2007] IEEE)

C. Embedded DC–DC Converter for Minimum Energy Operation

This section talks about the design of the DC–DC converter that enables minimum energy operation. Since the minimum energy operating voltage usually falls in the sub-threshold regime of operation, the DC–DC converter is designed to deliver load voltages from 250mV to around 700mV. The power consumed by digital circuits at these sub-threshold voltages is exponentially smaller and hence the DC–DC converter needs to deliver efficiently load power levels of the order of micro-watts. This demands extremely simple control circuitry design with minimal overhead power to get good efficiency. The DC–DC converter shown in Figure 5.18 is a synchronous rectifier buck converter with off-chip filter components and operates in the discontinuous conduction mode (DCM). It employs a pulse frequency modulation (PFM) [31] mode of control in order to get good efficiency at the ultra-low load power levels that the converter needs to deliver. The PFM mode control also helps in seamlessly disabling the converter when energy sensing takes place, thereby making it feasible to use the energy-sensing technique described in Section 5.3.3.1A.

The reference voltage to the converter is set digitally by the minimum energy tracking loop and is converted to an analog value by an on-chip DAC before it is fed to the comparator. The comparator compares V_{DD} with this reference voltage and when V_{DD} is found to be smaller generates a pulse of fixed width to turn the PMOS power transistor ON and ramp up the inductor current. A variable pulse-width generator to achieve zero-current switching is used for the NMOS power transistor. The comparator is clocked by a divided and level-converted version of the system clock which feeds the load FIR filter.

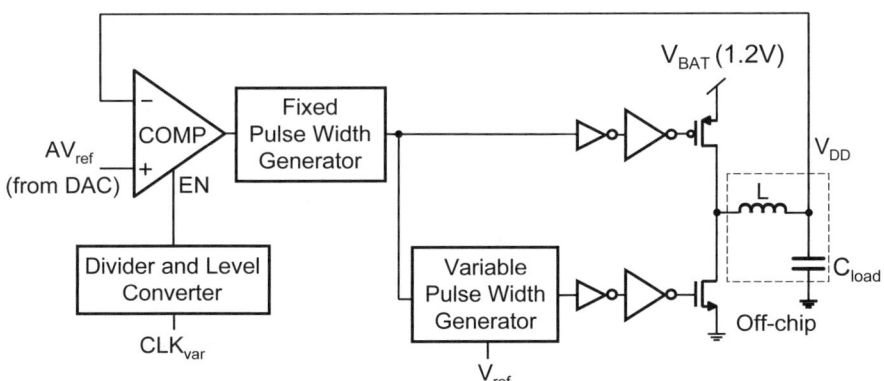

Figure 5.18 DC–DC Converter architecture. (© [2007] IEEE)

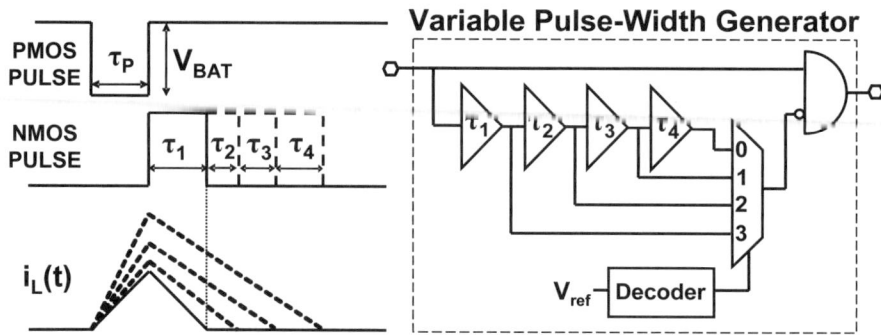

Figure 5.19 Approximate zero-current switching block. (© [2007] IEEE)

The ultra-low load power levels demand extremely simple control circuitry to achieve good efficiency. This precludes the usage of high-gain amplifiers to detect zero-crossing and thereby do zero-current switching [31]. In order to keep the control circuitry simple and consume little overhead power, an all-digital open-loop control as shown in Figure 5.19 is used to achieve zero-current switching. The variable pulse-width generator block which accomplishes this functions as follows: When the comparator senses that V_{DD} has fallen below the reference voltage, a PMOS ON pulse of fixed pulse width τ_P is generated. This ramps up the inductor current from zero. Once the PMOS is turned OFF, the NMOS power transistor is turned ON after a fixed delay. This ramps down the inductor current. Ideally, in the discontinuous conduction mode (DCM) used in this implementation, the NMOS has to be turned OFF just when the inductor current reaches zero. The amount of time it takes for the inductor current to reach zero is dependent on the reference voltage set, and in steady state, the ratio of the NMOS to PMOS ON-times is given by the following equation:

$$\frac{\tau_N}{\tau_P} = \frac{V_{BAT} - V_{DD}}{V_{DD}} \tag{5.4}$$

where τ_N and τ_P are the NMOS and PMOS ON-times and V_{BAT} is the battery voltage. Thus, by fixing τ_P, the values of τ_N for specific load voltages can be predetermined. The variable pulse-width generator block then suitably multiplexes these predetermined delays depending on the reference voltage set to achieve approximate zero-current switching. Increasing the number of these delay elements and the complexity of the multiplexer block gives a better approximation to zero-current switching. Since only the ratios of the NMOS and PMOS ON-time pulse widths need to match, this scheme is independent of absolute delay values and any tolerance in the inductor value. Furthermore, it consumes very little overhead power.

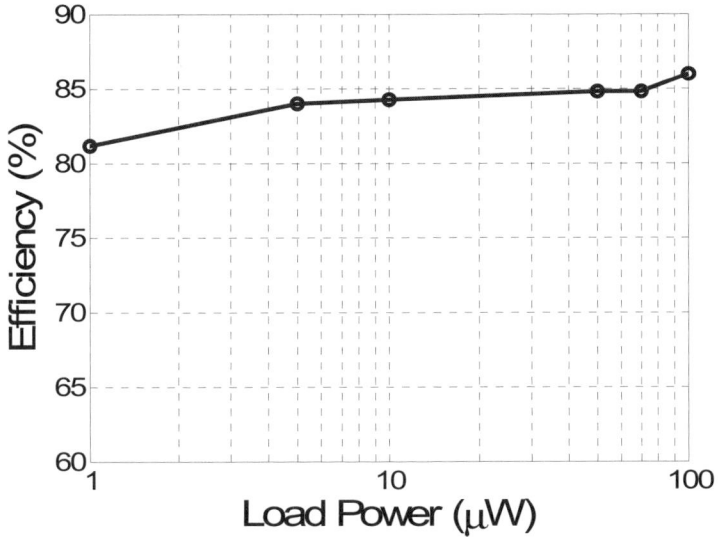

Figure 5.20 Efficiency of the inductor-based switching regulator embedded within the minimum energy tracking loop. (© [2007] IEEE)

With the help of the above-mentioned efficiency improvement techniques, the DC–DC converter was able to achieve an efficiency >80% at an extremely low load power level of 1μW as shown in Figure 5.20. While the switching and conduction losses bring down efficiency at load power levels of 100μW and above, the leakage losses kick in at lower load levels bringing the efficiency further down. The simplicity of the control blocks helps to maintain good efficiency at these ultra-low load power levels.

The proposed minimum energy tracking loop is *non-intrusive*, thereby allowing the load circuit to operate without being shut down. At the same time, it computes the energy per operation of the actual circuit and not of any replica. This eliminates the problems of designing a replica circuit that can track the energy behavior of a load circuit over varying operating conditions. The tracking methodology is independent of the size and type of digital circuit being driven and the topology of the DC–DC converter.

5.4 Conclusion

Dynamic voltage scaling is a popular method to minimize power consumption in digital circuits given a performance constraint. By introducing the capability of sub-threshold operation, DVS systems can be made to operate at their minimum energy operating voltage in periods of very little activity, leading to further savings in total energy consumed. These U-DVS systems provide energy savings by either reducing the supply voltage to just meet performance or operating at the minimum energy operating voltage in periods of very little activity.

The challenges involved in designing logic and memory circuits suitable for sub-threshold operation and the methodology to overcome these challenges have been described in this chapter. Furthermore, a control circuit to track the optimum energy point of digital circuits was presented. The DC–DC converter used within the control loop was designed to provide sub-threshold output voltages at very high efficiencies. The overall design methodology and the control circuit help in saving energy consumed in highly energy-critical applications leading to enhanced battery lifetimes and the ability to operate out of scavenged energy.

References

[1]   V. Gutnik and A. Chandrakasan, "Embedded power supply for low-power DSP," *IEEE Trans. VLSI Syst.*, vol. 5, no. 4, pp. 425–435, Dec. 1997.

[2]   A. Sinha and A. Chandrakasan, "Dynamic power management in wireless sensor networks," *IEEE Design and Test of Computers*, vol. 18, no. 2, pp. 62–74, March 2001.

[3]   B. Zhai *et al.*, "A 2.6pJ/Inst subthreshold sensor processor for optimal energy efficiency," in *Symp. VLSI Circuits Tech. Dig.*, pp. 192–193, June 2006.

[4]   O. Soykan, "Power sources for implantable medical devices," *Medical Device Manufacturing and Technology*, 2002.

[5]   S. Roundy, P. K. Wright, and J. Rabaey, "A study of low level vibrations as a power source for wireless sensor nodes," *Computer Communications*, vol. 26, no. 11, pp. 1131–1144, July 2003.

[6]   A. Wang and A. Chandrakasan, "A 180-mV Sub-threshold FFT processor using a minimum energy design methodology," *IEEE J. Solid-State Circuits*, vol. 40, no. 1, pp. 310–319, Jan. 2005.

[7]   A. Wang, B. H. Calhoun, and A. P. Chandrakasan, "Sub-Threshold Design for Ultra Low-Power Systems," New York, Springer, pp. 75–102, 2006.

[8]   M. J. M. Pelgrom, A. C. J. Duinmaijer, and A. P. G. Welbers, "Matching properties of MOS transistors," *IEEE J. Solid-State Circuits*, vol. 24, no. 5, pp. 1433–1439, Oct. 1989.

[9]   J. Kwong, A. P. Chandrakasan, "Variation-driven device sizing for minimum energy sub-threshold circuits," *IEEE Intl. Symp. on Low Power Electronics and Design*, 2006. pp. 8–13.

[10]   A. Srivastava, D. Sylvester, D. Blaauw, "Statistical analysis and optimization for VLSI: timing and power," New York, Springer, pp. 79–132, 2005.

[11] B. Zhai, S. Hanson, D. Blaauw, and D. Sylvester, "Analysis and mitigation of variability in subthreshold design," *IEEE Intl. Symp. on Low Power Electronics and Design*, pp. 20–25, 2005.

[12] J. Pille et al., "Implementation of the CELL broadband engine in a 65nm SOI technology featuring dual-supply SRAM arrays supporting 6GHz at 1.3V," *IEEE ISSCC Dig. Tech. Papers*, pp. 322–323, Feb. 2007.

[13] N. Verma and A. Chandrakasan, "A 65nm 8T sub-V_t SRAM employing sense-amplifier redundancy," *IEEE ISSCC Dig. Tech. Papers*, pp. 328–329, Feb. 2007.

[14] E. Seevinck, F. List and J. Lohstroh, "Static noise margin analysis of MOS SRAM cells," *IEEE J. Solid-State Circuits*, vol. SC-22, no. 5, pp. 748–754, Oct. 1987.

[15] M. Agostinelli, et al., "Erratic fluctuations of SRAM cache Vmin at the 90nm process technology node," *IEDM Dig. Tech. Papers*, pp. 671–674, Dec. 2005.

[16] R. Rodriguez, et al. "The impact of gate-oxide breakdown on SRAM stability," *IEEE Electron Device Letters*, vol. 23, no. 9, pp. 559–561, Sept. 2002.

[17] L. Chang, et al., "A 5.3GHz 8T-SRAM with operation down to 0.41V in 65nm CMOS," *Symp. VLSI Circuits*, pp. 252–253, June 2007.

[18] B. Calhoun and A. Chandrakasan, "A 256kb sub-threshold SRAM in 65nm CMOS," *IEEE ISSCC Dig. Tech. Papers*, pp. 628–629, Feb. 2006.

[19] T. Pering, T. Burd and R. Brodersen, "The simulation and evaluation of dynamic voltage scaling algorithms," *IEEE Intl. Symp. Low Power Electronics and Design*, pp. 76–81, 1998.

[20] B. H. Calhoun and A. P. Chandrakasan, "Ultra-dynamic voltage scaling using sub-threshold operation and local voltage dithering in 90nm CMOS," *IEEE ISSCC Dig. Tech. Papers*, pp. 300–301, Feb. 2005.

[21] G-Y.Wei and M. Horowitz, "A fully digital, energy-efficient, adaptive power-supply regulator," *IEEE J. Solid-State Circuits,* vol. 34, no. 4, pp. 520–528, Apr. 1999.

[22] P. Hazucha et al., "Area efficient linear regulator with ultra-fast load regulation," *IEEE J. Solid-State Circuits*, vol. 40, no. 4, pp. 933–940, Apr. 2005.

[23] J. Xiao, A. Peterchev, J. Zhang and S. Sanders, "A 4µA-quiescent-current dual-mode buck converter IC for cellular phone applications," *IEEE ISSCC Dig. Tech. Papers*, pp. 280–281, Feb. 2004.

[24] A. Rao, W. McIntyre, U. Moon and G. C. Temes, "Noise-shaping techniques applied to switched capacitor voltage regulators," *IEEE J. Solid-State Circuits*, vol. 40, no. 2, pp. 422–429, Feb. 2005.

[25] G. Patounakis, Y. Li and K. L. Shepard, "A fully integrated on-chip DC–DC conversion and power management system," *IEEE J. Solid-State Circuits,* vol. 39, no. 3, pp. 443–451, Mar. 2004.

[26] Y. K. Ramadass and A. Chandrakasan, "Voltage scalable switched capacitor DC–DC converter for ultra-low-power on-chip applications," *IEEE Power Electronics Specialists Conference,* pp. 2353–2359, June 2007.

[27] Y. K. Ramadass and A. P. Chandrakasan, "Minimum energy tracking loop with embedded DC–DC converter delivering voltages down to 250mV in 65nm CMOS," *IEEE ISSCC Dig. Tech. Papers*, pp. 64–65, Feb. 2007.

[28] H. P. Forghani-zadeh and G. A. Rincón-Mora, "Current-sensing techniques for DC–DC converters," *Proc. 2002 Midwest Symp. Circuits and Systems (MWSCAS)*, vol. 2, pp. 577–580, Aug. 2002.

[29] G. Bonfini *et al.*, "An ultralow-power switched opamp-based 10-B integrated ADC for implantable biomedical applications," *IEEE Trans. Circuits Syst. I, Reg. Papers*, vol. 51, no. 1, pp. 174–177, Jan. 2004.

[30] Y. K. Ramadass and A. P. Chandrakasan, "Minimum energy tracking loop with embedded DC–DC converter enabling ultra-low-voltage operation down to 250mV in 65nm CMOS," *to be published, IEEE J. Solid-State Circuits* vol. 43, Issue 1, pp. 256–265, Jan. 2008.

[31] A. J. Stratakos, "High-efficiency low-voltage DC–DC conversion for portable applications," University of California, Berkeley, *Ph.D. Thesis*, 1998.

Chapter 6 Dynamic Voltage Scaling with the XScale Embedded Microprocessor

Lawrence T. Clark [1], Franco Ricci [2], William E. Brown [3]

[1]Arizona State University, [2]Marvell Semiconductor Inc.,[3]Ellutions, LLC

6.1 The XScale Microprocessor

The XScale microprocessors [1] were intended as a follow-on to the StrongARM microprocessors [2] developed at Digital Equipment Corp. The XScale work began in 1998 to design a microprocessor that would be embedded in high-performance "tethered," i.e., line-powered, as well as handheld (battery-powered) system-on-chip (SOC) ICs. The ability of the processor core to operate over a wide range of supply voltages (V_{DD}) is key to achieving both high-performance and low power consumption across such a wide application range. Using the same microprocessor core in many, diversely targeted ICs, maximizes the core development return on investment.

Dynamically scaling the power supply to different voltages (V_{DD}) to fit the application that is presently running maximizes both overall performance vs. power and energy efficiency. It was thus deemed critical to the XScale effort. Such a capability had been suggested by [3] and had been a topic of university research [4] before the XScale processor development began. Around the same time, notebook computers introduced static voltage scaling schemes, e.g., "Speed-Step," whereby the processor power is minimized when running on battery power by using a lower V_{DD} and clock frequency, compared to operation when powered from a wall socket. As of 2007, it is a commonly available commercial capability, and the body of academic work investigating circuits and scheduling algorithms has become quite large.

A. Wang, S. Naffziger (eds.), *Adaptive Techniques for Dynamic Processor Optimization*,
DOI: 10.1007/978-0-387-76472-6_6, © Springer Science+Business Media, LLC 2008

Due to its large market size and rapid growth, which includes cell phones, handheld devices became the primary market for the XScale processors. DVS improves upon the static ability to operate over a very wide range of V_{DD} and performance in achieving the best battery lifetime. Portable devices are diverse both in purpose, e.g., personal digital assistants (PDAs), sub-notebooks, and cell phones, and have greatly varying usage models, which range from simple text-messaging to surfing web pages using a broadband connection. The same device may be used for many of these diverse applications, therefore DVS is very beneficial.

6.1.1 Chapter Overview

This chapter discusses the implementation and usage of DVS on the XScale microprocessor cores implemented on 180 nm fabrication technologies. Obviously, DVS requires that the processor supports a wide V_{DD} operating range, which is essentially a circuit-design problem. However, it is made more effective by additional processor support ranging from the circuit to the architectural level.

The XScale micro-architecture provides a performance-monitoring unit (PMU) to allow software, presumably the operating system (OS), to determine the processor throughput and efficiency in real time. This improves the DVS control considerably over merely knowing that the processor is busy. These monitors and their use in DVS control are discussed using example code that runs on an XScale microprocessor demonstration board supporting DVS.

Increased transistor variation in highly scaled manufacturing processes has made SRAM read stability problematic when operating with low V_{DD}. This chapter then discusses this issue and how it is addressed in XScale SOCs that utilize DVS.

The chapter concludes by discussing clock generation schemes used in some XScale implementations. In the original 180 nm prototype/product, i.e., the 80200 design, the processor can continue to run while the V_{DD} is adjusted, but a performance penalty is incurred due to the PLL relock time. In the 90 nm XScale processor prototype [5], an improved PLL and clock-generation scheme is used that allows true on-the-fly DVS, with essentially no time penalty for speed changes. Here, the PLL runs at a constant frequency on a separately regulated power supply, requiring no relock time. Processor clock changes are handled completely digitally, and frequency changes are made in one bus clock cycle.

6.1.2 XScale Micro-architecture Overview

The XScale block diagram comprises Figure 6.1. The processor uses a seven-stage (eight-stage cache access) pipeline [1]. The pipeline depth, which at the time was more than usual for an embedded processor, allows higher performance at low V_{DD}, by shifting the maximum operating frequency (F_{max}) curve upward at all voltages. To support a wide range of operating voltages, as well as DVS, two separate timing databases were constructed as part of the performance validation. One was at the nominal target V_{DD} of 1.3 V and one was at 0.7 V. The low V_{DD} timing database allowed specific circuits, whose performance scaled poorly with reduced voltage, to be identified, and appropriate design changes to be made.

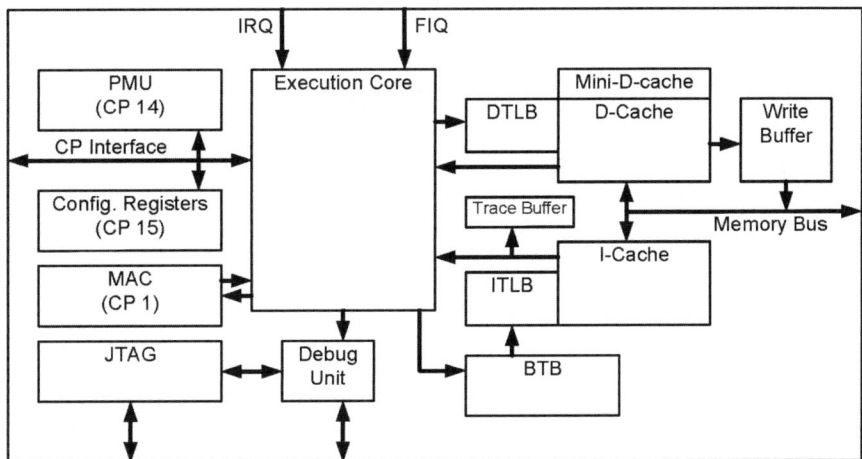

Figure 6.1 The 180 nm XScale microprocessor micro-architecture. The PMU is accessed through the coprocessor (CP14) interface. Frequency and V_{DD} controls reside in the CP15 configuration registers.

In particular, the differential cascade voltage-switched logic (DCVSL) circuit style was often problematic at low V_{DD}. DCVSL has poor delay vs. V_{DD} scaling properties due to its ratioed nature, where the input pull down transistors must overpower the cross-coupled PMOS load transistors. DCVSL was also incompatible with the static timing analysis tool and therefore required increased engineering effort. Pulse-clocked latches replaced master-slave latches in much of the design. These allowed about a 45% decrease in the sequential circuit energy per clock and reduced the path delays due to sequential elements.

6.1.3 Dynamic Voltage Scaling

DVS scales the processor performance by adjusting the frequency at which the processor operates to an estimate of the future workload. Scaling frequency down delivers a linear power savings, while simultaneously scaling V_{DD} with the frequency changes allows quadratically reduced power dissipation, which is compounded with the linear power dissipation savings due to reduced operating frequency. On the 180 nm XScale, the maximum operating frequency (F_{max}) at each V_{DD} scales with $V_{DD}^{1.75}$ as shown in Figure 6.2. The actual F_{max} vs. V_{DD} behavior in the figure differs from the ideal V^2 since the submicron transistors are velocity saturated, and interconnect RC has an impact. Transistor currents scale at a reduced exponent [5], while metal RC is constant with V_{DD}. The net result is an approximately cubic reduction in power dissipation, as shown in Figure 6.2. Note that the F_{max} approaches 0 MHz at a V_{DD} greater than 0 V.

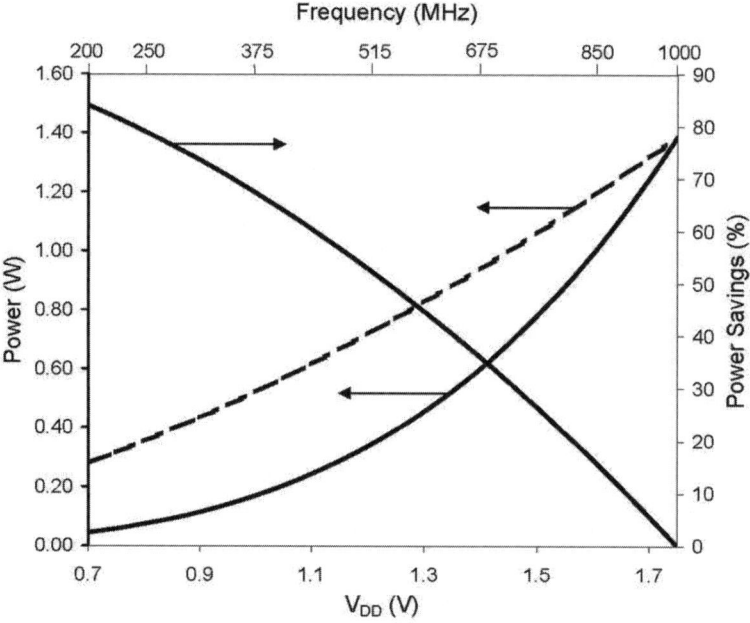

Figure 6.2 80200 XScale processor power dissipation vs. operating frequency with constant V_{DD} (dashed line) and scaled V_{DD} (solid line) at the F_{max} for each voltage. The savings due to DVS is also shown.

CPUs on modern semiconductor fabrication processes dissipate a considerable portion of their power from transistor-leakage currents [6]. Process scaling into the deep submicron region has resulted in exponential increases in leakage currents. Reducing V_{DD} scales these leakage components more rapidly than even the V^2 dependency of the active power component. Consequently, V_{DD} scaling is also desirable in managing leakage power dissipation and will become more so in the future [7].

When operating with DVS, the future workload must be estimated by the OS from present operations and hints about future needs. The key is to avoid missed deadlines, i.e., when scaling back the processor performance to reduce the system power usage, the required tasks should still finish in time. Examples of tasks that have deadlines are MPEG or audio decode and playback, where each block must be delivered in time to the screen or speakers. If a block is decoded late, the user experience suffers. Ideally, the OS schedules tasks so that the processor is kept continually busy, i.e., there is no idle time, but so that no deadlines are missed. In reality, there is always some idle time, since the scheduling must avoid degrading the user experience, so the power savings is not as high as the ideal case. It is also important to maintain the overall system responsiveness.

The XScale microprocessor PMU is critical to effective DVS use. The PMU allows real-time determination of not just whether tasks are keeping the processor busy, but whether they are being executed *efficiently*.

6.1.4 The Performance Measurement Unit

A performance measurement capability is necessary to effectively use DVS in practical applications. Since the actual mix of applications and their interaction with the OS cannot be known a priori, whether or not the processor is running efficiently must be measured in real time. This application mix-dependent behavior implies the need for some form of hardware counting support to minimize power dissipation while ensuring adequate quality of service. The additional hardware allows the OS to estimate the future workload from the present one, ideally with hints from the applications about priorities [3].

The XScale micro-architecture includes a performance measurement unit (PMU) [8, 9] that supports this need. The monitors are accessed through coprocessor registers (specifically CP14). The basic counting mechanism is provided by a dedicated 32-bit clock counter and two programmable 32-bit performance counters, PMN0 and PMN1. The counters can trigger interrupts on rollovers under software control. The performance monitor control register (PMNC) controls the monitored events, resets counters, determines which counters have events, and enables and disables interrupts. Table 6.1 lists the events that can be monitored by the XScale PMU.

Table 6.1 Performance monitoring events supported by the XScale PMU [8]. The numbers refer to the counters as chosen by the CP14 enables.

0	Instruction cache miss caused external memory access
1	Instruction not delivered by I-cache—I-cache or I-TLB miss
2	Data dependency stall
3	I-TLB miss
4	D-TLB miss
5	Branch instruction executed
6	Mispredicted branch
7	Instruction executed
8	Stall due to full data cache buffers (once per clock cycle)
9	Stall due to full data cache buffers (once per stall sequence)
10	D-cache access
11	D-cache miss
12	D-cache write back (for each four words written back)
13	Software-controlled PC change with no mode change
16	Bus memory request from core
17	Bus memory request queue full
18	Bus queues drained
20	Unlogged bus ECC error
21	Single-bit bus error
22	ECC required read–modify–write cycle for narrow write

The performance monitors can be used to determine the number of actual clocks per instruction (CPI), bus activity, translation lookaside buffer (TLB), and cache efficiency, for the code being run. The latter are determined easily by counting TLB or cache misses. The performance counters can also be used to determine average fetch latency, by counting the stall cycles waiting for memory.

For DVS applications, the PMU is used to distinguish intervals where the processor is continually busy, i.e., those where there is no idle time, vs. those where it is actually accomplishing useful work. Consider an application that is memory bandwidth limited. In this case, if the working set does not fit in the D-cache, there is no idle time between tasks, but a significant amount of the processor cycles are spent with the pipeline stalled and waiting for bus operations, resulting in a high number of clocks per instruction (CPI). In this case, lowering the voltage and frequency can provide significant power savings with no impact on performance. When the processor is running at a lower core voltage and frequency, there are fewer stall cycles and hence a lower CPI.

As a concrete example, assume the processor is running at 1 GHz and V_{DD} = 1.75 V. If half of the cycles are stalls waiting for the bus, as determined by a combination of the total clock count, instructions executed, and data dependency stall or bus request counts, the V_{DD} can be adjusted to 1.2 V (see Figure 6.2) and the core frequency reduced to 500 MHz. Useful work is then performed in a greater number of the (fewer overall) core clock cycles. Referring to Figure 6.2, the power savings is nearly 50% with the same work finished in the same amount of time.

6.2 Dynamic Voltage Scaling on the XScale Microprocessor

This section describes experimental results running DVS on the 180 nm XScale microprocessor. The value of DVS is evident in Figure 6.3. Here, the 80200 microprocessor is shown functioning across a power range from 10 mW in idle mode, up to 1.5 W at 1 GHz clock frequency. The idle mode power is dominated by the PLL and clock generation unit. The processor core includes the capacity to apply reverse-body bias and supply collapse [10, 11] to the core transistors for fully state-retentive power-down. The microprocessor core consumes 100 μW in the low standby "Drowsy" mode [12]. The PLL and clock divider unit must be restarted when leaving Drowsy mode. When running with a clock frequency of 200 MHz, the V_{DD} can be reduced to 700 mV, providing power dissipation less than 45 mW.

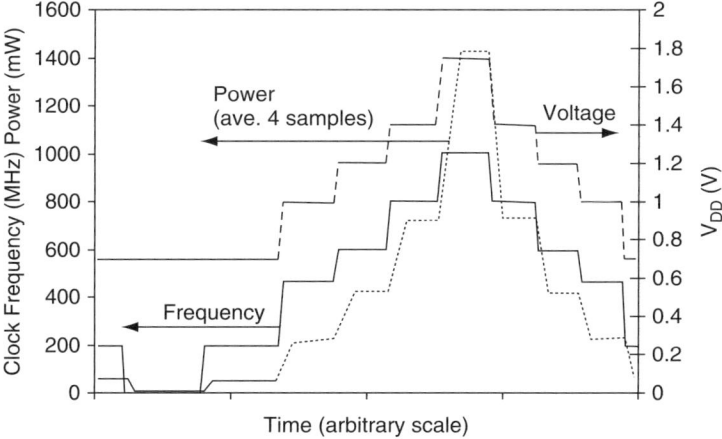

Figure 6.3 The value of dynamic voltage scaling is evident from this plot of the 80200 power and V_{DD} voltage over time. The power lags due to the latency of the measurement and time averaging.

6.2.1 Running DVS

To demonstrate DVS on the XScale, a synthetic benchmark programmed using the LRII demonstration board is used here. The onboard voltage regulator is bypassed, and a daughter-card using a Lattice GAL22v10 PLD controller and a Maxim MAX1855 DC-DC converter evaluation kit is added. The DC–DC converter output voltage can vary from 0.6 to 1.75 V. The control is memory mapped, allowing software to control the processor core V_{DD}.

The synthetic benchmark loops between a basic block of code that has a data set that fits entirely in the cache (these pages are configured for write-back mode) and one that is non-cacheable and non-bufferable. The latter requires many more bus operations, since the bus frequency of 100 MHz is lower than the core clock frequency, which must be at least 3× the bus frequency on the demonstration board.

The code monitors the actual operational CPI using the processor PMU. The number of executed instructions as well as the number of clocks, since the PMU was initialized and counting began, are monitored. The C code, with inline assembly code to perform low-level functions is

```
unsigned int count0, count1, count2;

int cpi() {
  int val;

  // read the performance counters
  asm("mrc p14, 0, r0, c0, c0, 0":::"r0");  // read the PMNC register
  asm("bic r1, r0, #1":::"r1");             // clear the enable bit
  asm("mcr p14, 0, r1, c0, c0, 0":::"r1");  // clear interrupt flag, disable counting

  // read CCNT register
  asm("mrc p14, 0, %0, c1, c0, 0" : "=r" (count0) : "0" (count0));
  asm("mrc p14, 0, %0, c2, c0, 0" : "=r" (count1) : "0" (count1));
  asm("mrc p14, 0, %0, c3, c0, 0" : "=r" (count2) : "0" (count2));

  return(val = count0);
}

int startcounters() {
  unsigned int z;

  // set up and turn on the performance counters
  z = 0x00710707;
  asm("mov r1, %0" :: "r" (z) : "r1");       // initialization value in reg. 1
  asm("mcr p14, 0, r1, c0, c0, 0" ::: "r1"); // write reg. 1 to PMNC
}
```

Note that the code to utilize the PMU is neither large nor complicated. It is also straightforward to implement the actual V_{DD} and core clock rate changes. To avoid creating a timing violation in the processor logic, the core voltage V_{DD} must always be sufficient to support the core operating frequency. This requires that the voltage be raised before the frequency is and conversely that the frequency be reduced before the voltage is. The XScale controls the clock divider ratio from the PLL through writes to CP14. The C code to raise the V_{DD} voltage is

```
int raisevoltage() {
  int i;

  // raise the voltage first
  if (voltage <= TOP_V) {              // leave it alone
    printf ("V at end of range--");
  }
  else {
    voltage--;
    *voltagep = voltage;

    // adjusting the frequency

    if (frequency < TOP_F) {        // do nothing

    frequency = uf[voltage];;
    asm("mov r1, %0" :: "r" (frequency) : "r1");
    asm("mcr p14, 0, r1, c6, c0, 0" ::: "r1");
    }
  }
  return(voltage);
}
```

The code to lower the voltage is very similar. The supported clock multipliers range from 3 to 11 [9]. The array `uf[]` is a lookup table of appropriate voltages for each frequency. The PLD is programmed so that the highest voltage of 1.7 V is programmed by setting the value to 0 and higher values increase the voltage by 50 mV (for the first four entries) or 100 mV increments. The constant `TOP_V` = 0. For the `lowervoltage()` function, an equivalent `BOTTOM_V` constant avoids setting the voltage too low. No delay is required, since the coprocessor register write forces the core clocks to be inactive for approximately 20 μs while the PLL relocks to the new clock fraction—this is handled automatically by the XScale core hardware. Excellent power supply rejection ratio (PSRR) in the 80200 PLL allows the relock to occur in parallel with the voltage movement. The code to lower the voltage is similar, but as mentioned, the frequency is reduced before lowering the voltage. Again, the PLL lock time, invoked before the MCR P14 instruction can finish, hides the latency of the voltage movement from the software.

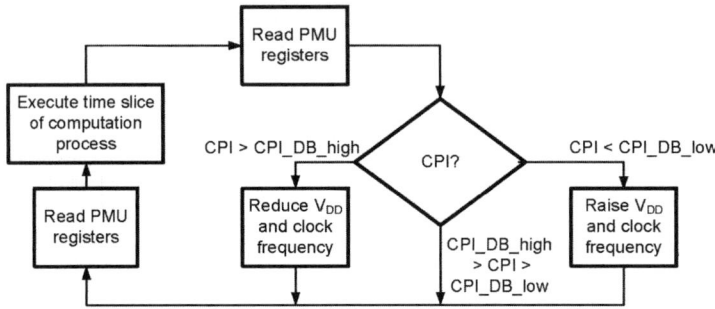

Figure 6.4 Simple DVS control heuristic using an estimate of the CPI as determined by the PMU. The CPI is estimated for each time slice and VDD adjusted if it is outside the dead-band parameters CPI_DB_high and CPI_DB_low. Otherwise the V_{DD} and clock frequency are unchanged.

Here, for illustration purposes, the control algorithm is very simple, as shown in Figure 6.4. All but the "Execute time slice…" block would be part of the OS. Behavior of the synthetic benchmark, using the code shown above, is shown in Figure 6.5(a). Many complicated and hence more optimal V_{DD} control algorithms have been developed but are application dependent and beyond the scope of this discussion. The frequency and voltage are increased by one increment if the measured CPI is below the predetermined value CPI_DB_high, and they are decreased by one increment if the CPI is above another predetermined value CPI_DB_low. It is left the same otherwise, i.e., the control dead-band is defined by the separation of the two values. Figure 6.5(b) shows the intervals more closely. The intervals running the bus limited data access code are marked by **A**, and the faster running (cacheable data) code is marked by **B**. The distinct V_{DD} voltage steps when the frequency and voltage are changed as the data accesses move from one behavior to the other are evident.

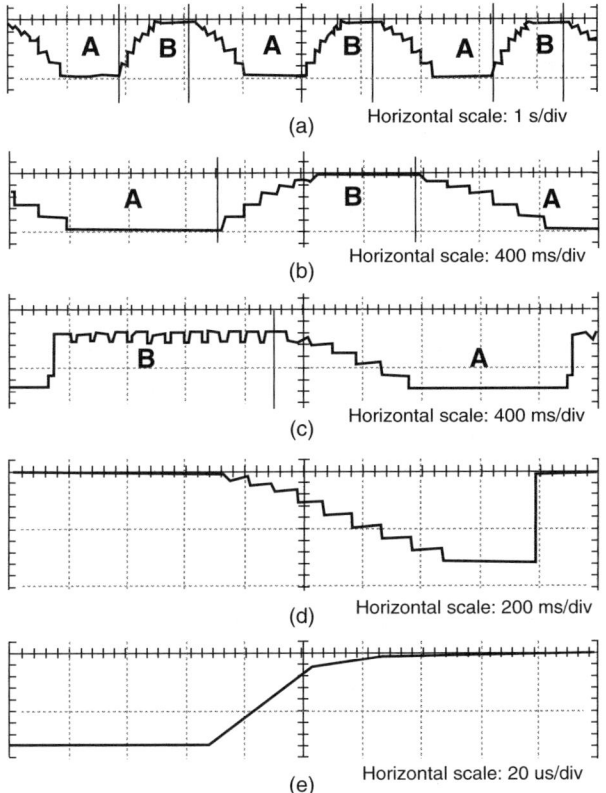

Figure 6.5 Oscilloscope traces of V_{DD} on the LRH test system. The system is running a synthetic benchmark that modifies V_{DD} based on the CPI as determined by the PMU (a)–(d). The distinct steps in voltage with each software-controlled clock rate and V_{DD} change are evident. The V_{DD} slew rate is shown in (e), where the supply ripple can also be seen.

Adjusting the size of the control heuristic dead-band to be too small causes the voltage to "hunt" when running the faster code, as evident in Figure 6.5(c) section **B** since a stable CPI value between that which causes an increase and that which causes a decrease is not found. This hunting behavior is not efficient, since the PLL lock time is wasted for each 50 mV V_{DD} movement. It is therefore important to define a large enough stable region and make DVS changes (monitor the CPI) infrequently enough to keep the total voltage change time insignificant compared to the total operating time. A further adjustment in the heuristic affects the minimum usable voltage, by allowing still slower operation for the bus limited code.

Figure 6.5(e) shows the maximum slew rate for the large voltage change from 1.0 to 1.7 V, which is the nearly vertical V_{DD} movement near the end of the trace in Figure 6.5(d). The core V_{DD} is slightly over-damped, as evident in Figure 6.5(e).

6.3 Impact of DVS on Memory Blocks

As mentioned in the introduction, some circuits may limit operation at low V_{DD}. Microprocessors and SOC ICs include numerous memories, usually implemented with six transistor SRAM cells. In future devices, it is expected that memory, and SRAM in particular, will dominate IC area [13]. Unfortunately, SRAM has diminishing read stability [14] as manufacturing processes are scaled down in size and transistor level variations increase [15]. Lower V_{DD} profoundly reduces SRAM read stability, making it a primary limiting circuit when applying DVS.

When the SRAM is read, the low storage node rises due to the voltage divider comprised of the two series NMOS transistors in the read current path, which includes one of the storage nodes. Monte Carlo simulations of SRAM static noise margin are shown in Figure 6.6. As V_{DD} is decreased, the static noise margin (SNM) as measured by the smallest side of the square with largest diagonal in the small side of the static voltage curves (see Figure 6.6(a)) decreases as well. The large transistor mismatch due to both systematic (intra-die) and random (within-die) variations cause asymmetry in the SNM plot as shown in Figure 6.6(a). An IC contains many SRAM cells, so the combination of worst-case systematic and random variations can cause some cells to fail, significantly impacting the manufacturing yield at low V_{DD}. The simulated behavior of the SRAM SNM vs. voltage, using Monte Carlo device variations to 5σ, is shown in Figure 6.6(b). It is evident that the SRAM read margins are strongly affected by the combination of transistor variation and reduced V_{DD}. Register file memory, which is also ubiquitous in microprocessors and SOC Ics, does not suffer from reduced SNM when reading since the read current path does not pass through the SRAM storage nodes. These memories can scale with any core logic and can in fact operate effectively well into subthreshold, i.e., they allow operation with $V_{DD} < V_{th}$. [16, 17].

6.3.1 Guaranteeing SRAM Stability with DVS

In the 180 nm process used for the XScale, the manufacturing yield is negligibly impacted by SRAM read stability, even at $V_{DD} = 0.7$ V when

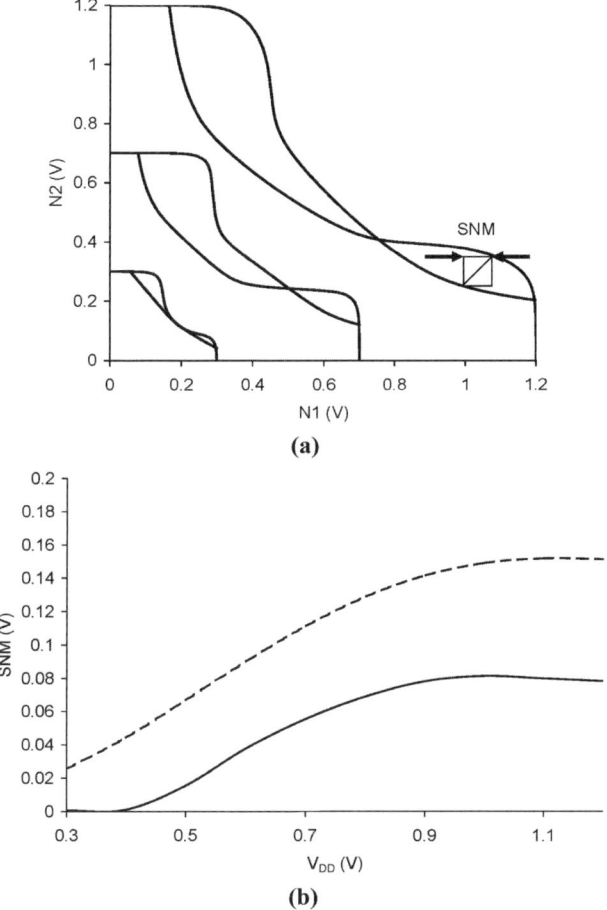

Figure 6.6 SRAM SNM at various voltages (a). The mean and 5σ SNM from Monte Carlo simulations (b) show vanishing SNM at low voltages. The XScale SOC logic level shifts SRAM input signals and operates the SRAMs at a constant voltage where SNM is maintained.

only the two 32kB caches are considered. However, adding large SOC SRAMs significantly affects the IC manufacturing yield at low V_{DD}. The solution used for the 180 nm "Bulverde" application processor SOC [18] is to scale the XScale cache circuits with the dynamically scaling core and SOC logic supply voltage, while operating the large SOC SRAM on a fixed supply [19]. The SRAMs and their voltage domains are shown in Figure 6.7. The SOC logic clock rate is 104 MHz or less depending on the DVS point, while the core clock frequency scales from 104 MHz to over

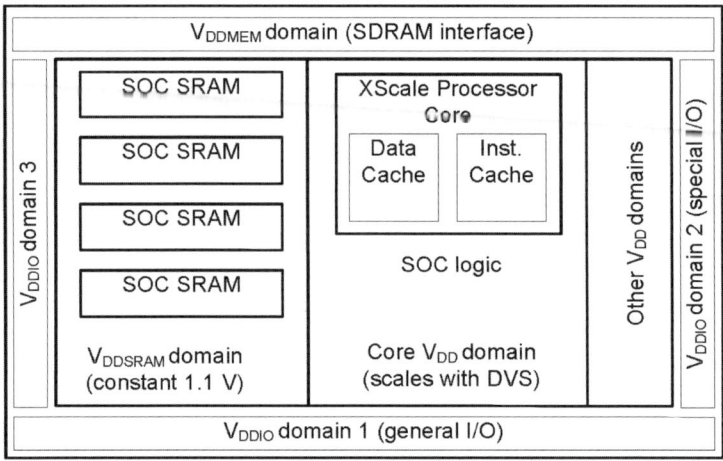

Figure 6.7 SRAMs and their voltage domains in the XScale core and in the Bulverde application processor [20]. This diagram is greatly simplified to emphasize the DVS vs. constant V_{DD} domains.

500 MHz [18, 20]. A constant 1.1 V SRAM power supply voltage (V_{DDSRAM}) provides adequate access times for the slower SOC logic. In this manner, the SOC and microprocessor core logic V_{DD} employ DVS, but the embedded SOC SRAM supply V_{DDSRAM} is fixed. The fixed, higher minimum V_{DD} for the additional SOC SRAMs assures high manufacturing yield with a low minimum V_{DD} for DVS. The fixed SRAM supply voltage also facilitates the low standby power Drowsy modes, which have a single optimal V_{DD} that must be sufficient to allow raising the NMOS transistor source nodes toward V_{DD} to apply NMOS body bias [11].

With two differing supply voltages, level shifting is required between the memories and the SOC logic. The added level shifters degrade the maximum performance, since they add delay. This is not an issue for low V_{DD} operation—the higher SRAM V_{DD} makes them fast compared to the surrounding logic operating at lower V_{DD}. The problem is that the level shifters slow the maximum clock rate of the design at high V_{DD} by injecting extra delay in the memory access path.

The Bulverde SOC memory level shifting scheme is shown in Figure 6.8(a). To minimize the number of level shifters and limit the complexity, the address ADD(1:m) and some control signal voltages are translated to the different V_{DDSRAM} power supply domain by the cross-coupled level shifting circuit evident at the decoder inputs. This scheme has the drawback that the word-line enable signal WLE, which is essentially a clock, and the array pre-charge signal PRECHN must be level shifted. The write and read column multiplexer control signals must also be level shifted—for clarity, these circuits are not shown in the figure. The differential sense amplifiers, which operate at the (potentially lower) DVS domain supply voltage, automatically shift the SRAM outputs OUTDATA to the correct voltage range. The sense timing signal SAE is also in the DVS domain.

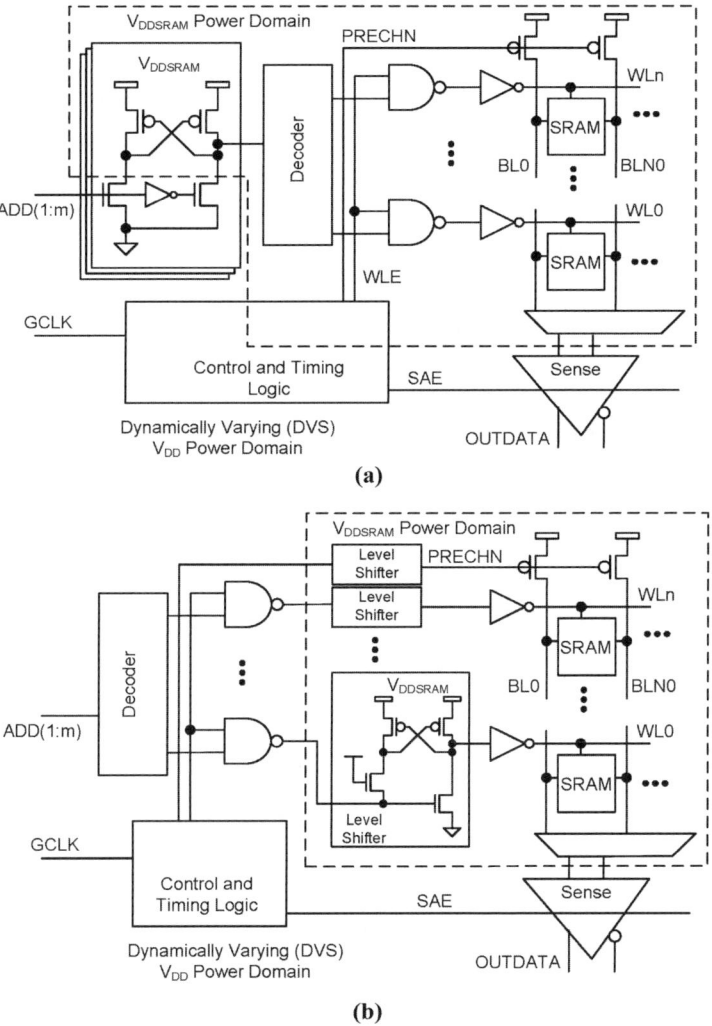

Figure 6.8 Level shifting paths to allow the SRAM supply voltage V_{DDSRAM} to remain constant while applying DVS to the surrounding logic. In (a) the level shifters are placed at the SRAM block interface, while in (b) the level shifters are at the storage array interface. In both cases, the sense amplifiers shift back to the DVS domain.

Additional power can be saved by the scheme shown in Figure 6.8(b), which shifts the voltage levels at the decoder outputs, i.e., the SRAM word-line drivers. Here, the decoders reside in the scaled V_{DD} domain and fewer control signals must be level shifted to the V_{DDSRAM} domain.

6.4 PLL and Clock Generation Considerations

In this section, the implications of DVS on microprocessor clocking are considered. In the original 180 nm implementation, a simple approach was taken—there are minimal changes to the PLL and clock generation unit to support DVS. The feedback from the core clock tree to the PLL requires a PLL relock time for each clock change. In the 90 nm prototype, the PLL and clock generation unit was explicitly designed to support zero latency clock frequency changes. Here, the PLL is derived from the I/O supply voltage via an internal linear regulator. Hence, the PLL power supply is not dynamically scaled with the processor core.

6.4.1 Clock Generation for DVS on the 180 nm 80200 XScale Microprocessor

The clock generation unit in the 80200 is shown in Figure 6.9. The ½ divider provides a high quality, nearly 50% duty cycle output. The feedback clock is derived from the core clock, to match the core clock (and I/O clock, which is not shown) phase to the reference clock. Experiments with PLL test chips showed that phase and frequency lock can be retained during voltage movements, if the PLL power supply rejection ratio is sufficient and the slew rate is well controlled [21,22]. This allows voltage adjustment while the processor is running, as mentioned. However, a change in the clock frequency changes the numerator in the 1/N feedback clock divider. This causes an abrupt change in the frequency of the signal Feedback Clk, which necessitates the PLL to relock to the new frequency. The PLL generates a lock signal, derived from the charge pump activity. Depending on the operating voltage, the PLL can achieve lock as quickly as a few microseconds. However, a dynamic lock time makes customer specification and testing more difficult—hence, a fixed lock time is used. Another scheme, which allows digital control of the clock divider ratio was developed for the 90 nm XScale prototype test chip.

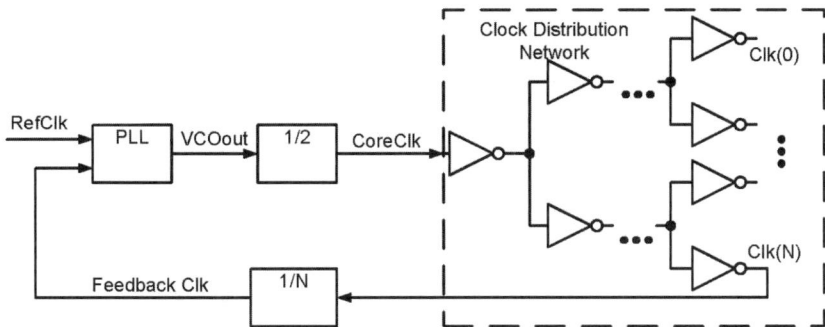

Figure 6.9 The 80200 PLL and clock generation scheme. The PLL must relock for each frequency change, since the feedback divider ratio is altered. The clock distribution network is much deeper than that indicated in the figure.

6.4.2 Clock Generation 90 nm XScale Microprocessor

The clock generation unit in the 90 nm XScale prototype microprocessor [5] is shown in Figures 6.10 and 6.11. Placing the PLL, which is implemented with the thin-oxide high-speed core transistors on the regulated power supply V_{DDPLL} reduces the jitter component caused by power supply noise, which allows the core to operate at higher frequencies. The low dropout (LDO) linear regulator is connected to the same voltage as the I/O supply, which may be 1.8 V to 3.3 V. The LDO reduces the V_{DDPLL} voltage to the nominal core voltage value, while allowing the PLL and clock generation unit to operate continuously through core voltage changes. It also greatly improves the PLL PSRR, since it acts as a low-pass filter as well as a level down-converter. The system bill of materials (BOM) cost is not increased, since the PLL supply is the same voltage as the I/O. It uses a separate supply pin, ideally with filtering for even greater power supply noise rejection. The PLLs used in both the 180 nm and 90 nm XScale prototype designs discussed are based on the self-biased PLL design presented in [23].

True on-the-fly DVS, with no time penalty for speed changes, is enabled by having multiple separate dividers of the VCO clock for each SOC function, as shown in Figure 6.11. Again, the VCO clock output of the PLL is run at the twice maximum desired frequency to ensure a high-quality clock with a 50% duty cycle, i.e., M in the 1/M divider is at least two. Changes in the core, I/O or SOC clock frequencies, derived from the 1/M, 1/X, and 1/Y dividers, respectively, are achieved without affecting the PLL lock since the PLL feedback is not derived from any of these counters. The feedback clock divider sets 1/N to the appropriate value for the required maximum VCOout frequency.

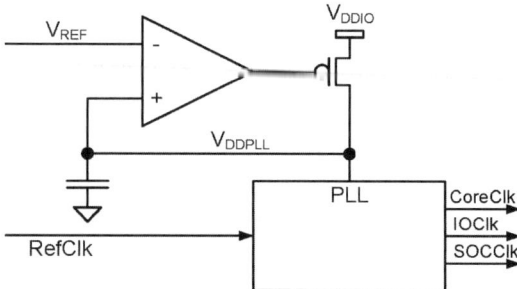

Figure 6.10 Regulated supply PLL used on the 90 nm XScale prototype [5]. The PLL power is supplied through a LDO regulator, which improves the PSRR and makes the PLL supply voltage independent of that of the processor core.

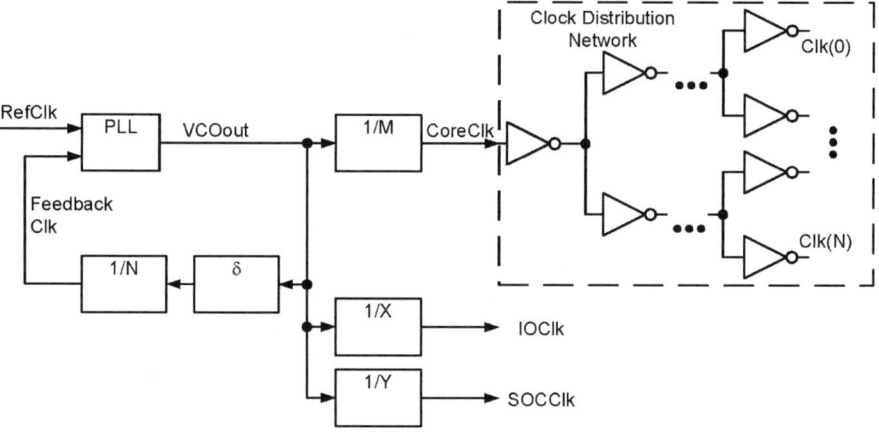

Figure 6.11 PLL and clock distribution network for the 90 nm XScale prototype allowing true on-the-fly clock speed changes. The δ delay in the feedback path is necessary to match insertion delays between the I/O and reference clock.

The 1/N, 1/M, 1/X, and 1/Y divider blocks in Figure 6.11 are digital counters of the PLL voltage controlled oscillator (VCO) output. These signals are compared to configuration bits to determine when output clock transitions are to be generated. The δ delay is programmable using the standard JTAG interface, to assure the edges of the I/O clock are aligned with those of the core clock.

The synchronization of a frequency change is achieved by first latching in the divider change request on the falling edge of the externally generated reference clock signal RefClk. Then, the new values are latched into the VCO domain by comparing the counter values in the VCO control latches to their inputs and enabling the VCO domain latch clocks to transition for one VCO clock cycle when a difference is detected. Finally, the comparison values are transferred to the divider comparators precisely one VCO cycle before the following synchronized rising edges of the M, X, and N dividers. This ensures that the derived clocks do not glitch and that the transition from one frequency to another occurs simultaneously in all domains. The on-the-fly frequency changes are shown in Figure 6.12. Here, the core frequency switches from 500 MHz (labeled A) down to 125 MHz (labeled B) and then back up to 250 MHz in the section labeled C. As evident in the figure, the transitions can occur closely spaced, and the PLL operation remains unchanged, i.e., the frequency changes occur without re-synching or relocking the PLL.

Since all the dividers are eventually synchronized on the final output generated by the 1/N divider, a global reset from the 1/N divider is logically OR-ed into the local reset of each of the other dividers. Loss of synchronization at high speeds, caused by critical timing paths in the clock generation logic that includes the reset signal, is thus prevented.

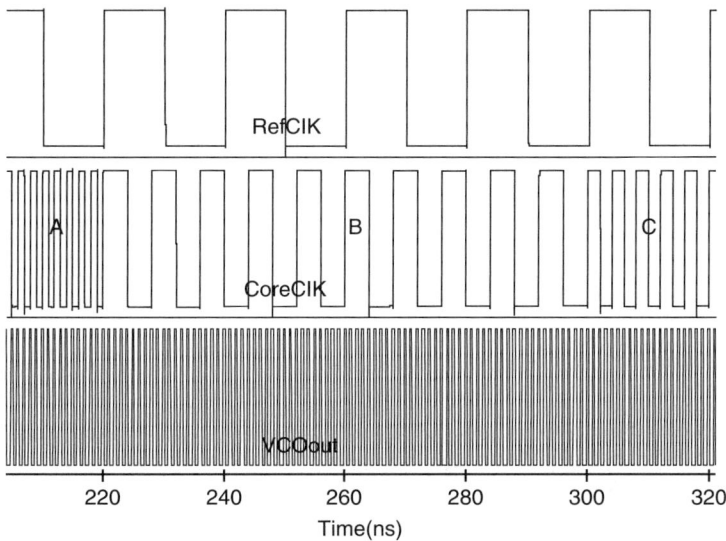

Figure 6.12 Simulated clock waveforms showing true on-the-fly clock frequency changes. Different frequencies are indicated by sections A, B, and C. Short clocks (glitches) in the generated clocks must be avoided while changing the divider ratios. The feedback clock (not shown) is a divided version of CoreClk, matching RefClk in frequency and phase.

6.5 Conclusion

This chapter has described DVS on the XScale micro-architecture. The XScale circuits, micro-architectural and architectural level supports for DVS have been described. Supporting DVS requires that the software running on the processor be able to predict the future workload to adjust the V_{DD} appropriately, based on past operations. DVS requires architectural support for not only dynamic frequency and voltage adjustments but also for real-time performance monitoring.

Increasing transistor mismatch, which is exacerbated by aggressive transistor scaling, has made low-voltage SRAM operation problematic. Consequently, ICs employing DVS must comprehend the SRAM yield impact. In this chapter, the methods used to provide SRAM stability, i.e., level-shifting and operation of the embedded SRAMs at higher single power supply voltage, in the XScale SOCs have been described. The relatively small first-level cache SRAMs maintain full high V_{DD} performance by their inclusion in the DVS domain. To support DVS on future scaled manufacturing processes, which exhibit even greater transistor variability, separate power supplies may be needed for all SRAMs.

DVS can be implemented with minimal clock level support, e.g., requiring the PLL to relock at each frequency change. Better performance and finer granularity clock changes can be obtained with an improved clocking scheme which does not place the core clocks in the PLL feedback loop. This implementation, used in the 90 nm XScale processor prototype, allows clock frequency changes with no wasted time for re-synchronization.

References

[1]   Clark, L, et al., "An embedded microprocessor core for high performance and low power applications," *IEEE Journal of Solid-State Circuits*, Vol. 36, No. 11, pp. 498–506, November 2001.

[2]   Montonarro, J, et al., "A 160 MHz, 32b 0.5W CMOS RISC microprocessor," *IEEE Journal of Solid-state Circuits*, vol. 31, pp. 1703–1714, November 1996.

[3]   Weiser, M, Welch, B, Demers, A, Shenker, S, "Scheduling for reduced CPU energy," *Proceedings of the Fisrt Symposium on Operating Systems Design and Implementation*, November, 1994.

[4]   Pering, T, Burd, T, Broderson, R, "The simulation and evaluation of dynamic voltage scaling algorithms," *Proceedings of International Symposium on Low Power Electronics*, pp. 76–81, August 1998.

[5]   Ricci, F, et al., "A 1.5 GHz 90-nm embedded microprocessor core," *VLSI Circuits Symposium on Technology Design*, pp. 12–15, June 2005.

[6]   Sakurai, T, Newton, A, "Alpha-power law MOSFET model and its applications to CMOS inverter delay and other formulas," *IEEE Journal of Solid-State Circuits*, Vol. 25, No. 2, pp. 584–594, April 1990.

[7]   Mudge, T, "Power: A first-class architectural design constraint," *Computer*, Vol. 34, No. 4, pp. 52–58, April 2001.

[8]   Intel 80200 Processor based on Intel XScale Microarchitecture Developers Manual, November 2000.

[9]   Intel XScale Core Developers Manual, December 2000.

[10] Clark, L, Deutscher, N, Ricci, F, Demmons, S, "Standby power management for a 0.18-μm microprocessor," *Proceedings of International Symposiums on Low Power Electronics*, pp. 7–12, August 2002.

[11] Clark, L, Morrow, M, Brown, W, "Reverse body bias for low effective standby power," *IEEE Transactions on VLSI Systems*, Vol. 12, No. 9, pp. 947–956, September, 2004.

[12] Morrow, M, "Micro-architecture uses a low power core," *Computer*, p. 55, April 2001.

[13] ITRS roadmap. Online at www.itrs.org.

[14] Seevinck, E, List, F, Lohstroh, J, "Static noise margin analysis of MOS SRAM cells," *IEEE Journal of Solid-State Circuits*, Vol. 22, No. 5, pp. 748–754, October 1987.

[15] Bhavnagarwala, A, Tang, X, Meindl, J, "The impact of intrinsic device fluctuations on CMOS SRAM cell stability," *IEEE Journal of Solid-State Circuits*, Vol. 36, No. 4, pp. 658–665, April 2001.

[16] Chen, J, Clark, L, Chen, T, "An ultra-low-power memory with a subthreshold power supply voltage," *IEEE Journal of Solid-State Circuits*, Vol. 41, No. 10, pp. 2344–2353, October 2006.

[17] Calhoun, B, Chandrakasan, A, "A 256-kb 65-nm Sub-threshold SRAM design for ultra-low-voltage operation," *IEEE Journal of Solid-State Circuits*, Vol. 42, No. 3, pp. 680–688, March 2007.

[18] Intel PXA27× Processor Family Developers Manual.

[19] US patent 6,650,589: "Low Voltage Operation of Static Random Access Memory," November 18, 2003.

[20] Intel PXA27× Processor Family Power Requirements.

[21] US patent 6,519,707: "Method and Apparatus for Dynamic Power Control of a Low Power Processor," February 11, 2003.

[22] US patent 6,664,775: "Apparatus Having Adjustable Operational Modes and Method Therefore," December 16, 2003.

[23] Maneatis, J, "Low-jitter process-independent DLL and PLL based on self-biased techniques," *IEEE Journal of Solid-State Circuits*, Vol. 31, No. 11, pp. 1723–1732, November 1996.

Chapter 7 Sensors for Critical Path Monitoring

Alan Drake

IBM

7.1 Variability and Its Impact on Timing

Modern processes are becoming more sensitive to noise [25]. In addition to technology parameters having larger variation with each new technology generation [1], [20], timing sensitivity to such environmental conditions as temperature [19], [31], aging [3], workload [19], [13], cross-talk noise in wires [18], NBTI [12], and many other effects is increasing.

Noise processes that effect timing are described as random or systematic, and they are measured from die to die and within die [6], [32]. Random noise is less dependent on the integrated circuit's design than systematic noise and it is characterized by a number of statistics such as its mean and standard deviation. Systematic noise results from characteristics of the manufacturing process or from the physical design and can be predicted once the underlying process causing the variation is understood. For example, the wire thickness in technologies that use copper metallization is dependent on the wire density and wire width [21]. Once the source of systematic variation is identified, designs can be adjusted or processing can be modified to reduce the variation [27], [1]. Die-to-die noise measurements measure the statistical difference between separate integrated circuit die and inter-die noise is measured within a single die [5], [4].

The increasing timing uncertainty due to noise processes as technology scales is creating a need for on-chip timing sensors that can be explained using Figure 7.1. During the early design stages of the integrated circuit, assuming it is a microprocessor, an iterative process is used to develop the architecture to meet the performance targets of the intended application. Once the architecture is defined, the microprocessor passes through logic,

A. Wang, S. Naffziger (eds.), *Adaptive Techniques for Dynamic Processor Optimization*,
DOI: 10.1007/978-0-387-76472-6_7, © Springer Science+Business Media, LLC 2008

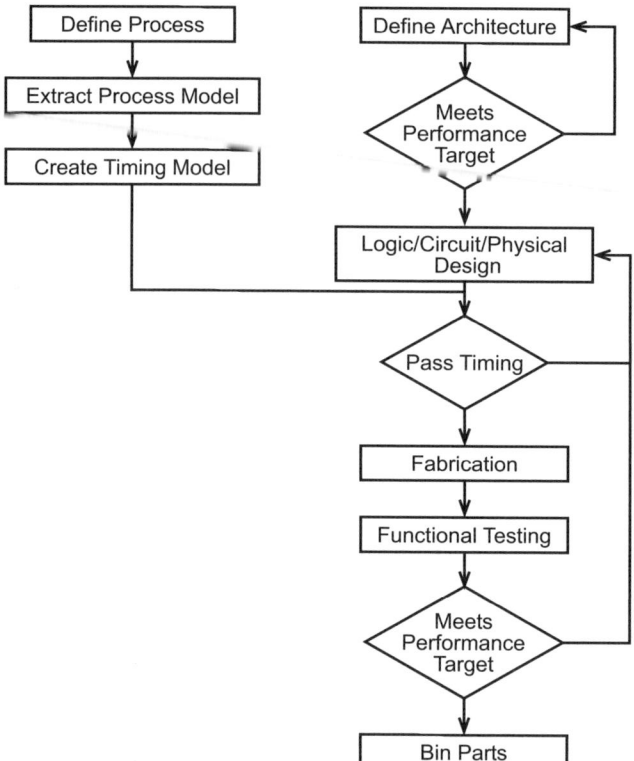

Figure 7.1 A simplified design flow is shown for a large-scale integrated circuit. Emphasis is placed on the development of the target performance and testing to ensure that performance is met.

circuit, and physical design. Models describing the timing of the target technology are used to predict the timing of the microprocessor during its design phases. Once timing is met, the processor is fabricated and tested. If performance targets are met, the microprocessor will be binned into performance categories and sold. If not, the design cycle must iterate at some point to fix the errors.

When the timing models can accurately predict the performance of the microprocessor, even when within-die variation is significant, adding a design margin and binning is sufficient for determining the performance of the microprocessor [32]. However, as sensitivity to environmental conditions increases, the needed margins to ensure functionality cause valuable performance to be lost. Because much of the timing variation (caused by such things as power supply noise and temperature shifts) is related to workload, it can be considered systematic noise and compensated for using dynamic voltage and frequency scaling (DVFS).

DVFS is typically used to optimize the power/performance of a microprocessor [7], [11], [26], [24], but if the DVFS system can sense a change in temperature, workload, etc., then it can compensate for environmental noise and recover some of the design margin [22]. For DVFS to be functional, it must have a means to determine the operating point of the microprocessor. This can be done using workload estimates and look-up tables, but this is usually expensive in terms of calibration time and complexity. Another solution, especially when dealing with fast environmental changes like supply voltage noise, is to use on-chip sensors to monitor the operating condition. Such sensors, typically called critical path monitors, can provide real-time performance information to DVFS systems with a simpler calibration. The critical path is used because it is the benchmark of timing and is most sensitive to environmental conditions. In addition to providing real-time timing analysis, critical path monitors are extremely useful as an aid in testing microprocessors. Since there is a cost overhead to including critical path monitors, they must provide better performance than just binning and margining by themselves. This chapter will describe in detail the design of critical path monitors as real-time sensors providing output to DVFS systems.

7.2 What Is a Critical Path

As discussed in the introduction, the timing for an integrated circuit is determined during the design phase based on performance needs, power dissipation, technology limitations, and design architecture. Once the cycle time has been determined and design begins, a number of timing paths within the integrated circuit emerge whose timing exceeds the cycle time. These paths, called critical paths, must be retimed to meet the cycle time. Part of the timing distribution of each path will exceed the cycle time as shown in Figure 7.2. An equation for the delay of a critical path, including sources of timing variation, is given in the following equation:

$$T_d + \delta T_d + \delta T_{\theta 1} + \delta T_{\theta 2} > T_\theta - \delta T_\theta - T_{setup}, \qquad (7.1)$$

where T_d is the path delay, $\delta T_{\theta 1}$ and $\delta T_{\theta 2}$ are the jitter in the sending and receiving clock edges, T_θ is the clock period, δT_θ is the clock jitter, and T_{setup} is the latch setup time [32]. To ensure that all critical paths meet timing requirements, T_θ must be increased to meet the following equation:

$$T_d + m\sigma(\delta T_d + \delta T_{\theta 1} + \delta T_{\theta 2}) < T_\theta - \delta T_\theta - T_{setup}, \qquad (7.2)$$

where $\sigma(\delta T_d + \delta T_{\theta1} + \delta T_{\theta2})$ is the standard deviation of the combined jitter of the critical path timing and m is a multiplier determining the number of standard deviations of error that are required for appropriate yield [32]. A large value of m (which is an expression of design margin) decreases not only the probability of timing failure but also the integrated circuit's performance. As shown by the process spreads that overlap the cycle time in Figure 7.2, the distribution of the critical path timing will exceed the cycle time for all but the largest m (increasing m moves the cycle time to the left on the graph). Because integrated circuits are sensitive to environmental conditions, at a given operating point, the timing of one or more of the paths may exceed the cycle time, causing a timing failure. Critical path monitors are a way to provide feedback to the integrated circuit when critical paths may be approaching the cycle time so that the DVFS system can respond appropriately to prevent a system failure.

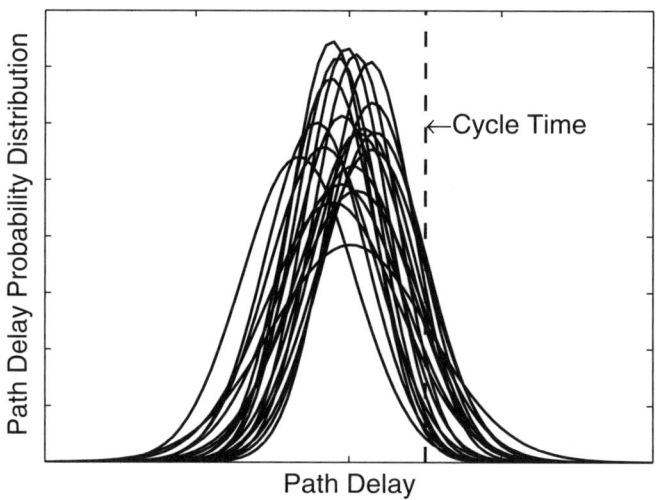

Figure 7.2 Representative probability distribution of critical paths in a microcircuit showing placement of the cycle delay. The location of the cycle time in the process spread is determined by the desired yield.

7.3 Sources of Path Delay Variability

There are a number of noise processes that cause timing variability in an integrated circuit. It is important to understand these processes and their effect on the number and location of critical path monitors.

7.3.1 Process Variation

Random, uncorrelated variations [1] cause two transistors carefully matched and sitting next to each other to operate differently. Because of this, a large-enough number of critical path monitors is needed to capture meaningful estimates of the mean and standard deviation of the random variations. Fortunately, the effect of random variation on delay is reported to be small [32] since these variations average out, so they are not a driver in determining the needed number of critical path monitors.

Systematic variations [20], which may also be random, are correlated and make a smooth transition across a die or a wafer. If there is little within die variation, then a single critical path monitor per die will be sufficient to capture the die-specific timing variation. For die that has significant inter-die variation, because the variation is correlated, critical path monitors located near the circuits being monitored should be sufficient to capture the systematic variation. The actual number of needed critical path monitors will depend on how the noise process varies, but for large microprocessors that only have on the order of tens of critical paths, only a few critical path monitors are needed as long as they are located close to the critical paths to capture the systematic variation.

7.3.2 Environmental Variation

Environmental noise processes are a function of the operating point of the microprocessor. Some timing variations, such as clock jitter, clock skew, path jitter, aging, and NBTI, have an uncorrelated, random component. A single critical path monitor can track these random changes if sampled over time as they will have a zero mean and constant variance across the integrated circuit. Random, uncorrelated variation accounts for a small portion of the variation in timing caused by environmental conditions [32], so most timing variation is a result of the workload of the integrated circuit [9], [33]. All of the environmental noise processes listed above with the addition of temperature and power supply noise correlate directly to the power consumed in the integrated circuit, which is a function of the workload of the circuit. In order to correctly monitor timing, critical path monitors must be located close enough to where the noise is occurring to detect its effect on critical path timing. Each of the environmental effects has a different time and spatial constant that determines how many sensors are needed to measure how the critical path's timing responds.

Temperature [34], [16] resolves on the order of milliseconds and has a spatial constant around 1mm: the temperature in any 1mm square is approximately equal. A critical path monitor located near high-power density circuits will track the temperature-induced timing changes of those circuits. The number of sensors is determined by the number of regions of high-power density on the integrated circuit.

Supply voltage [19], [31] variation has a much shorter time constant. The initial depth of a voltage droop, ΔV, is determined by the effective decoupling capacitance, C_{dc}, and the amount of current drawn, I, over a time period, Δt, as given by

$$\Delta V = \frac{I\Delta t}{C_{dc}}.$$

(7.3)

The duration of the voltage droop is a function of the RLC characteristics of the power supply network and its ability to provide enough current to boost the power supply backup to its nominal value. In integrated circuits where decoupling capacitance is insufficient, but a robust power supply distribution exists, voltage droops will be large, but short lived. Adding additional decoupling capacitance will slow down and reduce the amplitude of voltage droops.

In a 65nm, dual-core processor designed to test the performance of the power supply distribution, large changes in the number of registers used in each cycle resulted in voltage droops around 150mV that lasted several nanoseconds. A voltage droop caused by activity changes in one core traveled to the second core on-chip in around 4ns where it was attenuated by the capacitive load of the second core. A large droop in both cores at simultaneous moments caused a large drop in the overall supply voltage [19].

Because power supply droop travels from where the current draw is highest to other parts of the integrated circuit, relatively few critical path monitors are needed to detect them, as even a single critical path monitor will eventually see the attenuated supply droop. Of more importance is how soon after its occurrence the droop needs to be detected. In DVFS systems that track the supply noise, more critical path monitors will be needed, and they will need to be located close to the circuits most responsible for dynamic current draw. For slower systems, fewer monitors are needed.

Clock jitter and skew are largely dependent on the power supply noise. The value of each of these noise processes depends on the stability of the switching points of the logic gates in the clock distribution and in the logic paths. As power supply noise increases, the switching point of the logic gates changes, injecting the power supply noise into the clock distribution [8].

Placing sufficient critical monitors to track power supply noise should also capture clock jitter and skew.

Aging [3] and NBTI [12] have long time constants, but their spatial constant can be quite small. General aging across a chip will be tracked by a single critical path monitor, but some aging processes may affect a single transistor. The best response to tracking these types of changes in timing is to locate the critical path monitors close to the most active circuitry, which sees the widest swing in environmental conditions.

7.4 Timing Sensitivity of Path Delay

In order to build an effective critical path monitor, it is essential to understand the sensitivity of path delay to noise. The typical logic path begins at a latch and ends at a latch: on receipt of a clock signal, the data is passed through the logic from the source latch to the final latch. SRAM critical paths are more complicated than logic paths because the control signal often crosses supply voltage boundaries and interfaces with analog sense-amps. Because of this, we will ignore the intricacies of SRAM and just deal with the timing of regular logic.

Figure 7.3 shows a simplified model of a critical path consisting of logic elements driving equal lengths of wire [17]. Most any logic path can be reduced, to the first order, to a buffer-driven delay-line model by converting any gate with multiple fan-in to an equivalent inverter. The wire length of each segment is adjusted to match the wire length between gates. Fan-out is added as additional gate capacitance load at a given stage. While these modifications can tailor the model in Figure 7.3 to most any logic path, for this analysis, it is simpler and sufficient to analyze the path as a simple buffered delay line.

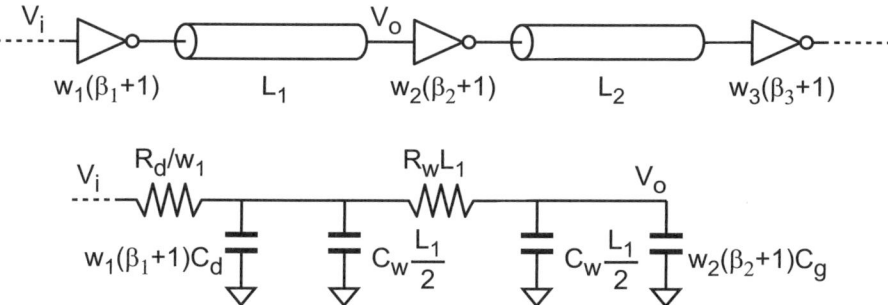

Figure 7.3 A simplified model of a delay line based on the theory developed in [17].

Some commonly used simplifications are helpful for estimating the delay per segment in the delay line in Figure 7.3. The on current of a drive transistor can be approximated to first order as

$$I_{on} = C \frac{dV_{out}}{dt}, \qquad (7.4)$$

where C is the load capacitance of the gate. Equation (7.4) can be rearranged, under the simplifying assumption that I_{on} is constant and that dV_{out} changes linearly from V_{DD} to 0, as

$$D = t = C \frac{V_{DD}}{I_{on}} = C \frac{R_d}{w}. \qquad (7.5)$$

Equation (7.5) is pessimistic because it is based on the property that all charge on the wire, gate, and drain capacitance is removed [17], but in reality, only part of the charge is removed before the load gate switches and the signal is passed to the next stage of logic.

Using Equation (7.5), a generalized expression for the delay in one section of the path in Figure 7.3 is given by [17]:

$$D = a\left\{ \frac{R_d}{w}\left[w(\beta+1)(C_g + C_d) + lC_w\right] + l^2 \frac{R_w C_w}{2} + lR_w w(\beta+1)C_g \right\}. \quad (7.6)$$

R_d is the equivalent resistance of the gate and is approximated to first order by V_{DD}/I_{sat} of the transistor and its units are $\Omega \cdot cm$. The width of the equivalent NFET is w, β is the pfet/nfet ratio, l is the length of the wire segment, C_g is the capacitance/width of the gate, C_d is the capacitance/width of the drain, and R_w and C_w are the resistance and capacitance per unit length of the wire. For buses, the value of l is large, while for dense logic, the value of l can be quite small. The coefficient a, which typically has a value of 0.7, is a factor that accounts for the non-ideality of the input slope and the pessimism of Equation (7.6).

If there are multiple delay stages in a path, each with a different equivalent inverter and wire length, the path delay can be approximated by

$$D_{path} = \sum_1^n a_n \left\{ \begin{array}{l} \dfrac{R_{dn}}{w_n}\left[w_n(\beta_n+1)(C_{gn+1}+C_{dn})+l_n C_{wn}\right]+ \\[2mm] l_n^2 \dfrac{R_{w1}C_{w1}}{2}+l_n R_{w1} w_{n+1}(\beta_{n+1}+1)C_{gn+1} \end{array} \right\}. \qquad (7.7)$$

Equation (7.7) is an Elmore approximation to the delay of the line. If there are n inverters of the same length driving the same wire load, Equation (7.7) reduces to

$$D_{path} = an\left\{\begin{array}{l} \dfrac{R_d}{w}\left[w(\beta+1)(C_g + C_d) + lC_w\right] + \\ l^2\dfrac{R_w C_w}{2} + lR_w w(\beta+1)C_g \end{array}\right\}. \qquad (7.8)$$

From Equation (7.8), the stage delay, D_{stage}, is simply D_{path}/n.

The general equation for the sensitivity of a parameter to small changes in one of its variables is given by

$$S_x^y = \frac{dy}{dx}\frac{x}{y}. \qquad (7.9)$$

To simplify the algebra needed to calculate sensitivity, the following variable substitutions can be made:

$$t_{wire} = l^2\frac{R_w C_w}{2} + lR_w w(\beta+1)C_g, \qquad (7.10)$$

$$C_{source} = w(\beta+1)(C_g + C_d) + lC_w, \qquad (7.11)$$

and

$$C_{load} = l^2\frac{C_w}{2} + lw(\beta+1)C_g. \qquad (7.12)$$

The value t_{wire} is the delay caused by the wire and the value C_{source} is the capacitance seen by the source driver. To further simplify, let the value of the wire delay equal some fraction of the delay of the driver,

$$l^2\frac{R_w C_w}{2} + lR_w w(\beta+1)C_g = \gamma\left(\frac{R_d}{w}\left[w(\beta+1)(C_g + C_d) + lC_w\right]\right), \qquad (7.13)$$

where γ is the proportionality constant of wire delay to gate delay.

Substituting Equations (7.11) and (7.13) into Equation (7.8) gives

$$D = an(1+\gamma)\frac{R_d}{w}C_{source}. \qquad (7.14)$$

Variations in a transistor's driver strength are manifest in changes in the output resistance, R_d. The derivative of Equation (7.14) with respect to R_d is

$$\frac{dD}{dR_d} = an\frac{1}{w}C_{source}\left(1 + \gamma \mid R_d \frac{d\gamma}{dR_d}\right),$$ (7.15)

From Equation (7.13), $d\gamma/dR_d$ is

$$\frac{d\gamma}{dR_d} = \frac{-\left[l^2\dfrac{R_wC_w}{2} + lR_ww(\beta+1)C_g\right]}{\dfrac{R_d^2}{w}C_{source}} = -\frac{\gamma}{R_d}$$ (7.16)

Combining Equations (7.9) and (7.13)–(7.15) gives the sensitivity of delay to transistor output resistance as a function of wire versus FET delay:

$$S_{R_d}^D = an\frac{1}{w}C_{source}\frac{R_d}{an(1+\gamma)\dfrac{R_d}{w}C_{source}} = \frac{1}{1+\gamma}.$$ (7.17)

The sensitivity of the delay to the effective output resistance of the driver is a function of the percentage of wire delay versus FET delay in the delay line. Path delay can be written as the sum of the FET delay and the RC delay:

$$D_{path} = D_{FET} + D_{RC}.$$ (7.18)

The percentage of delay in the wires, D_{RC}, is often given as a percentage, η, of path delay:

$$D_{RC} = \eta D_{path}.$$ (7.19)

Rewriting Equation (7.13) as

$$D_{RC} = \gamma D_{FET},$$ (7.20)

then substituting Equations (7.19) and (7.20) into Equation (7.18) allows us to calculate γ and R_d when the percentage of wire delay is known:

$$\gamma = \frac{\eta}{1-\eta}.$$ (7.21)

Equation (7.18) is only valid for values of η that are realistic (Equation (7.18) predicts an infinite γ for η approaching 1, a path composed completely of RC delay).

The maximum percentage of path delay in the wires (RC delay) is 50% in repeated lines and less than 25% in pipeline stages [32]. The 50% RC delay limit, even for long repeater driven paths, is due to two primary factors: first, design rules limit the length of wire driven by each repeater to minimize noise; second, all paths are latch to latch, so there is significant FET delay in the launching and capturing of data signals. If 50% of the delay is in the wires, then the wire and FET delay are equal and $\gamma = 1$. If 25% of the delay is RC delay, $\gamma = 1/3$. The delay sensitivity due to R_d varies from 0.5 to 1 as shown in Figure 7.4.

Using a similar derivation as above, the sensitivities of delay to other parameters can be computed. Without giving the steps in the derivation, some of these are as follows. The path delay sensitivity to length is given as

$$S_l^D = \frac{\dfrac{R_d}{w} C_w l + R_w C_w l^2 + R_w w (\beta + 1) C_g l}{\dfrac{R_d}{w} \left[w(\beta + 1)(C_g + C_d) + C_w l \right] + \dfrac{R_w C_w}{2} l^2 + R_w w (\beta + 1) C_g l}, \quad (7.22)$$

which has a value that ranges between 0 and 1. While Equation (7.19) is messy, some assumptions can be made to simplify its analysis. The

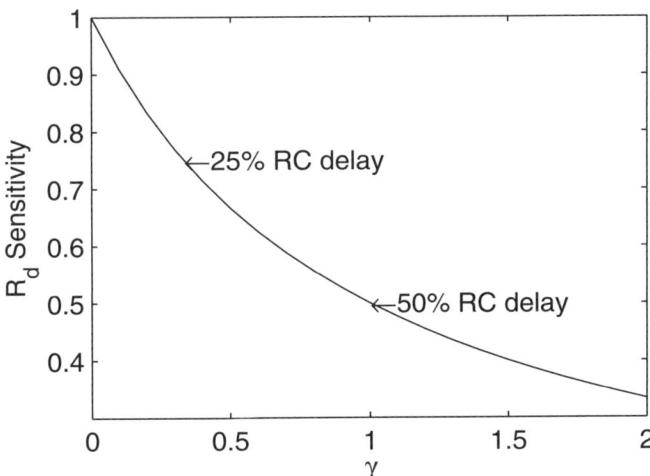

Figure 7.4 Delay sensitivity to R_d as a function of γ, the ratio of wire to gate delay in a delay path.

denominator of Equation (7.19) is the stage delay without the correction factor a in Equation (7.8). The numerator consists of the following Elmore delay components: the output driver resistance and the total wire capacitance, two times the delay of the unloaded wire, and the wire resistance and the load capacitance. If it is assumed, as most design rules stipulate, that the rise times at the receiving end are reasonable, then the wire delay is no more than 50% of the path delay, and the numerator can be approximated as two times the RC delay of the path. The uncertainty arises from the fact that the third term in the numerator is not multiplied by 2 as it would be if the second and third terms equaled exactly twice the wire delay. The addition of the first term makes up for some of the uncertainty. Using these approximations,

$$S_l^D \approx \frac{2D_{RC}}{D_{stage}},\tag{7.23}$$

so the path delay sensitivity to length ranges between 0.5 and 1 for RC delays 25–50% of the path delay.

The path delay sensitivity due to width is given by

$$S_w^D = \frac{-\dfrac{R_d}{w}C_w l + R_w w(\beta+1)C_g l}{\dfrac{R_d}{w}\left[w(\beta+1)(C_g+C_d)+C_w l\right]+\dfrac{R_w C_w}{2}l^2 + R_w w(\beta+1)C_g l},\tag{7.24}$$

which has a value that ranges from –1 to 1 and has a value of zero when

$$\frac{R_d}{w}C_w l = R_w w(\beta+1)C_g l.\tag{7.25}$$

The numerator of Equation (7.24) consists of the following Elmore delays: driver resistance and wire capacitance and wire resistance and load capacitance. The denominator is simply the stage delay without the factor a in Equation (7.8). A repeater stage designed with equal wire delay and FET delay has 60% of the capacitance in the wires [17]. If the output resistance and wire resistance are also equal, which is often a design goal, then the numerator of Equation (7.24) is $-0.4RC+0.6RC$, for a path delay sensitivity of approximately $0.2RC/D_{stage}$. Notice that the length term, l, falls out of Equation (7.25), so the sensitivity of the path delay depends mostly on the ratio of driver resistance to wire resistance. For delay paths with little wire delay, the second term of the numerator falls out and the sensitivity is approximately

$$S_w^D \approx \frac{-\dfrac{R_d}{w} C_w l}{D_{stage}} .$$ (7.26)

Because the numerator of Equation (7.26) is a component of the delay, and a small one for short wires, the path delay sensitivity to width changes will be small.

The path delay sensitivity to temperature is found by replacing the resistors by a simplified linear resistance model,

$$R_T = R_o + \alpha T ,$$ (7.27)

which is used to represent changes in resistance to small changes in temperature. Using Equations (7.11), (7.12), and (7.25) in Equation (7.8) gives

$$D = an \left\{ \frac{R_d + \alpha_d T}{w} C_{source} + (R_w + \alpha_w T) C_{load} \right\} .$$ (7.28)

The variables α_d and α_w are the temperature coefficients for the driver and wire resistance, respectively. The sensitivity of Equation (7.27) with respect to temperature is

$$S_T^D = \frac{\left(\dfrac{C_{source}}{w} \alpha_1 + C_{load} \alpha_2 \right) T}{\left(\dfrac{C_{source}}{w} \alpha_1 + C_{load} \alpha_2 \right) T + \dfrac{R_d}{w} C_{source} + R_w C_{load}} ,$$ (7.29)

which ranges between 0 for low temperatures and 1 for high temperatures. The numerator of Equation (7.29) consists of two Elmore delay components: the change in resistance due to temperature of the driver and the driver capacitance, and the change in resistance due to temperature of the wire resistance and the load capacitance. Notice that path delay sensitivity increases as temperature increases.

This analysis indicates that, to first order, the sensitivity of the delay to small changes in any of its parameters is never greater than 1. This information is important when determining what type of circuit and path is most important when deciding how to monitor critical paths.

7.5 Critical Path Monitors

Critical path monitors are generally used as part of a closed loop DVFS control system. A number of critical path monitors in association with DVFS systems have been reported in the literature [2], [9], [10], [24] While the specific details of the implementations vary, they all share a basic structure similar to the block diagram shown in Figure 7.5. The operation of a critical path monitor is as follows: the system clock triggers the launch of a timing signal into a delay path; after the delay of the clock period, the phase of the timing signal and the system clock is captured by some time-to-digital conversion and compared to the expected phase; the difference between the captured and the expected phase indicates the amount of slack available in the timing. A block of logic is added to control the critical path monitor for operation and testing, and calibration data is maintained to provide the needed sensor accuracy. Each of these components will now be discussed.

7.5.1 Synchronizer

The first component in Figure 7.5 is a synchronizer that times the launch of the timing signal to coincide with the system clock. The synchronizer is most often a latch or a pulse generator. Since critical paths exist from latch to latch, it is advantageous for the timing signal to be generated by a latch, to capture the clock-to-data timing variance accurately.

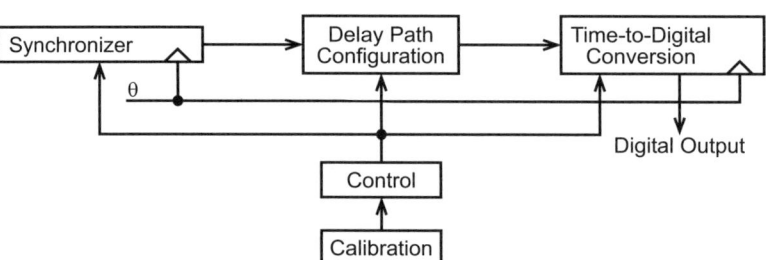

Figure 7.5 A simplified block diagram showing the basic building blocks inherent in most published critical path monitors.

7.5.2 Delay Path Configuration

The second component in Figure 7.5 is the delay path configuration which is used to synthesize the critical path of the integrated circuit. Several path types are used in the literature, but they all have one of the two forms shown in Figure 7.6. The parallel paths type uses multiple paths which can be individually selected or selected in parallel. Because the critical path can change with the operating point of the integrated circuit, selecting paths in parallel allows different paths to be combined for a synthesized path that would be difficult to design by itself. For example, the two paths may include a wire-dominated path and a FET-dominated path that when combined (selecting the slowest path using an AND gate) provide a mixed path. The serial delay paths use a multiplexing scheme to change the percentage of FET and RC delay in the delay path. While the most accurate approach to critical path selection is to place the critical path monitor in the critical paths themselves [2], the critical path can be synthesized using a delay line that varies the amount of RC versus FET delay [10], [24], or by a small group of representative paths in parallel [9].

The largest timing sensitivity in delay paths is to voltage (R_d as a function of γ in Equation (7.13)). Figure 7.7 shows a graph of how path delay changes as a function of the ratio of RC versus FET delay while following the strict design rules used for a microprocessor. Eight paths were simulated: Path 1 consisted mostly of RC delay, with RC delay decreasing from Path 1 to Path 8. As predicted by Equation (7.13), Path 1 has the least delay change with voltage change due to its high wire delay content. However, instead of having continuously varying slopes as wire delay is

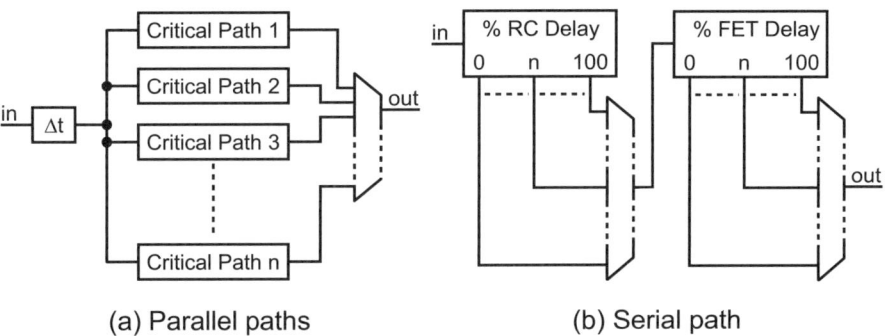

(a) Parallel paths   (b) Serial path

Figure 7.6 Block diagrams of the two basic path types used for critical path synthesis. In (a), parallel paths, where each path has different timing characteristics, are selected as the synthesis path. In (b), a serial mix of RC and FET delay are combined to synthesize the critical path.

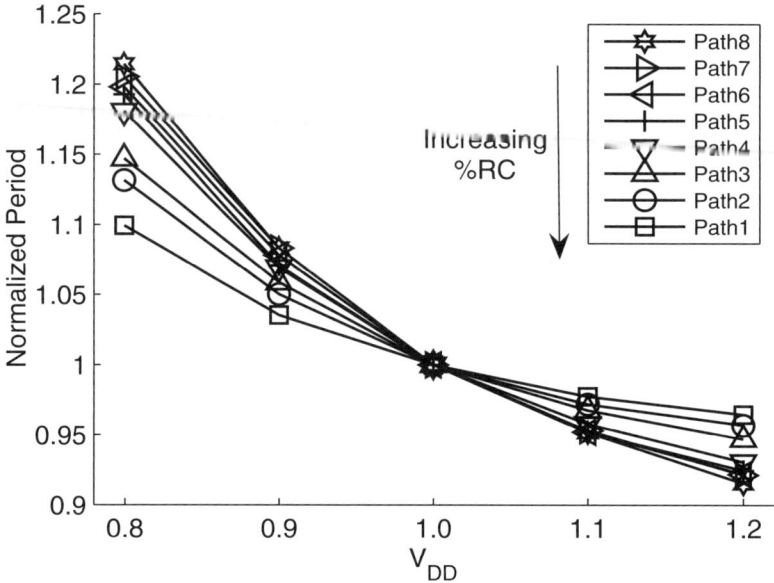

Figure 7.7 Normalized path delay of a 65nm process as the ratio of wire versus FET delay is changed. Path wire length varies from 10μm to 560μm. Each path is simulated separately with no programming overhead. Simulations were at the nominal corner and at 85°C with a target path delay of 200 ps.

reduced, the slopes bunch themselves into three groups. The large gain of each added inverter in the delay path quickly reduces the percentage of wire delay in the path. This is most apparent in low FO4 designs.

A variety of gate types is used in any design. Figure 7.8 shows the normalized delay of five different delay paths: an adder path consisting of a mix of XOR, NOR, and NAND gates; a wire path consisting of a series of buffers separated by long wires; a pass-gate path consisting of a series of buffers separated by a number of pass-gates in series: essentially an FET wire; and NAND and NOR gate paths consisting of a series of 4-high NAND and 3-high NOR gates respectively. Simulations were performed at two frequencies, F and $F/3$ where F was 4.5 GHz, to demonstrate the changes in sensitivity with increasing clock frequency. There are three distinct slopes due to the wire, gate, and pass-gate sensitivities to voltage, however; the NAND and NOR gates seem to be no more sensitive than the adder gates. In fact, gates, as long as they have sufficient gain, have remarkably similar sensitivities, regardless of stacking and arrangement. Reducing the frequency brings out sensitivity differences between the paths, for example, the pass-gate path delay increases by 32% at 0.8 V when the

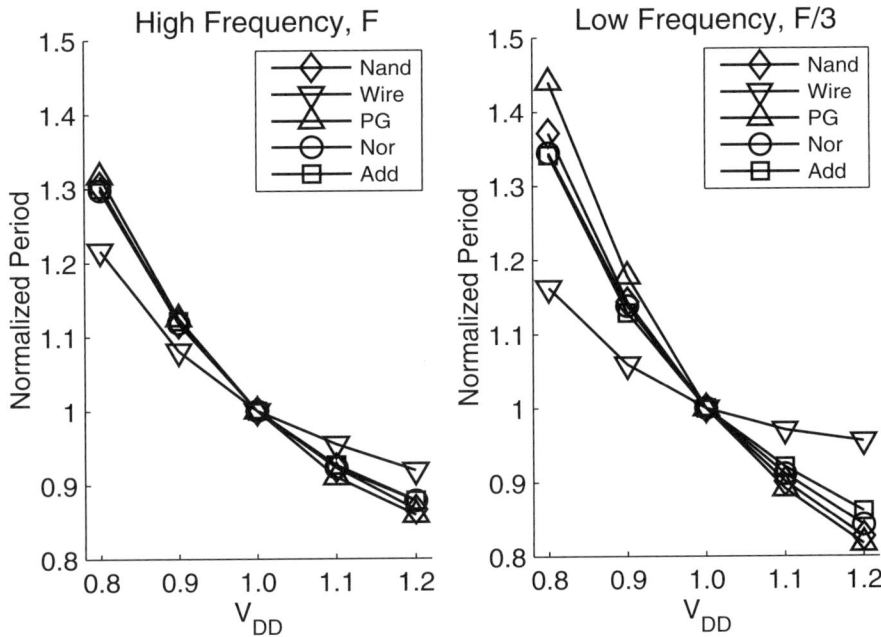

Figure 7.8 The normalized delay of five different gate types at two different frequencies is shown. Each path was normalized to its delay value at 1V to show the percentage change of delay as voltage changes. Paths were simulated separately without any programming overhead. Simulations were performed at the nominal process corner and 85°C for a 65nm process.

frequency is *F*, but 45% when the frequency is *F*/3. This is because there are more gates "seeing" the change in voltage. This is not meant to imply that high fan-in gates are just as fast as inverters, just that the difference in sensitivity in high fan-in gates versus inverters is small.

The paths simulated in Figure 7.7 and Figure 7.8 do not include the setup and hold times of the latch-to-latch timing, nor the required muxes and gates needed to configure the timing paths. The launching, capturing, and programming overhead can reduce the amount of time spent in the delay paths by as much as 8–10 FO4, which reduces the sensitivity differences between paths composed of differing delay elements. Figure 7.9 plots the measured normalized path delay of the critical path monitor described in [9] at the two target frequencies. Since FET delay is similar, only the wire and NOR path delays are shown, along with the normalized minimum cycle time of the microprocessor tested. The wire path tracks the critical path very closely in the high-frequency core regions; there is little

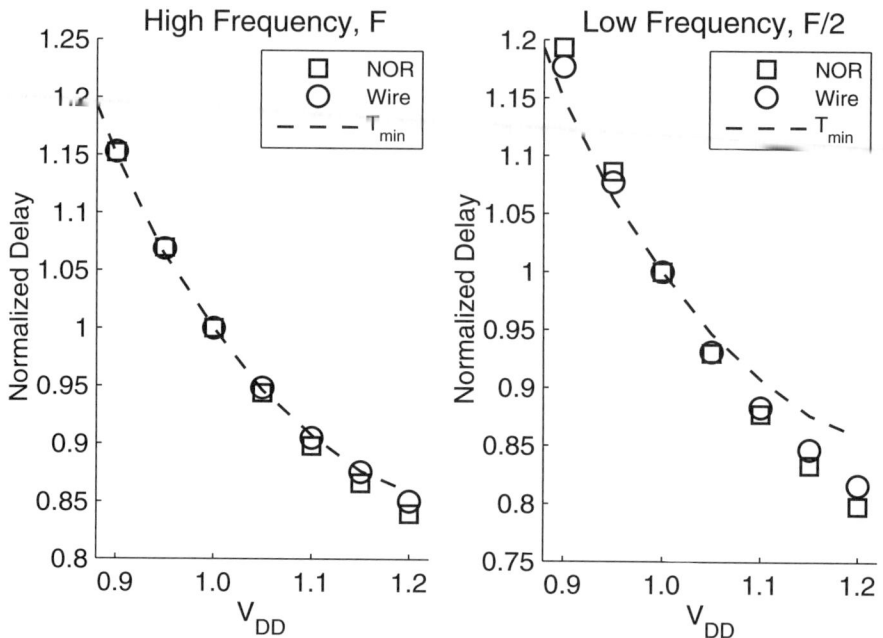

Figure 7.9 Measured path delay in critical path monitor designed for a high-performance microprocessor in 65nm technology [9]. The delays are normalized to the delay at 1V to show the percent change in delay for each path type. Two frequency domains were measured: the high-speed core and low-speed non-core. The minimum period of the microprocessor, T_{min}, is also plotted as a dashed line.

difference between the wire path and the NOR path; the slope differences seen in the simulations in Figure 7.8 disappear once all the overhead circuits are added, even for the much lower frequency non-core regions. In this particular process, extra slack was required for the non-core regions, explaining the deviation of the path delay from the critical path at high voltages.

More complexity has been added to selecting the gate types needed to correctly synthesize the critical path delay than is warranted. For custom microprocessors, there may only be tens of critical paths while ASICs may have up to hundreds of critical paths. Each of these paths will have a fixed skew depending on the gate type, but that can be calibrated out. Simulations and measurements indicate that some combination of FET and RC delay is needed, but the resolution or how much of each need not be very fine, especially for low FO4 designs. Fortunately, design rules limit the extreme gate types that would cause the most significant varition and require that edge rates and wire lengths be eliminated to reduce noise. Following the design rules allows for simpler critical path monitors to be designed.

7.5.3 Time-to-Digital Conversion

The third component of Figure 7.5 is the time-to-digital conversion that turns the path delay into a digital signal using the system clock as a reference. The system clock is the most accurate signal on most integrated circuits, especially for microprocessors, and provides the most accurate reference for timing conversion: 2GHz clocks may have skew as low as 9ps and a timing variability (combined jitter and skew) less than 25ps (5% of the clock) [30]. Conveniently, the system clock is also the signal that determines what the timing is, so jitter-caused delay changes in the critical paths will correlate to delay changes in the critical path monitors. In most types of measurement circuits, such as a digital-to-analog converter, it is preferable for the conversion circuitry to not be sensitive to the noise processes that affect the signal being measured; however, for critical path monitors, the reverse is desirable. Timing margin is the quantity desired from a critical path monitor, and it needs to include the variance of the clock as well as the variance of the capture latches. Making the measurement circuit immune to the process variation reduces the sensitivity of the critical path monitor and will cause it to not track the critical path circuits.

The time-to-digital conversion used in critical path monitors is a phase comparison between a clock-synchronized signal (typically the clock itself) and a delayed signal sent through the critical path synthesis circuit. The comparison is made once per cycle, providing very high bandwidth measurements.

All latches are phase comparators. Figure 7.10 catalogs a number of latch configurations that can be used for phase comparison. The first, Figure 7.10a, is a simple D flip-flop where the two input signals have similar phases. If φ_1 arrives before φ_2, the output is one, otherwise, it is zero. This phase comparator is simple to design and small. It is, however, subject to metastability problems as will be discussed later, and it often has different timing for latching rising and falling edges which must be taken into account when designing the critical path monitor. Metastability can be reduced by reducing edge rates and by adding a second flip-flop stage to give the metastable flip-flop time to resolve.

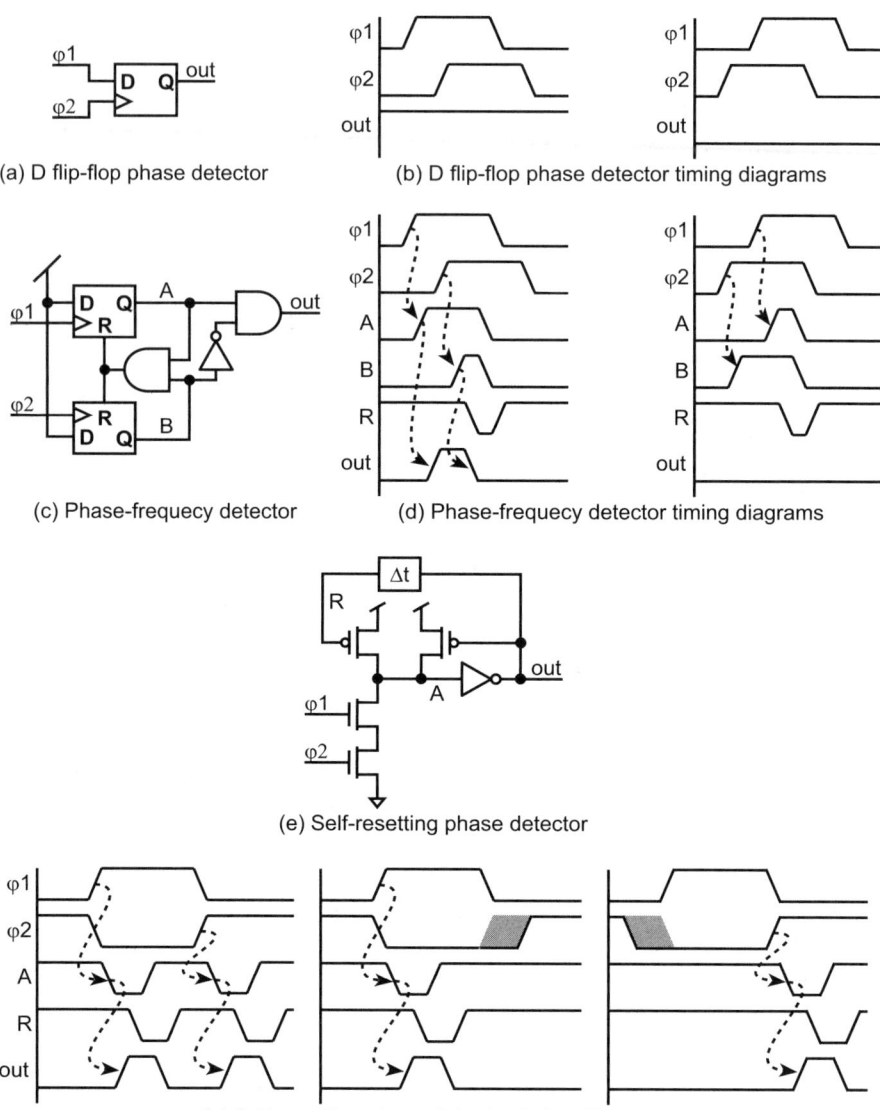

(a) D flip-flop phase detector

(b) D flip-flop phase detector timing diagrams

(c) Phase-frequecy detector

(d) Phase-frequecy detector timing diagrams

(e) Self-resetting phase detector

(e) Self-resetting phase detector timing diagrams

Figure 7.10 A catalog of phase comparison latches and their respective timing diagrams. The first two phase detectors in this figure were derived from circuits described in [29].

Figure 7.10c is a phase frequency detector that is a self-resetting circuit. When signal φ_1 arrives, it latches a 1 into its latch, which remains until φ_2 arrives. When both latches are set, it sends a reset signal to clear the latches to await the next input. An AND gate must be used to suppress the pulse that occurs when φ_1 arrives after φ_2. This phase comparator is useful when both timing signals are of the same frequency. Because the comparison is always based on the same clock edge (rising or falling), it eliminates zero versus one latching differences. Metastability problems are eliminated because the latch input is always stable when the clock signal arrives.

Figure 7.10e is a self-resetting phase detector based on a dynamic latch that allows a phase comparison to be made on both edges of the timing signal. The phase comparison is made by determining whether the rising edge of φ_1 arrives before the falling edge of φ_2. If it does, a one is latched. The same comparison is then made on the falling edge of φ_1 and the rising edge of φ_2. The reset time is determined by Δt. Because the rising and falling edges are both compared, in practice it is difficult to tell which pair of edges has caused the phase comparison to trigger since it is difficult to track to current edge. If one of the signals being phase compared is chopped, then the comparison is only made on one of the edge sets. This allows the phase detector to have a multiplexer built into it. Metastability is eliminated in this phase comparator because the dynamic node does not have a metastable state. Because both of its inputs are full-swing signals, the pull-down network will either discharge or not, so no metastable state is created. The width of the output pulse of this latch indicating a phase mismatch is determined by the length of the delay added between the latch output and the pre-charge transistors. This delay must be sufficient to create a pulse wide enough to meet the timing requirements of any latches located after the phase comparator.

The basic phase comparators described above can be combined in a number of ways to perform time-to-digital conversion. Figure 7.11 shows a single bit design where the clock signal is delayed between two flip-flops that receive the same data. If the delay of the critical path is long compared to the clock cycle and the clock delay is long enough to cover all required design margins, the contents of the two latches will differ and an error is signaled to the system. The clock delay may be small to capture fine

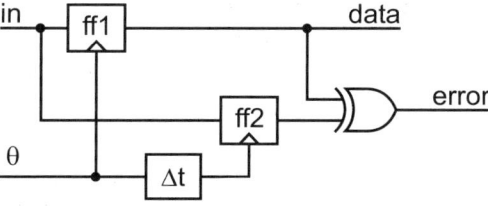

Figure 7.11 Block diagram of a delayed clock phase comparator used as a building block in the Razor latch [2]. A master flip-flop and a secondary flip-flop both capture the data, but the second flip-flop does so using a delayed clock. If the flip-flop contents differ, an error is sent to the system.

timing differences when the timing is on the edge, or large to cover all possible margins for robust timing. The Razor latch uses the phase comparator in Figure 7.11 as a basis for its design.

If a delay is added to one of the phase inputs of the phase comparator, then several comparators can be combined in series to form a more complicated time-to-digital converter similar to the one shown in Figure 7.12. This configuration is often called an edge detector because the location of the edge with respect to the synchronizing signal within a bank of latches determines the phase difference. The timing difference between latch bits is fixed by the delay line driving the latches. The example shown in Figure 7.12 utilizes a level scheme: the level of the timing signal inverts with each clock cycle. The converter holds two timing edges at any given time. A pulsed timing signal can also be used with this edge detector. The output of the bank of latches can be converted to a thermometer code by comparing the latch output to the expected value of the level, or a bit-wise XOR can be used to convert the output into an edge position. This type of time-to-digital conversion is convenient for DVFS systems that respond to timing changes over several thousand clock cycles.

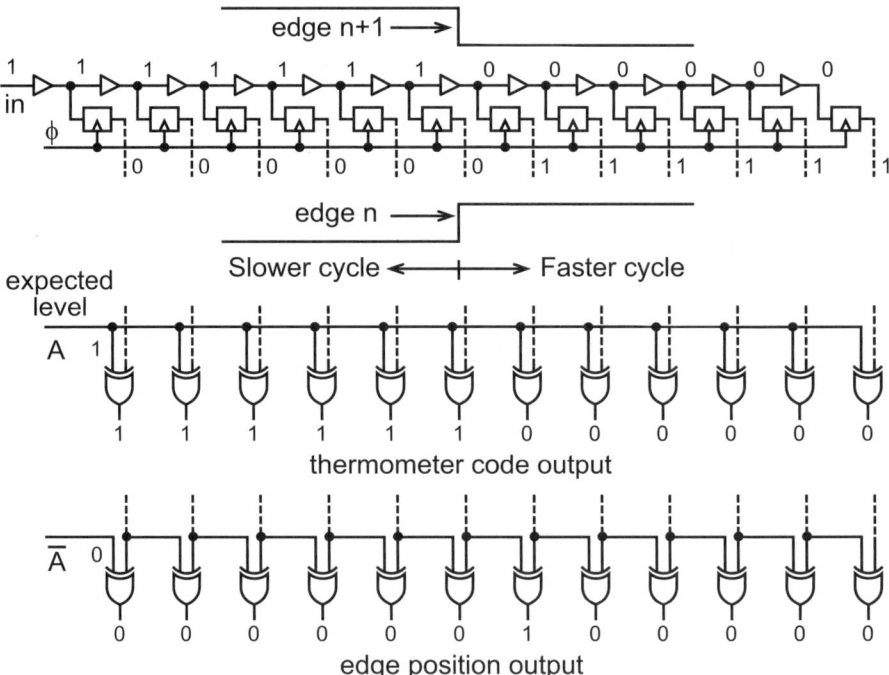

Figure 7.12 Multiple-bit time-to-digital converter. Derived from figure in [9].
[©IEEE 2007]

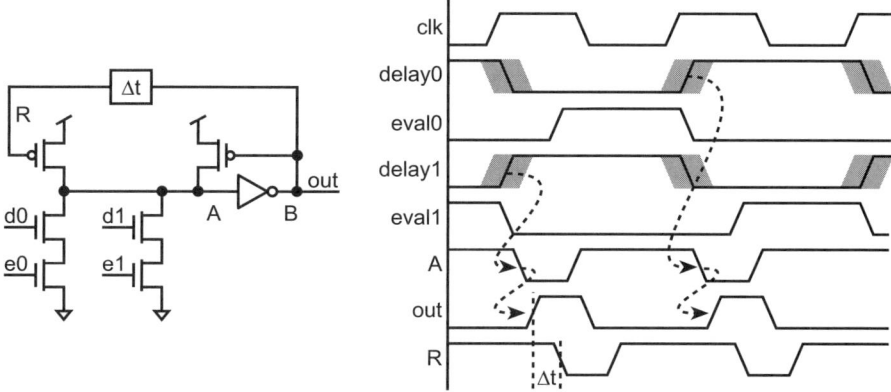

Figure 7.13 A multiplexing time-to-digital converter for multiple timing signals.

A multiplexing time-to-digital converter can be created by adding a second pull-down path in the self-resetting phase comparator in Figure 7.10e. If the *eval* signals are chopped as shown in the timing diagram of Figure 7.13, then two different sets of signals can be phase detected on alternating clock cycles. A technique similar to this was used in [14]. The advantage of this converter is it allows two delay paths to be used: the first can be pre-set, or cleared, while the second has its delay measured, and then the first is measured and the second is cleared on the following cycle.

7.5.3.1 Sensitivity

The sensitivity of the critical path monitor is largely determined by the time-to-digital converter. The per-bit sensitivity is the amount of delay between each phase of the inputs to the phase comparator as well as the amount of delay change built up in the delay line. For example, in a 65nm critical path monitor recently developed using a thermometer-code converter [9], the sensitivity per bit was 20mV AC voltage, 10mV DC voltage, 1 FO2 of clock jitter, and about 10°C of temperature change. Any noise that caused a delay change less than the 1 FO2 inverter delay in the delay line would be missed. Using an interpolating delay line such as that shown in Figure 7.14 would improve the sensitivity, but there is a noise floor caused by the jitter of the clock and the random variations in timing delay in the inverters and latches themselves. Time-to-digital converters that use a Razor-type approach are as sensitive as the delay to the slave latch.

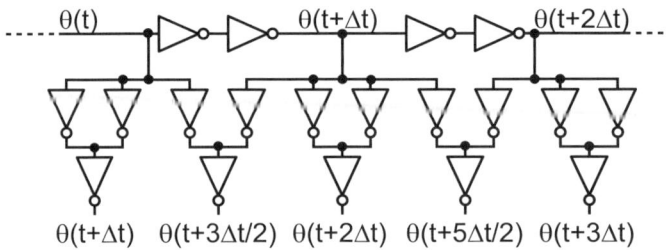

Figure 7.14 A schematic for an interpolating delay line useful for producing phase differences less than the delay of an inverter [15].

7.5.4 Control and Calibration

The control and calibration blocks in Figure 7.5 hold the necessary data for the critical path monitor to function. The control block may also have data collection capabilities. Some critical path monitors provide data once per cycle as part of a high-speed feedback loop, but others are only read once every million cycles. Lower speed monitors must maintain some record of the critical path such as worst delay, or average delay. If sufficient area is available, significant statistical data about the noise processes can be accumulated in the critical path monitor.

The calibration can be a simple fuse bank that indicates the appropriate delay setting to use, or may be a complex look-up table. Simpler calibration is preferred because calibration time is expensive.

In an ideal critical path monitor, the critical path will track exactly with the maximum frequency of the microprocessor and only one calibration measurement would need to be taken. However, in practice, there will be some skew between the critical path monitor and the critical paths that shows up as error that must be accounted for in the timing margins. One advantage of thermometer-coded critical path monitors is that a multiple-point calibration can be used to measure the failure position at environmental extremes, and interpolation used to calculate the failure bit position in between these readings, reducing the error. Single bit monitors do not have this advantage and whatever error occurs between the critical path monitor and the critical paths must be accounted for in the timing margin. Figure 7.15 is a graph of the change in the failure bit position of a thermometer-code-based critical path monitor as a function of voltage. There is a nearly linear increase in failure bit position as voltage increases. A two-point calibration would provide for accuracy of the critical path timing measurement of 1.5 bits in this design.

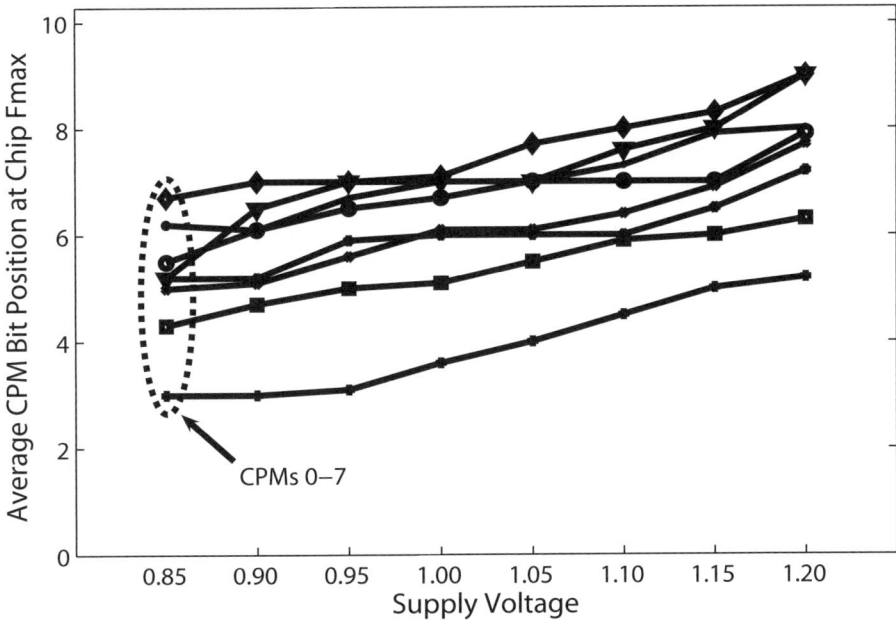

Figure 7.15 This graph shows the failure bit position of the critical path monitors in one core as well as the normalized frequency as described in [9] [©IEEE 2007]. Notice the nearly linear increase with failure position as voltage increases.

7.6 Conclusion

Table 7.1 lists the characteristics of a number of critical path monitors described in the literature. Column order is by date of publication.

The Razor latch is the smallest of the monitors and utilizes the critical paths themselves as the timing path. Because of this, it is the most accurate in its "emulation" of the critical path. As a small monitor, it can be distributed extensively throughout the system, but it adds a multiplexer and some gate loading to the critical path which may not be acceptable in high-performance designs.

The critical path monitor in [24] is a single monitor, although it could probably be distributed more extensively as it does not appear, from the description, to be large. It utilizes a thermometer-code output.

The critical path monitor in [14] is used as part of a voltage following feedback loop. While it is described as a regional voltage detector, it uses critical path timing to sense voltage changes. This monitor provides high-speed 2-bit data to a control loop that adjusts frequency to respond to voltage changes.

The critical path monitor in [9] is part of an out-of-band control loop that adjusts voltage and frequency to respond to the operating point of a microprocessor. This monitor has a 12-bit thermometer-code output. One advantage of a thermometer-code output is the ability to quantify noise processes in test and debug, as well as during operation. This monitor provides a sampling function as well as maintaining the worst-case delay since it was last read.

Table 7.1 Comparison chart of different critical path monitors in the literature. Column order is by date of publication.

	[10]	[9]	[14]	[24]	[2]
Application	16×16 multiplier	IBM POWER6	Intel Montecito	MPEG4 decoder	64-bit alpha
Synchronizer	Flip-flop	Flip-flop	Finite state machine	Pulse generator	Flip-flop or Razor latch
Delay path	One serial path	Five parallel paths	Two synthesis paths: each has two serial paths in parallel	1 serial path	Embedded into actual critical path
Time-to-digital conversion	Flip-flop: 1 bit	Flip-flop: 12-bit thermometer code	Multiplexing latch: 2 bits	Flip-flop: n-bit thermometer code	Razor latch: 1 bit
Monitor density	1	8/core	12/core	1/chip	192/chip
Technology	0.18 μm	65 nm	90 nm	0.18 μm	0.18 μm
Approximate area[1]	>1000 flip-flops	215 flip-flops	> 100 flip-flops	>100 flip-flops	2–3 flip-flops
Target Frequency	90 MHz	4–5 GHz	2–2.5 GHz	8–123 MHz	200 MHz

[1] The monitor area is the approximate area as a multiple of the area of a single flip-flop, not the number of flip-flops in the monitor. This metric is used to allow comparisons to be made independently of technology. Target frequency has a large impact on area as it impacts the length of the delay lines used to synthesize the critical path. Area is based on published descriptions and, except for [9], does not include configuration, control, and test logic not described.

The critical path monitor in [10] has a clever self-calibrating scheme that adjusts the critical path settings based on the output of a process sensitive ring oscillator. This monitor is very large and could not be widely distributed around an integrated circuit without significant area penalties.

Voltage and temperature sensitivity are increasing with process scaling, making processor timing very susceptible to workload. Because workload is a systematic noise, critical path monitors allow DVFS systems (which have begun to multiply in recent years) to respond to the workload and improve the efficiency of the microprocessors. These circuits tend to be small, having the same area of roughly 100–200 flip-flops. They can be made sensitive to voltage, process, temperature, aging, NBTI, and workload while allowing the system to respond and adapt to these noise processes. Critical path monitors have been reported to be sensitive to changes in delay as small as an FO2 inverter delay ([14] showed a sensitivity of 1.5% of a clock period). In addition to providing accurate timing measurements, critical path monitors can be valuable tools in testing and debugging new integrated circuit designs. In order to make a critical path monitor worthwhile, it must provide enough accuracy to reduce the design margins allocated for environmental changes in the integrated circuit.

Acknowledgments

The contribution made by each of the following individuals is gratefully acknowledged: Robert Senger, Harmander Deogun, Gary Carpenter, Tuyet Nguyen, Jeremy Schaub, Soraya Ghiasi, Norman James, Michael Floyd, Phillip Restle, Scott Taylor, Kevin Nowka, Sani Nassif, Fadi Gebara, Robert Montoye, and Hung Ngo. Funding was provided in part under DARPA contract number NBCH30390004.

References

[1]   K. Agarwal and S. Nassif, "Characterizing Process Variation in Nanometer CMOS," *DAC*, 4–8 June 2007, pp. 396–399.

[2]   T. Austin, D. Blaauw, T. Mudge, and K. Flautner, "Making Typical Silicon Matter with Razor," *Computer*, vol. 27, no. 3, Mar 2004, pp. 57–65.

[3]   J. Blome, S. Feng, S. Gupta, and S. Mahlke, "Self-Calibrating Online Wear-out Detection," *MICRO*, 1–5 Dec 2007.

[4]   S. Borkar, T. Karnik, S. Narendra, J. Tschanz, A. Keshavarzi, and V. De, "Parameter Variations and Impact on Circuits and Microarchitecture," *DAC*, 2–6 June 2003, pp. 338–342.

[5]   K. Bowman, S. Duvall, and J. Meindl, "Impact of Die-to-Die and Within-Die Parameter Fluctuations on the Maximum Clock Frequency Distribution for Gigascale Integration," *IEEE J. Solid-State Circuits*, vol. 37, no. 2, Feb 2002, pp 183–190.

[6]   K. Bowman, S. Samaan, and N. Hakim, "Maximum Clock Frequency Distribution Model with Practical VLSI Design Considerations," *Integrated Circuit Design and Technology*, 17–20 May 2004, pp. 183–191.

[7]   Cool 'n' Quiet Technology Installation Guide for AMD Athlon 64 Processor Based Systems. AMD, Corp. CA [Online] 0.04, 2004, June. www.amd.com/us-en/assets/content_type/DownloadableAssets/Cool_N_ Quiet_Installation_Guide3.pdf.

[8]   A. Drake, (2005), *Power Reduction in Digital Systems Through Local Resonant Clocking and Dynamic Threshold MOS*, Ph.D. Dissertation, University of Michigan.

[9]   A. Drake, R. Senger, H. Deogun, G. Carpenter, S. Ghiasi, T. Nguyen, N. James, M. Floyd, and V. Pokala, "A Distributed Critical-Path Monitor for a 65nm High-Performance Microprocessor," *ISSCC*, 11–15 Feb 2007, pp. 398–399.

[10] M. Elgebaly and M. Sachdev, "Variation-Aware Adaptive Voltage Scaling System," *IEEE Transactions on VLSI Systems*, vol. 15, no. 5, May 2007, pp. 560—571.

[11] *Enhanced Intel SpeedStep Technology for the Intel Pentium M Processor*, Order No. 301170-001. Intel, Corp. OR. [Online]. 2004, March. www.intel.com/ technology/silicon/power/chipdesign.htm.

[12] M. Ershov, S. Saxena, H. Karbasi, S. Winters, S. Minehane, J. Babcock, R. Lindley, P. Clifton, M. Redford, and A. Shibkov, "Dynamic Recovery of Negative Bias Temperature Instability in p-type Metal-Oxide-Semiconductor Field-Effect Transistors," *Applied Physics Letters*, vol. 83, no. 8, 25 Aug 2003, pp. 1647–1649.

[13] E. Fetzer, "Using Adaptive Circuits to Mitigate Process Variations in a Microprocessor Design," *IEEE Design and Test of Computers*, vol. 23, no. 6, Nov/Dec 2006, pp. 476–483.

[14] T. Fischer, J. Desai, B. Doyle, et al., "A 90-nm Variable Frequency Clock System for Power-Managed Itanium Architecture Processor," *IEEE J. Solid-State Circuits*, vol. 41, no. 1, Jan 2006, pp. 218–228.

[15] B. Garlepp, K. Donnelly, J. Kim, P. Chau, J. Zerbe, C. Huang, C. Tran, C. Portmann, D. Stark, Y.-F. Chan, T. Lee, and M. Horowitz, "A Portable Digital DLL for High-Speed CMOS Interface Circuits," *IEEE J. Solid-State Circuits*, vol. 34, no. 5, May 1999, pp. 632–644.

[16] H. Hamann, A. Weger, J. Lacey, Z. Hu, P. Bose, E. Cohen, and J. Wakil, "Hotspot-Limited Microprocessors: Direct Temperature and Power Distribution Measurements," *IEEE J. Solid-State Circuits*, vol. 42, no. 1, Jan 2007, pp. 56–65.

[17] R. Ho, K. Mai, and M. Horowitz, "The Future of Wires," *Proceedings of the IEEE*, vol. 89, no. 4, April 2001, pp. 490–504.

[18] R. Ho, K. Mai, and M. Horowitz, "Managing Wire Scaling: A Circuit Perspective," *International Technology Conference*, 2–4 June 2003, pp. 177–179.

[19] N. James, P. Restle, J. Friedrich, B. Huott, and B. McCredie, "Comparison of Split-Versus Connected-Core Supplies in the POWER6 Microprocessor," *ISSCC*, 11–15 Feb 2007, pp. 298–604.

[20] H. Mahmoodi, S. Mukhopadhayay, and K. Roy, "Estimation of Delay Variations Due to Random-Dopant Fluctuations in Nanoscale CMOS Circuits," *IEEE J. Solid-State Circuits*, vol. 40, no. 9, Sept 2005, pp. 1787–1796.

[21] V. Mehrotra, S. L. Sam, D. Boning, A. Chandrakasan, R. Vallishayee, and S. Nassif, "A Methodology for Modeling the Effects of Systematic Within-Die Interconnect and Device Variation on Circuit Performance," *DAC*, 5–9 June 2000, pp. 172–175.

[22] S. Naffziger, B. Stackhouse, T. Grutkowski, D. Josephson, J. Desai, and M. Horowitz, "The Implementation of a 2-Core, Multi-Threaded Itanium Family Processor," *IEEE J. Solid-State Circuits*, vol. 41, no. 1, Jan 2006, pp. 197–209.

[23] M. Nakai, S. Akui, K. Seno, N. Makai, T. Meguro, T. Seki, T. Kondo, A. Hashiguchi, H. Kawahara, K. Kumano, and M. Shimura, "Dynamic Voltage and Frequency Management for a Low-Power Embedded Microprocessor," *IEEE J. Solid-State Circuits*, vol. 40, no. 1, Jan 2005, pp. 28–35.

[24] M. Nakai, S. Akui, K. Seno, T Meguro, T. Seki, T. Kondo, A. Hashiguchi, H. Kawahara, K. Kumano, and M. Shimura, "Dynamic Voltage and Frequency Management for a Low-Power Embedded Microprocessor," *IEEE J. Solid-State Circuits*, vol. 40, no. 1, Jan 2005, pp. 28–35.

[25] S. Nassif, "Delay Variability: Sources, Impacts and Trends," *ISSCC*, 7–9 Feb 2000, pp. 368–369.

[26] K. Nowka, G. Carpenter, and B. Brock, "The Design and Application of the PowerPC 405LP Energy-Efficient System-on-a-Chip," *IBM Journal of Research and Development*, vol. 47, no. 5/6, Sept/Nov 2003, pp. 631–639.

[27] S.-I. Ochkawa, M. Aoki, and H. Masuda, "Analysis and Characterization of Device Variations in an LSI Chip Using an Integrated Device matrix Array," *IEEE Transactions on Semiconductor Manufacturing*, vol. 17, no. 2, May 2004, pp. 155–165.

[28] R. Rao, A. Srivastava, D. Blaauw, and D. Sylvester, "Statistical Analysis of Subthreshold Leakage Current for VLSI Circuits," *IEEE Transactions on VLSI Systems*, vol. 12, no. 2, Feb 2004, pp. 131–139.

[29] B. Razavi, *Design of Analog CMOS Integrated Circuits*, McGraw Hill, Boston, 2001, pp. 550–556.

[30] P. Restle, R. Frach, N. James, W. Huott, T. Skergan, S. Wilson, N. Schwartz, and J. Clabes, "Timing Uncertainty Measurements on the Power5 Microprocessor," *ISSCC*, 15–19 Feb 2004, pp. 354–355.

[31] M. Saint-Laurent and M. Swaminathan, "Impact of Power-Supply Noise on Timing in High-Frequency Microprocessors," *IEEE Transactions on Advanced Packaging*, vol. 27, no. 1, Feb 2004, pp. 135–144.

[32] S. Samaan, "The Impact of Device Parameter Variations on the Frequency and Performance of VLSI Chips," *ICCAD*, 7–11 Nov 2004, pp. 343–346.

[33] A. Strak and H. Tenhunen, "Investigation of Timing Jitter in NAND and NOR Gates Induced by Power-Supply Noise," *ICECS*, 10–13 Dec 2006, pp. 1160–1163.

[34] H. Su, F. Liu, A. Devgan, E. Acar, and S. Nassif, "Full Chip Leakage-Estimation Considering Power Supply and Temperature Variations," *ISLPED*, 25–27 Aug 2003, pp. 78–83.

Chapter 8 Architectural Techniques for Adaptive Computing

[1,2]Shidhartha Das, [2]David Roberts, [2]David Blaauw, [1]David Bull, [2]Trevor Mudge

[1]ARM Ltd., UK, [2] University of Michigan

8.1 Introduction

As critical geometries shrink to the 45nm region and beyond, lithographic limitations have led to rising intra- and inter-die process variations. Increased variability makes it significantly difficult to accurately model transistor behavior on silicon, and often probabilistic methods are required [1]. The consequent loss in silicon predictability implies that design uncertainties become severe and are made even worse at the lower supply voltages used for future technologies [2].

In addition to process variability, deep sub-micron technologies also suffer from increased power consumption which compromises structural reliability of processors. Indeed, as current densities have increased, chip failure through effects like electro-migration [3] and time-dependent dielectric breakdown (TDDB) [4] has become major challenge, especially for high-end processors. Furthermore, at lower supply voltages, noise margins for sensitive circuits significantly reduce. Consequently, signal integrity concerns assume greater relevance. Smaller noise margins enhance susceptibility to capacitive and inductive coupling, thereby adversely affecting computational robustness. Robustness is further aggravated by resistive voltage drops and inductive overshoots in the supply voltage network. As such, it will be exceedingly difficult to sustain the current rate of technology scaling unless power and robustness concerns are suitable addressed [5].

A. Wang, S. Naffziger (eds.), *Adaptive Techniques for Dynamic Processor Optimization*,
DOI: 10.1007/978-0-387-76472-6_8, © Springer Science+Business Media, LLC 2008

The traditional approach of fabricating robust circuits has been to design for the worst-case scenario. In this approach, circuits are built with sufficient safety margins such that they operate correctly even under the worst-case combination of process, voltage and temperature conditions. As design uncertainties worsen, it is expected that safety margins will increase at future technology nodes. At these nodes, the worst-case transistor performance is likely to vary widely from that under typical conditions. This limits the operating frequency of processors, thereby reducing the performance improvements that technology scaling traditionally afforded. Furthermore, safety margins typically require the use of wider devices, higher operating voltage and thicker interconnects, all of which have the undesirable effect of increased power consumption. Thus, while design margining ensures robust operation, unfortunately, it also leads to reduced performance and increased power consumption.

A key observation is that robust computing and low power are fundamentally at odds with each other. Low-power methodologies typically sacrifice robustness for lower power consumption and vice versa. This trade-off is especially significant in the mobile and battery-operated world where meeting robustness and performance targets under restrictive power budgets makes design closure difficult. For example, an effective low-power technique is dynamic voltage scaling (DVS), which enables quadratic power savings by scaling supply voltage during low CPU utilization periods. However, low voltage operation causes signal integrity concerns by reducing the static noise margins for sensitive circuits. Furthermore, sensitivity to threshold voltage variation also increases at low voltages [2] which can lead to circuit failure. Another popular technique for low power relies on downsizing off-critical paths [6]. This balances path delays in the design leading to the so-called *timing wall*. In a delay-balanced design, the likelihood of chip failure significantly increases because more paths can now fail setup requirements. Conversely, most robust design techniques, such as hardware redundancy and conservative margining, hurt power consumption. Thus, the traditional design paradigm leads to a very complex optimization space where design closure by simultaneously meeting power, performance, and robustness objectives can be exceedingly difficult.

In order to effectively address the issue of design closure, it is helpful to analyze and categorize the sources of design uncertainties, depending on their spatial reach and temporal rate of change.

8.1.1 Spatial Reach

Based on spatial reach, design uncertainties can be further subdivided as follows:

- *Global uncertainties*

   Those that affect all transistors on the die are *global* in nature. For example, global supply voltage variations affect the entire die and could be due to voltage fluctuations onboard or within the package. Other examples of such global phenomena are inter-die process variations and ambient temperature.

- *Local uncertainties*

   Local effects are limited to a few transistors in the immediate vicinity of each other. Voltage variations due to resistive drops in the power grid and temperature hot spots in regions of high switching activity have local effects. Cross-coupling noise events are extremely local and are restricted to a few signal nets near the aggressor. Other examples of local effects are intra-die process variations.

8.1.2 Temporal Rate of Change

Based on their rate of change with time, design uncertainties can be broadly divided under the following categories.

- *Slow-changing effects*

   Design uncertainties that have time constants of the order of millions of cycles or more can be categorized as slow-changing. Thus, they could be

   (a) Invariant with time: Effects such as intra- and inter-die process variations are fixed after fabrication and remain effectively invariant over the lifetime of the processor.

   (b) Extremely slow-changing, spread over the lifetime of the die: Wear-out mechanisms such as negative bias temperature instability [7], TDDB [4] and electro-migration are typical examples of such effects that gradually degrade processor performance over its lifetime.

   (c) Moderately slow-changing, spread over millions of cycles: Temperature fluctuations fall under this category.

- *Fast-changing effects*

   Such effects develop over thousands of cycles or less. They could be

   (a) Moderately fast-changing, spread over thousands of cycles: Supply voltage uncertainties attributed to the Voltage Regulation Module or board-level parasitics can cause supply voltage variations on-die. Such effects develop over a range of few microseconds or thousands of processor cycles.

(b) Fast changing, spread over tens of cycles: Inductive overshoots due to package inductance can cause supply voltage noise with time constants of the order of tens of processor cycles.

(c) Extremely fast-changing, spread over a few cycles or less: IR drops in the on-chip power supply network develop over a few cycles. Coupling noise effects exist for even shorter durations; typically for less than a cycle.

In addition to process and silicon conditions, input vector dependence of circuit delay is another major source of variation which cannot be captured easily in the above categories. Circuits exhibit worst-case delay for very specific instruction and data sequences [8]. Consequently, most input vectors do not sensitize the critical path, thereby aggravating the pessimism due to overly conservative safety margins.

Addressing the issue of excessive margins requires a fundamental departure from the traditional technique of operating every dice at a single, statically determined operating point. Adaptive design techniques seek to mitigate excessive margining by dynamically adjusting system parameters (voltage and frequency) to account for variations in environmental conditions and silicon grade. Thus, a significant portion of worst-case safety margins is eliminated leading to improved energy efficiency and performance over traditional methods. Broadly speaking, adaptive techniques can be divided into two main categories.

- **"Always-correct" techniques**

The key idea of "always-correct" techniques is to predict the point of failure for a die and to tune system parameters to operate near this predicted point. Typically, safety margins are added to the predicted failure point to guarantee computational correctness.

- **"Error detection and correction" techniques**

Such approaches rely on scaling system parameters to the point of failure. Computation correctness is ensured by detecting timing errors and suitably recovering from them.

Table 8.1 compiles a list of different adaptive design techniques discussed in literature and the margins eliminated by each of them. We survey these techniques in detail in Sections 8.2 and 8.3, respectively. In Section 8.4, we discuss "Razor" as a special case study of error detection and correction approaches. In this section, we introduce the basic concepts of Razor. We follow it with measurement results on a test chip using Razor for adaptive voltage control in Section 8.5. Section 8.6 deals with the recent research related to Razor. Finally, Section 8.7 concludes the chapter with few re-marks on the future direction of research on adaptive techniques.

Table 8.1 Adaptive techniques landscape.

Category	Technique	Data	Process		Ambient (V,T)				General-purpose computing?
			Intra-die	Inter-die	Local		Global		
					Fast	Slow	Fast	Slow	
Always correct	Table look-up [Section 8.2.1]	N	N	N	N	N	N	N	Y
	Canary circuits [Section 8.2.2]	N	N	Y	N	N	N	Y	Y
	In situ triple-latch monitor [Section 8.2.3]	N	Y	Y	N	Y	N	Y	Y
	Typical delay adder structures [Section 8.2.4]	Y	N	N	N	N	N	N	Y
	Non-uniform cache architectures [Section 8.2.4]	Y	N	N	N	N	N	N	Y
Error detection and correction	Self-calibrating interconnects [Section 8.3.1]	Y	Y	Y	Y	Y	Y	Y	N
	ANT [Section 8.3.1]	Y	Y	Y	Y	Y	Y	Y	N
	Razor [Section 8.4]	Y	Y	Y	Y	Y	Y	Y	Y

8.2 "Always-Correct" Techniques

As mentioned before, "always-correct" techniques predict the operational point where the critical path fails to meet timing and to guarantee correct-ness by adding safety margins to the predicted failure point. The conven-tional approach toward predicting this point of failure is to use either a look-up table or the so-called *canary* circuits.

8.2.1 Look-up Table-Based Approach

In the look-up table-based approach [9][10][11], the maximum obtainable frequency of the processor is characterized for a given supply voltage. The voltage–frequency pairs are obtained by performing traditional timing analysis on the processor. Typically, the operating frequency is decided based on the deadline under which a given task needs to be completed. Accordingly, the supply voltage corresponding to the frequency requirement is "dialed in". The table look-up approach is able to exploit periods of low CPU utilization by dynamically scaling voltage and frequency, thereby leading to energy savings. Furthermore, owing to its relative simplicity, this approach can be easily deployed in the field. However, its reliance on conventional timing analysis performed at the combination of worst-case process, voltage and temperature corners implies that none of the safety margins due to uncertainties are eliminated.

8.2.2 Canary Circuits-Based Approach

An alternative approach relies on the use of the so-called canary circuits to predict the failure point [12], [13]–[17], [38]. Canary circuits are typically implemented as inverter chains which approximate the critical path of the processor. They are designed to track the critical-path delay across process, voltage and temperature corners. Voltage and frequency are scaled to the extent that this replica-delay path fails to meet timing. The key requirement for this approach is that the main processor operates correctly even when the replica-delay path fails to meet timing. In order to ensure this, worst-case safety margins are added to the replica path to account for *local* variations due to temperature hot spots, cross-talk noise, power supply droops and intra-die process variation. Furthermore, the replica path cannot respond to the *fast-changing* effects for which worst-case safety margins need to be budgeted. Margins also need to account for mismatches in the scaling characteristics of the replica path and the critical path.

There are several systems reported in literature based on canary circuits. One approach uses the replica path as a delay reference for a voltage-controlled oscillator (VCO) unit. The VCO monitors the delay through the chain at a given supply voltage and scales the operating frequency to the point of failure of the replica path. An example of such an approach is Uht's TEATime which is illustrated in Figure 8.1.

A toggle flip-flop initiates a new transition through the replica path every cycle. The transition is correctly captured at the receiving flip-flop only if the clock period is greater than the propagation delay through the replica path. A simple up–down counter is used to control the VCO frequency output via a digital-to-analog converter (DAC). IBM's PowerPC System-on-Chip design reported in [14] and the Berkeley Wireless Research Center's [16][12] low-power microprocessor are all based on a similar concept. An alternative approach developed by Sony and reported

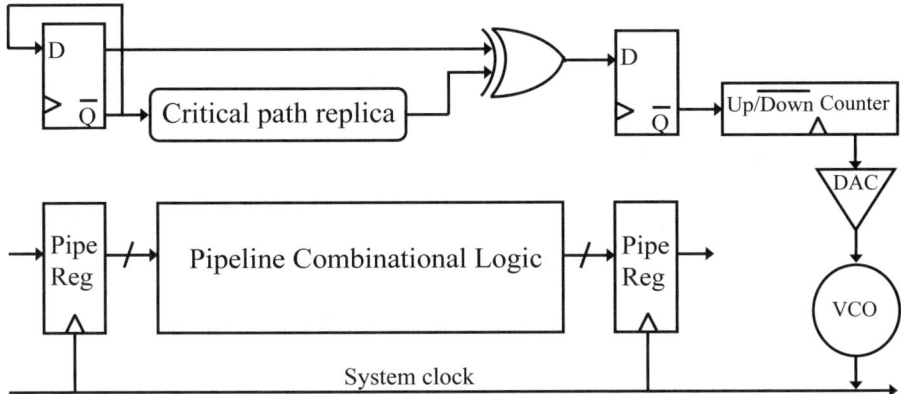

Figure 8.1 Uht's TEATime: A canary circuit-based approach.

in [17] uses a delay synthesizer unit consisting of several delay chains which selects a safe frequency depending on the maximum propagation delay through the chains. Typically, canary circuits enable better energy efficiency than the table look-up approach because unlike the latter, they are able to eliminate margins due to inter-die process variations and *global* fluctuations in voltage and temperature that are essentially *slow-changing* effects [18].

8.2.3 In situ Triple-Latch Monitor

Kehl's triple-latch monitor is similar to the canary circuits-based techniques, but utilizes in situ monitoring of circuit delay [19]. Using this approach, all monitored system states are sampled at three different latches with a small delay interval between each sampling point, as shown in Figure 8.2(a). The value in the latest-clocked latch which is allowed the most time is assumed correct and is always forwarded to later logic. The system is considered "tuned" (Figure 8.2b) when the first latch does not match the second and third latch values, meaning that the logic transition was very near the critical speed, but not dangerously close. If all latches see the same value, the system is running too slowly and frequency should be increased. If the first two latches see different values than the last, then the system is running dangerously fast and should be slowed down.

Because of the in situ nature of this approach, it can adjust to *local* variations such as intra-die process and temperature variations. However, it still cannot track *fast-changing* conditions such as cross-coupling and voltage noise events. Hence, the delay between the successive samples has to be sufficiently separated to allow for margins for such events. In addition,

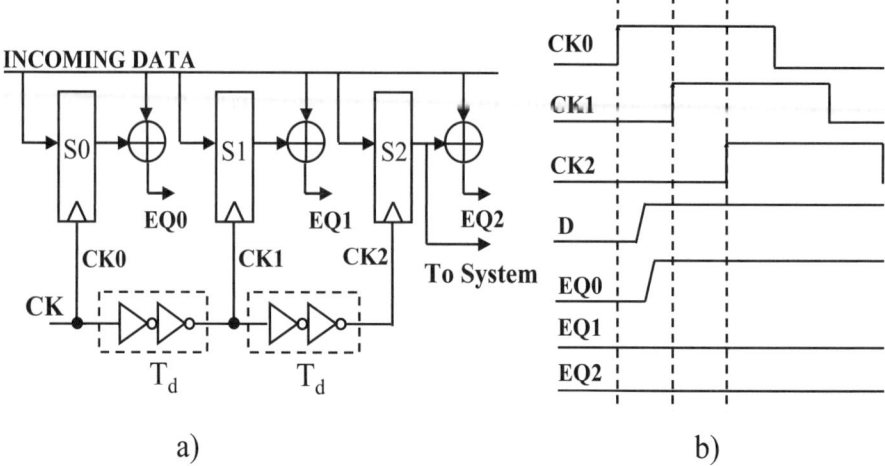

a)   b)

Figure 8.2 In situ monitoring: Kehl's triple-latch monitor. (a) In situ monitoring of delay. (b) Timing diagrams showing when system is "tuned".

to avoid overly aggressive clocking, evaluations of the latch values must be limited to tests using worst-case latency vectors. Kehl suggests that the system should periodically stop and test worst-case vectors to determine whether the system requires tuning. This requirement severely limits the general applicability of this approach since writing vectors that account for the worst-case delay and coupling noise scenario are difficult to generate and exercise for general-purpose processors.

8.2.4 Micro-architectural Techniques

A potential shortcoming of all the techniques discussed above is that they seek to track variations in the critical-path delay and, consequently, cannot adapt to input vector-dependent delay variations. The processor voltage and frequency are still limited by the worst-case path, even if it is not being sensitized. This issue is addressed by several micro-architectural techniques discussed in literature, specifically related to adder architectures [8] [20]. Such designs exploit the fact that the worst-case carry-chain length is rarely sensitized. This allows them to operate the adder block at a higher frequency than what is dictated by the worst-case carry path. If a latency-intensive add operation is detected, then the clock frequency is halved to allow it to complete without errors. An example of such a design is the stutter adder [8] which uses a low-overhead circuit for a priori determination of the carry-chain length. If the carry-chain length in a cycle exceeds a certain number of bits, then a "stutter" signal is raised which clock-gates the next cycle. Thus a "long" adder computation is effectively given two cycles to execute. Recent studies [21] on SPECInt 2000 benchmarks have shown that the maximum carry-chain length for 64-bit additions rarely exceeds 24 bits. In fact, in [8], the authors report that in 95% of cases, the adder required only one cycle to compute. Lu [20] proposes a similar technique where an "approximate" but faster implementation of a functional unit is used in conjunction with a slow but always-correct checker to exploit typical latencies and clock the system a higher rate.

Data-dependent delay variations are also exploited by non-uniform cache architectures (NUCA) [22][23]. In aggressively scaled technologies, interconnect delay can become a significant portion of the cache access time. This causes wide variations in the fetch latencies of data words located near the access port versus those located further off. In traditional cache designs, the worst-case latency limits the cache access time. However, NUCA allows early access times for addresses near the access port, thereby achieving throughput improvement. Additional throughput can be achieved by mapping frequently accessed data to banks located nearest to the access port. Thus, in the context of NUCA, data dependence of delay relates to the frequency with which an address in the cache is accessed.

While the stutter adder and the NUCA architectures adapt to data-dependent variations, they still require margins to account for slow silicon grade and worst-case ambient conditions. On the contrary, error detection and correction approaches seek to achieve both, i.e., eliminate worst-case safety margins for all types of uncertainties and adapt to data-dependent variations as well. However, they are more complex and incur additional overhead in their implementation. Such approaches are discussed in detail in the next section.

8.3 Error Detection and Correction Approaches

The key concept of these schemes is to scale the system parameters (e.g., voltage and frequency) until the point where the processor fails to meet timing. A detection block flags the occurrence of a timing error after which a correction block is engaged to recover the correct state. To ensure that the system does not face persistent errors, an additional controller monitors the error rate and tunes voltage and frequency to achieve a targeted error rate.

Allowing the processor to fail and then recover helps eliminate worst-case safety margins. This enables significantly greater performance and energy efficiency over "always-correct" techniques. Furthermore, by tuning system parameters based on the error rate, it is possible to exploit the input vector dependence of delay as well. Instead of safety margins, such systems rely on successful detection and correction of timing errors to guarantee computational correctness. The net energy consumption of the system is essentially a trade-off between the increased efficiency afforded by the elimination of margins and the additional overhead of recovery. Of course, the overhead of recovery can make sustaining a high error rate counterproductive. Hence, these systems typically rely on restricting operation to low error rate regimes to maximize energy efficiency.

Their relative complexity makes the general applicability of such systems difficult. However, they are naturally amenable for certain applications areas such as communications and signal processing. Communication systems require error correction to reliably transfer information across a noisy channel. Therefore, it is relatively easier to overload the existing error correction infrastructure to enable adaptivity to variable silicon and ambient conditions. Self-calibrating interconnects by Worm et al. [24] and algorithmic noise tolerance by Shanbhag et al. [25] are examples of applications of such techniques to on-chip communication and signal processing architectures.

8.3.1 Techniques for Communication and Signal Processing

Self-calibrating interconnects address the problem of reliable on-chip communication in aggressively scaled technologies. Signal integrity concerns require on-chip busses to be strongly buffered which consumes a significant portion of the total chip power. Hence, it is desirable to transfer bits at the lowest possible operating voltage while still guaranteeing the required performance and the targeted bit error rate (BER). Worm [24] addresses this issue by encoding the data words with the so-called self-synchronizing codes [24] before transmission. The receiver is augmented with a checker unit that decodes the received code word and flags timing errors. Correction occurs by requesting re-transmission through an automatic repeat request (ARQ) block, as shown in Figure 8.3. Furthermore, an additional controller obtains feedback from the checker and accordingly adjusts the voltage and the frequency of the transmission. By reacting to the error rates, the controller is able to adapt to the operating conditions and thus eliminate worst-case safety margins. This improves the energy efficiency of the on-chip busses with negligible BER degradation.

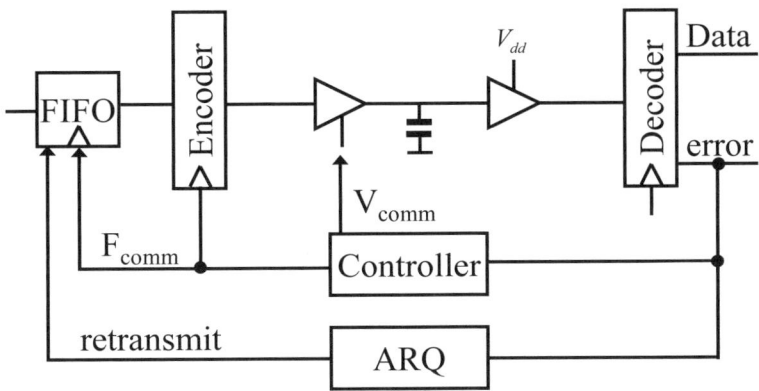

Figure 8.3 Self-calibrating interconnects.

Algorithmic noise tolerance (ANT) by Shanbhag et al. [25] uses a similar concept for low-power VLSI signal processing architectures. As conceptually illustrated in Figure 8.4, the main processor block is augmented with an estimator block. The main block is voltage scaled beyond the point of failure, thereby leading to intermittent timing errors. The result of the main block is validated against the result of the estimator block which computes correct result, based on the previous history. The estimator block is significantly cheaper in terms of area and power as compared to the main block which is being voltage scaled. At low error rates, the energy savings of aggressive scaling on the main block compensates for the overhead of correction, leading to significant energy savings. Error detection occurs when the difference in results of the main block and the estimator block exceeds a certain threshold. Error correction occurs by overwriting the result of the main block with that of the estimator block.

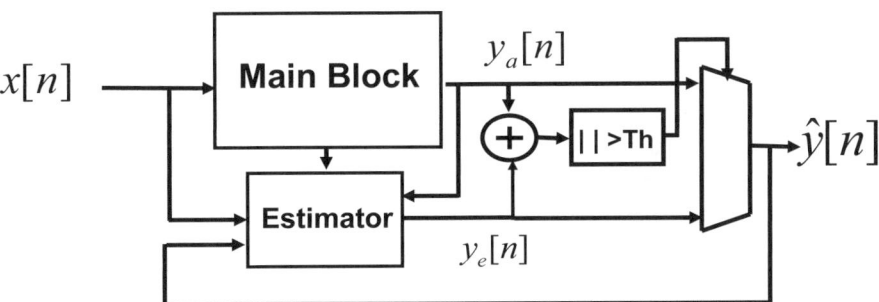

Figure 8.4 Algorithmic noise tolerance [25].

Since the estimator block depends on past history of correct results to make its prediction, its accuracy reduces as more errors are experienced. This adversely affects the BER of the entire block. In addition, the overhead of error correction also increases with increased error rate. Hence, it is desirable to keep the rate of timing errors low for maintaining a low BER and high energy efficiency. The authors built a FIR block in 0.35 micron technology [25] to demonstrate the efficacy of this technique. They obtained at least 70% savings over an error-free design for a 1% reduction in the signal-to-noise (SNR) ratio of the final output.

By reacting to error rates, both of the above techniques are able to exploit data-dependent delay variations because even under aggressively scaled voltage and frequency conditions, it is possible to maintain a low error rate as long as the critical paths are not being sensitized.

8.3.2 Techniques for General-Purpose Computing

Communication applications are inherently suited for error detection and correction techniques. Unfortunately, the same cannot be said for general-purpose computing. The key requirement for general-purpose computing is that the committed architectural state is always correct. Therefore, all timing errors that can possibly propagate to the architectural state have to be flagged and corrected. This is not an issue in communication because it does not affect the correct functionality of the system and leads to a negligible degradation of the BER, at worst. Unlike in communication, a timing error in the architectural state in general-purpose computing is a catastrophic failure and needs to be avoided at all costs. It is for this reason that there have been only a few examples of error detection and correction techniques in the area of general-purpose computing. Typically, such techniques rely on temporal redundancy for error detection. It was shown by Roberts et al. [26] that multi-bit bidirectional bit-flips occur in multiplier outputs under aggressive voltage scaling. Application of error-correcting codes for processor circuits to detect and correct such failures is infeasible due to the prohibitive area overhead incurred [26].

Using temporal redundancy for timing error detection requires two different samples of the monitored signal. The earlier speculative sample is validated against the latter "always-correct" version which is sampled according to conventional worst-case assumptions. The idea of temporal redundancy for error detection has been used extensively in the design and test community for at-speed delay testing. Anghel and Nicolaidis [27] use a similar concept for detecting SEU failures in combinational logic. A cosmic particle strike in the combinational logic manifests itself as a pulse which can get captured by downstream flip-flops. The authors detect such an event by re-sampling the flip-flop input after the pulse has died down. A discrepancy between the two samples indicates a SEU event in the combinational logic.

Razor [28] uses temporal redundancy for general-purpose computing. In Razor, a critical-path signal is speculatively sampled at the rising edge of the regular clock and is compared against a shadow latch which samples at a delayed edge. A timing error is flagged when the speculative sample does not agree with the delayed sampled. State correction involves overwriting the shadow latch data into the main flip-flop and engaging microarchitectural recovery features to recover correct state. As is common with most error detection and correction techniques, Razor is able to eliminate worst-case safety margins by allowing errors to occur and recovering from them. We discuss Razor in greater detail in the next section onward.

8.4 Introduction to Razor

Razor [28] is a circuit-level timing speculation technique based on dynamic detection and correction of speed path failures in digital designs. As mentioned in the previous section, a critical-path signal in Razor is sampled twice. The earlier sample is speculatively consumed by the pipeline downstream logic. A timing error is flagged by comparing the speculative sample against the correct, later sample. In such an event, suitable recovery mechanisms are engaged to achieve correct state. In situ detection and recovery ensures correct operation and allows for the elimination of worst-case safety margins. Thus, with Razor, it is possible to tune the supply voltage to the level where first delay errors happen. In addition, voltage can also be scaled below this first point of failure into the sub-critical regime, thereby deliberately tolerating a targeted error rate. Due to the strong data dependence of circuit delay, only a few critical instructions are expected to fail while a majority will operate correctly. Razor automatically exploits this by tuning the supply voltage to obtain a small, but non-zero error rate. Of course, error correction adds energy overhead but this is minimal at low error rates. The extra voltage scaling headroom enabled by sub-critical operation enables substantial energy savings.

The fundamental trade-off that exists between power overhead of error correction against additional power savings from operating at a lower supply voltage is qualitatively illustrated in Figure 8.5. The point of first failure of the processor (V_{ff}) and the minimum allowable voltage of traditional DVS techniques (V_{margin}) are also labeled in the figure. V_{margin} is much

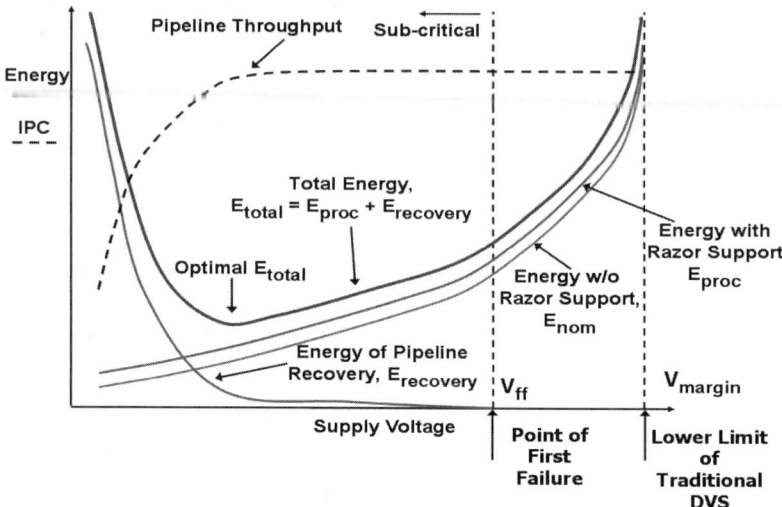

Figure 8.5 Qualitative relationship of energy and IPC as a function of supply voltage. (© IEEE 2005)

higher than V_{ff} under typical conditions, since safety margins need to be included to accommodate for worst-case operating conditions. In situ error detection and correction capability enables Razor to operate at V_{ff}, rather than at V_{margin}. The total energy of the processor (E_{tot}) is the sum of the energy required to perform standard processor operations (E_{proc}) and the energy consumed in recovery from timing errors ($E_{recovery}$). Of course, implementing Razor incurs power overhead due to which the nominal processor energy (E_{nom}) without Razor technology is slightly less than E_{proc}.

As the supply voltage is scaled, the processor energy (E_{proc}) reduces quadratically with voltage. However, as voltage is scaled below the first failure point (V_{ff}), a significant number of paths fail to meet timing. Hence, the error rate and the recovery energy ($E_{recovery}$) increase exponentially. The processor throughput also reduces due to the increasing error rate because the processor now requires more cycles to complete the instructions. The total processor energy (E_{tot}) shows an optimal point where the rate of change of $E_{recovery}$ and E_{proc} offsets each other.

8.4.1 Razor Error Detection and Recovery Scheme

Error detection in Razor occurs by augmenting the standard, positive edge-triggered critical-path flip-flops with a so-called shadow latch that samples off the negative edge of the clock. Figure 8.6 shows the conceptual representation of a Razor flip-flop, henceforth referred to as the RFF. The input data is given additional time, equal to the duration of the positive clock phase, to settle down to its correct state before being sampled by the shadow latch. To ensure that the shadow latch always captures the correct data, the minimum allowable supply voltage needs to be constrained during design time such that the setup time at the shadow latch is never violated, even under worst-case conditions. A comparator flags a timing error when it detects a discrepancy between the speculative data sampled at the main flip-flop and the correct data sampled at the shadow latch.

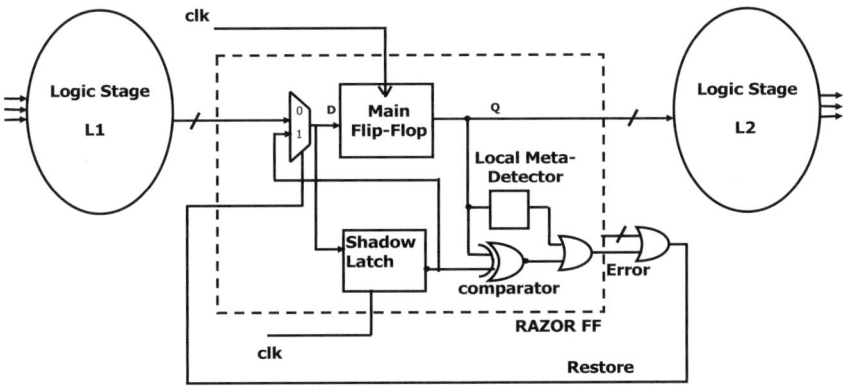

Figure 8.6 Razor flip-flop conceptual schematic. (© IEEE 2005)

Since setup and hold constraints at the main flip-flop input (D) are not respected, it is possible that the state of the flip-flop becomes metastable. A metastable signal increases critical-path delay which can cause a shadow latch in the succeeding pipeline stage to capture erroneous data, thereby leading to incorrect execution. In addition, a metastable flip-flop output can be inconsistently interpreted by the error comparator and the downstream logic. Hence, an additional detector is required to correctly flag the occurrence of metastability at the output of the main flip-flop. The outputs of the metastability detector and the error comparator are OR-ed to generate the *error* signal of the RFF. Thus, the system reacts to the occurrence of metastability in exactly the same way as a conventional timing failure.

A key point to note is the fact that metastability *need not be resolved correctly* in the RFF and that just the *detection* of such an occurrence is sufficient to engage the Razor recovery mechanism. However, in order to prevent potentially metastable signals from being committed to memory, at least two successive non-critical pipeline stages are required immediately before storage. This ensures that every signal is validated by Razor and is effectively double-latched in order to have a negligible probability of being metastable, before being written to memory. In our design, data accesses in the memory stage were non-critical and hence we required only one additional pipeline stage to act as a dummy stabilization stage.

Error signals of individual RFFs are OR-ed together to generate the pipeline *restore* signal which overwrites the shadow latch data into the main flip-flop, thereby restoring correct state in the cycle following the erroneous cycle. Thus, an erroneous instruction is guaranteed to recover with a single cycle penalty, without having to be re-executed. This ensures that forward progress in the pipeline is always maintained. Even if every instruction fails to meet timing, the pipeline still completes, albeit at a slower speed. Upon detection of a timing error, a micro-architectural recovery technique is engaged to restore the whole pipeline to its correct state.

8.4.2 Micro-architectural Recovery

The pipeline error recovery mechanism must guarantee that, in the presence of Razor errors, register and memory state is not corrupted with an incorrect value. In this section, we highlight two possible approaches to implementing pipeline error recovery. The first is a simple but slow method based on clock-gating, while the second method is a much more scalable technique based on counter-flow pipelining [29].

8.4.2.1 Recovery Using Clock-Gating

In the event that any stage detects a Razor error, the entire pipeline is stalled for one cycle by gating the next global clock edge, as shown in Figure 8.7(a). The additional clock period allows every stage to recompute its result using the Razor shadow latch as input. Consequently, any previously forwarded erroneous values will be replaced with the correct value from the Razor shadow latch, thereby guaranteeing forward progress. If all stages produce an error each cycle, the pipeline will continue to run, but at half the normal speed. To ensure negligible probability of failure due to metastability, there must be two non-speculative stages between the last Razor latch and the writeback (WB) stage. Since memory accesses to the data cache are non-speculative in our design, only one additional stage labeled ST (stabilize) is required before writeback (WB). In the general case, processors are likely to have critical memory accesses, especially on the read path. Hence, the memory sub-system needs to be suitably designed such that it can handle potentially critical read operations.

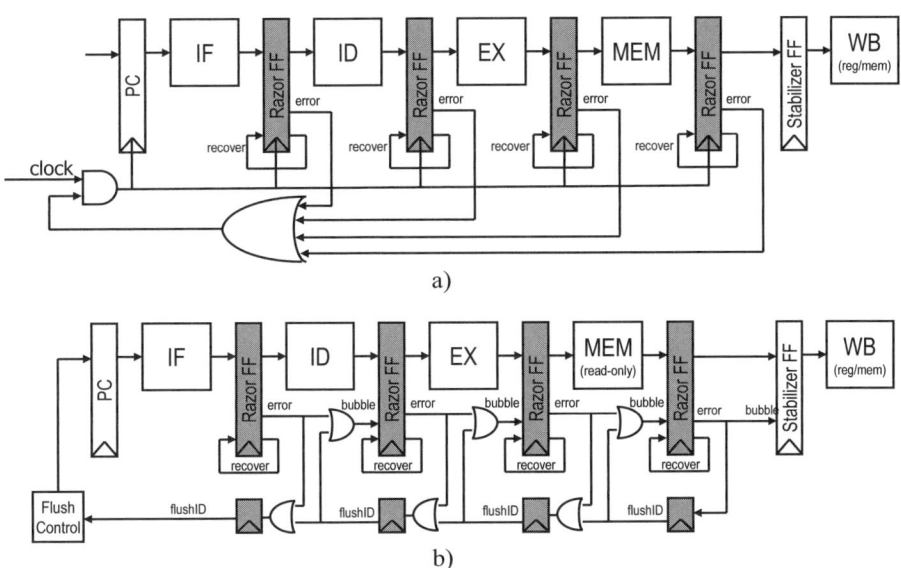

Figure 8.7 Micro-architectural recovery schemes. (a) Centralized scheme based on clock-gating. (b) Distributed scheme based on pipeline flush.
(© IEEE 2005)

8.4.2.2 Recovery Using Counter-Flow Pipelining

In aggressively clocked designs, it may not be possible to implement single cycle, global clock-gating without significantly impacting processor cycle time. Consequently, we have designed and implemented a fully pipelined error recovery mechanism based on counter-flow pipelining techniques [29]. The approach illustrated in Figure 8.7(b) places negligible timing constraints on the baseline pipeline design at the expense of extending pipeline recovery over a few cycles. When a Razor error is detected, two specific actions must be taken. First, the erroneous stage computation following the failing Razor latch must be nullified. This action is accomplished using the bubble signal, which indicates to the next and subsequent stages that the pipeline slot is empty. Second, the flush train is triggered by asserting the stage ID of failing stage. In the following cycle, the correct value from the Razor shadow latch data is injected back into the pipeline, allowing the erroneous instruction to continue with its correct inputs. Additionally, the flush train begins propagating the ID of the failing stage in the opposite direction of instructions. When the flush ID reaches the start of the pipeline, the flush control logic restarts the pipeline at the instruction following the erroneous instruction.

8.4.3 Short-Path Constraints

The duration of the positive clock phase, when the shadow latch is transparent, determines the sampling delay of the shadow latch. This constrains the minimum propagation delay for a combinational logic path terminating in a RFF to be at least greater than the duration of the positive clock phase and the hold time of the shadow latch. Figure 8.8 conceptually illustrates this minimum delay constraint. When the RFF input violates this constraint and changes state before the negative edge of the clock, it corrupts the state of the shadow latch. Delay buffers are required to be inserted in those paths which fail to meet this minimum path delay constraint imposed by the shadow latch.

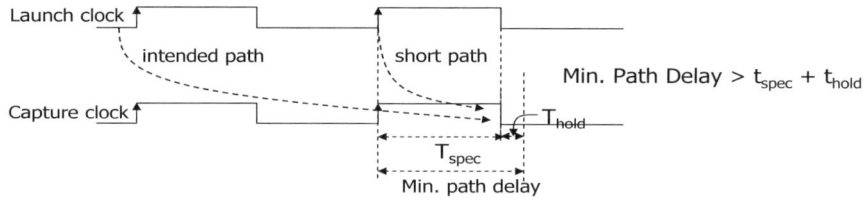

Figure 8.8 Short-path constraints.

The shadow latch sampling delay represents the trade-off between the power overhead of delay buffers and the voltage margin available for Razor sub-critical mode of operation. A larger value of the sampling delay allows greater voltage scaling headroom at the expense of more delay buffers and vice versa. However, since Razor protection is only required on the critical paths, overhead due to Razor is not significant. On the Razor prototype subsequently presented, the power overhead due to Razor was less than 3% of the nominal power overhead.

8.4.4 Circuit-Level Implementation Issues

Figure 8.9 shows the transistor level schematic of the RFF. The error comparator is a semi-dynamic XOR gate which evaluates when the data latched by the slave differs from that of the shadow in the negative clock phase. The error comparator shares its dynamic node, *Err_dyn,* with the metastability detector which evaluates in the positive phase of the clock when the slave output could become metastable. Thus, the RFF *error* signal is flagged when either the metastability detector or the error comparator evaluates.

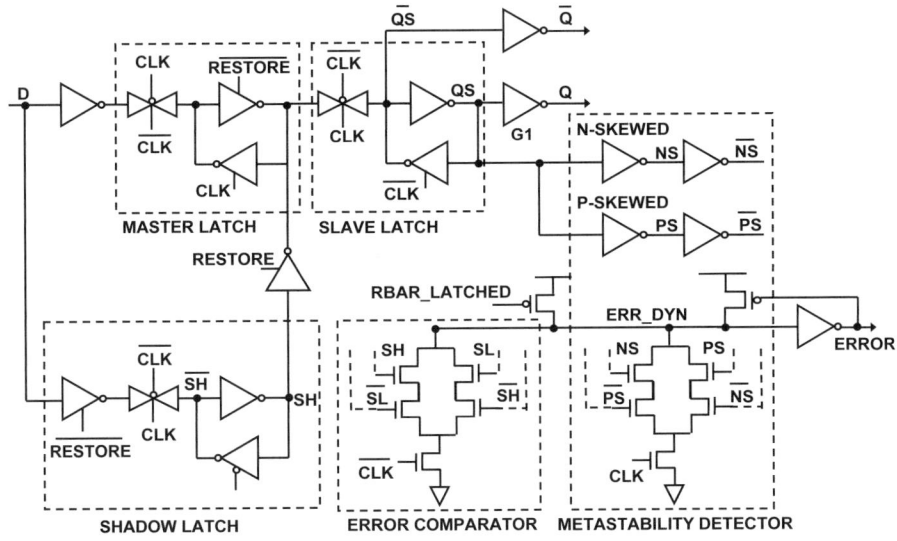

Figure 8.9 Razor flip-flop circuit schematic. (© IEEE 2005)

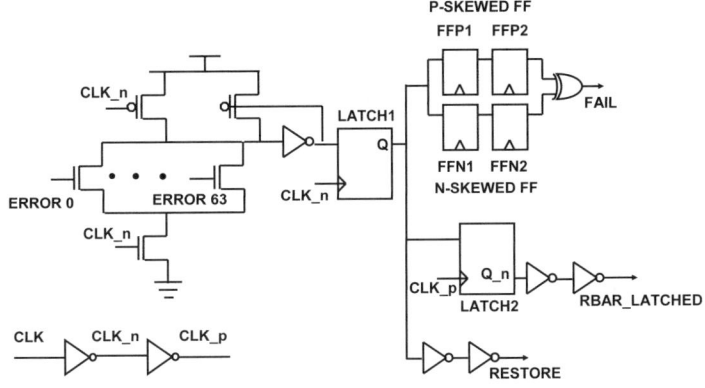

Figure 8.10 Restore generation circuitry. (© IEEE 2005)

This, in turn, evaluates the dynamic gate to generate the *restore* signal by "OR"-ing together the error signals of individual RFFs (Figure 8.10), in the negative clock phase. The *restore* needs to be latched at the output of the dynamic OR gate so that it retains state during the next positive phase (recovery cycle) during which it disables the shadow latch to protect state. The shadow latch can be designed using weaker devices since it is required only for runtime validation of the main flip-flop data and does not form a part of the critical path of the RFF.

The *rbar_latched* signal, shown in the restore generation circuitry in Figure 8.10, which is the half-cycle delayed and complemented version of the *restore* signal, precharges the *Err_dyn* node for the next errant cycle. Thus, unlike standard dynamic gates where precharge takes place every cycle, the *Err_dyn* node is conditionally precharged in the recovery cycle following a Razor error.

Compared to a regular DFF of the same drive strength and delay, the RFF consumes 22% extra (60fJ/49fJ) energy when sampled data is static and 65% extra (205fJ/124fJ) energy when data switches. However, in the processor, only 207 flip-flops out of 2388 flip-flops, or 9%, could become critical and needed to be RFFs. The Razor power overhead was computed to be 3% of nominal chip power.

The metastability detector consists of p- and n-skewed inverters which switch to opposite power rails under a metastable input voltage. The detector evaluates when input node SL can be ambiguously interpreted by its fan-out, inverter G1 and the error comparator. The DC transfer curve (Figure 8.11a) of inverter G1, the error comparator and the metastability detector show that the "detection" band is contained well within the ambiguously interpreted voltage band. Figure 8.11(b) gives the error detection and ambiguous interpretation bands for different corners. The probability that metastability propagates through the error detection logic and causes metastability of the restore signal itself was computed to be below 2e-30 [30]. Such an event is flagged by the fail signal generated using double-skewed flip-flops. In the rare event of a fail, the pipeline is flushed and the supply voltage is immediately increased.

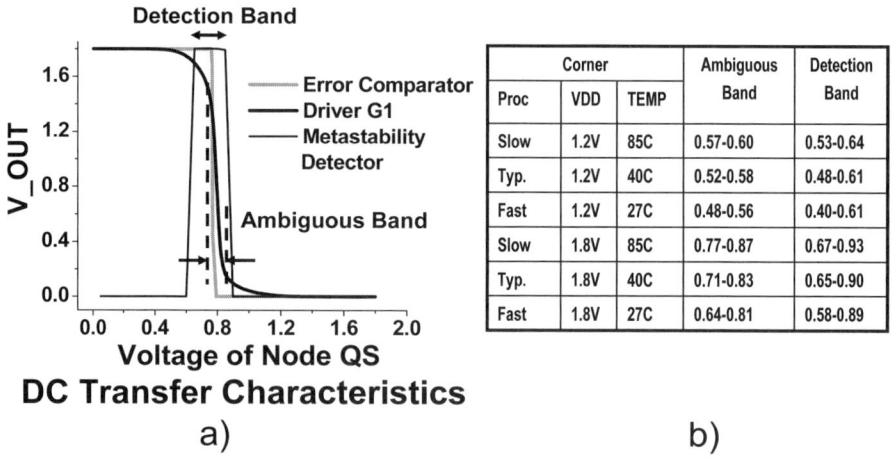

Figure 8.11 Metastability detector characteristics. (a) Principle of operation. (b) Metastability detector: corner analysis. (© IEEE 2005)

8.5 Silicon Implementation and Evaluation of Razor

A 64b processor which implements a subset of the Alpha instruction set was designed and built as an evaluation vehicle for the concept of Razor.   The chip was fabricated with MOSIS [31] in an industrial 0.18 micron technology. Voltage control is based on the observed error rate and power savings are achieved by (1) eliminating the safety margins under nominal operating and silicon conditions and (2) scaling voltage 120mV below the first failure point to achieve a 0.1% targeted error rate. It was tested and measured for savings due to Razor DVS for 33 different dies from two different lots and obtained an average energy savings of 50% over the worst-case operating conditions by operating at the 0.1% error rate voltage at 120MHz. The processor core is a five-stage in-order pipeline which implements a subset of the Alpha instruction set. The timing critical stages of the processor are the Instruction Decode (ID) and the Execute (EX) stages. The distributed pipeline recovery scheme as illustrated in Figure 8.7(b) was implemented. The die photograph of the processor is shown in Figure 8.12(a), and the relevant implementation details are provided in Figure 8.12(b).

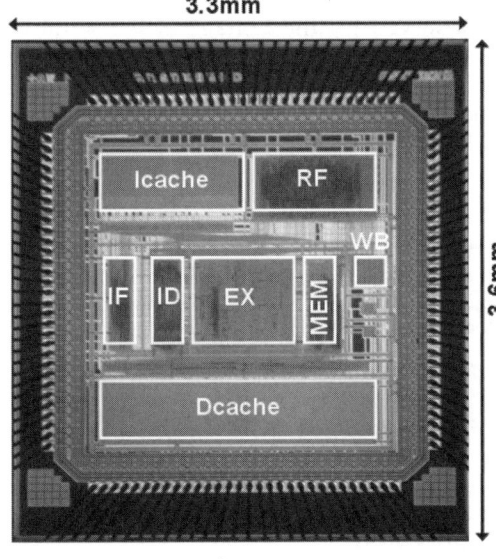

Technology Node	0.18μm
Max. Clock Frequency	140MHz
DVS Supply Voltage Range	1.2-1.8V
Total Number of Transistors	1.58million
Die Size	3.3mm*3.6 mm
Measured Chip Power at 1.8V	130mW
Icache Size	8KB
Dcache Size	8KB
Total Number of Flip-Flops	2388
Total Number of Razor Flip-Flops	207
Number of Delay Buffers Added	2801
Error Free Operation (Simulation Results)	
Standard FF Energy (Static/Switching)	49fJ/124fJ
RFF Energy (Static/Switching)	60fJ/205fJ
Total Delay Buffer Power Overhead	3.7mW
% Total Chip Power Overhead	2.9%
Error Correction and Recovery Overhead	
Energy of a RFF per error event	260fJ

a)   b)

Figure 8.12 Silicon evaluation of Razor. (a) Die micrograph. (b) Processor implementation details. (© IEEE 2005)

8.5.1 Measurement Results

Figure 8.13 shows the error rates and normalized energy savings versus supply voltage at 120 and 140MHz for one of the 33 chips tested, henceforth referred to as chip1. Energy at a particular voltage is normalized with respect to the energy at the point of first failure. For all plotted points, correct program execution with Razor was verified. The Y-axis on the left shows the percentage error rate and that on the right shows the normalized energy of the processor.

From the figure, we note that the error rate at the point of first failure is very low and is of the order of 1.0e-7. At this voltage, a few critical paths that are rarely sensitized fail to meet setup requirements and are flagged as timing errors. As voltage is scaled further into the sub-critical regime, the error rate increases exponentially. The IPC penalty due to the error recovery cycles is negligible for error rates below 0.1%. Under such low error rates, the recovery overhead energy is also negligible and the total processor energy shows a quadratic reduction with the supply voltage. At error rates exceeding 0.1%, the recovery energy rapidly starts to dominate, offsetting the quadratic savings due to voltage scaling. For the measured chips, the energy optimal error rate fell at approximately 0.1%.

The correlation between the first failure voltage and the 0.1% error rate voltage is shown in the scatter plot of Figure 8.14. The 0.1% error rate voltage shows a net variation of 0.24V from 1.38V to 1.62V which is approximately 20% less than the variation observed for the voltage at the point of

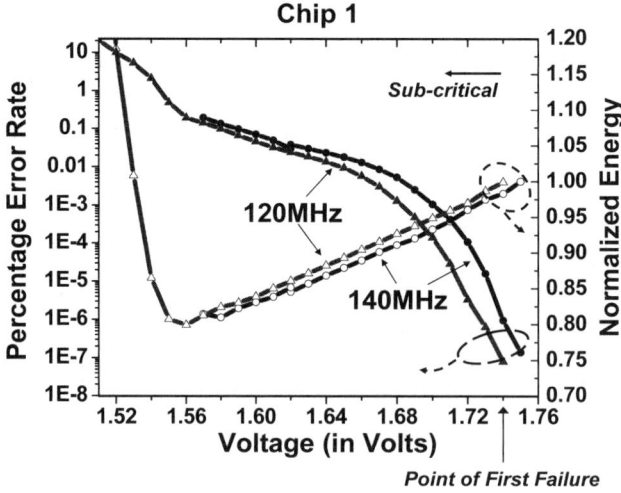

Figure 8.13 Measured error rate and energy versus supply voltage. (© IEEE 2005)

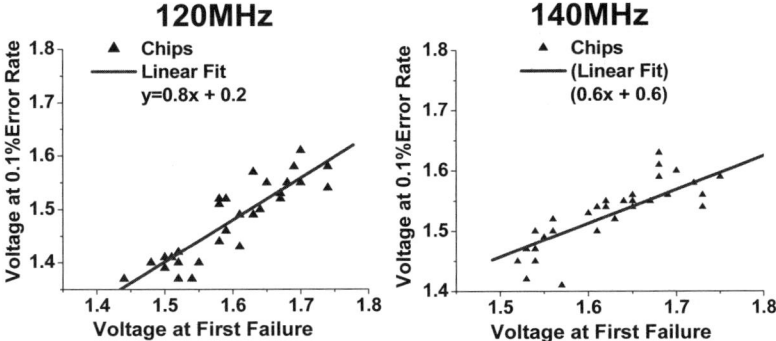

Figure 8.14 Scatter plot showing the point of 0.1% error rate versus the point of first failure. (© IEEE 2005)

first failure. The relative "flatness" of the linear fit indicates less sensitivity to process variation when running at a 0.1% error rate than at the point of first failure. This implies that a Razor-enabled processor, designed to operate at the energy optimal point, is likely to show greater predictability in terms of performance than a conventional worst-case optimized design. The energy optimal point requires a significant number of paths to fail and statistically averages out the variations in path delay due to process variation, as opposed to the first failure point which, being determined by the single longest critical path, shows higher process variation dependence.

8.5.2 Total Energy Savings with Razor

The total energy savings was measured by quantifying the savings due to elimination of safety margins and operation in the sub-critical voltage regime. Table 8.2 lists the measured voltage margins for process, voltage and temperature uncertainties for 2 out of the 33 chips tested, when operating at 120MHz. The chips are labeled as chip 1 and chip 2, respectively. The first failure voltage for chips 1 and 2 are 1.74V and 1.63V, respectively, and hence represent slow and typical process conditions, respectively.

Table 8.2 Measurement of voltage safety margins.

Chip (point of first failure)	Margins		
	Process	Voltage	Temperature
Slowest chip (1.76V)	0mV	180mV	100mV
Chip 1 (1.73V)	30mV	180mV	100mV
Chip 2 (1.63V)	130mV	180mV	100mV

The point of first failure of the slowest chip at 25°C is 1.76V. For this chip to operate correctly in the worst-case, voltage and temperature margins are added over and above the first failure voltage. The worst-case temperature margin was measured as the shift in the point of first failure of this chip when heated from 25°C to 10°5C. At 105°C, this chip fails at 1.86V, an increase of 100mV over the first failure voltage at 25°C. The worst-case voltage margin was estimated to be 10% of the nominal supply voltage of 1.8V (180mV). The margin for inter-die process variations was measured as the difference in the point of first failure voltage of the chip under test and the slowest chip. For example, chip 2 fails at 1.63V at 25°C when compared with the slowest chip which fails at 1.76V. This translates to 130mV process margin. Thus, with the incorporation of 100mV temperature margin and 180mV voltage margin over the first failure point of the slowest chip, the worst-case operating voltage for guaranteed correct operation was obtained to be 2.04V.

Figure 8.15 lists the energy savings obtained through Razor for chips 1 and 2. The first set of bars shows the energy when Razor is turned off and the chip under test is operated at the worst-case operating voltage at 120MHz, as determined for all the chips tested. At the worst-case voltage of 2.04V, chip 2 consumes 160.5mW of which 27.3mW is due to 180mV margin for supply voltage drop, 11.2mW is due to 100mV temperature margin and 17.3mW is due to 30mV process margin.

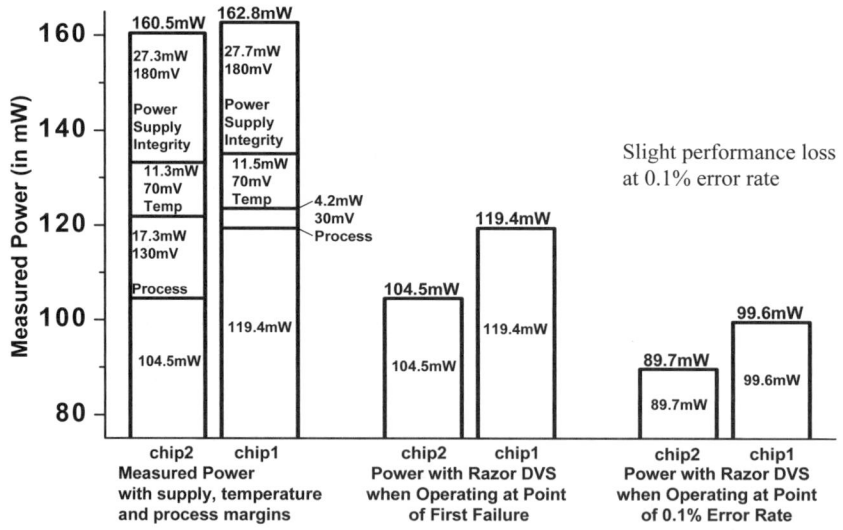

Figure 8.15 Total energy savings. (© IEEE 2005)

The second set of bars shows the energy when operating with Razor enabled at the point of first failure with all the safety margins eliminated. At the point of first failure, chip 2 consumes 104.5mW, while chip 1 consumes 119.4mW of power. Thus, for chip 2, operating at the first failure point leads to a saving of 56mW which translates to 35% saving over the worst case. The corresponding saving for chip 1 is 27% over the worst case.

The third set of bars shows the additional energy savings due to subcritical mode of operation of Razor. With Razor enabled, both chips are operated at the 0.1% error rate voltage and power measurements are taken. At the 0.1% error rate, chip 1 consumes 99.6mW of power at 0.1% error rate which is a saving of 39% over the worst case. When averaged over all die, we obtain approximately 50% savings over the worst case at 120MHz and 45% savings at 140MHz when operating at the 0.1% error rate voltage.

8.5.3 Razor Voltage Control Response

Figure 8.16 shows the basic structure of the hardware control loop that was implemented for real-time Razor voltage control. A proportional integral algorithm was implemented for the controller in a Xilinx XC2V250 FPGA [32]. The error rate was monitored by sampling the on-chip error register at a conservative frequency of 750KHz. The controller reacts to the error rate that is monitored by sampling the error register and regulates the supply voltage through a DAC and a DC–DC switching regulator to achieve a targeted error rate. The difference between the sampled error rate and the targeted error rate is the error rate differential, E_{diff}. A positive value of E_{diff} implies that the CPU is experiencing too few errors and hence the supply voltage may be reduced and vice versa.

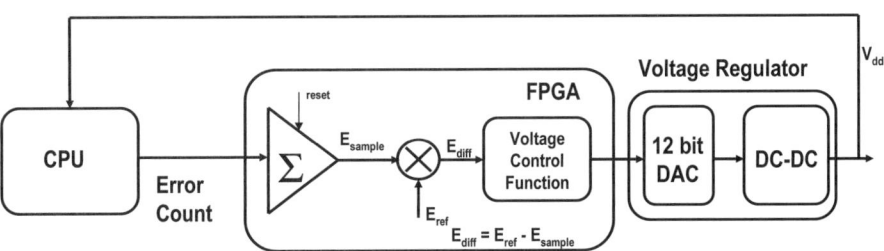

Figure 8.16 Razor voltage control loop. (© IEEE 2005)

The voltage controller response for a test program was tested with alternating high and low error rate phases. The targeted error rate for the given trace is set to 0.1% relative to CPU clock cycle count. The controller response during a transition from the lowerror rate phase to the high-error rate phase is shown in Figure 8.17(a). Error rates increase to about 15% at the onset of the high-error phase. The error rate falls until the controller reaches a high enough voltage to meet the desired error rate in each millisecond sample period. During a transition from the high-error rate phase to the low-error rate phase, shown in Figure 8.17(b), the error rate drops to zero because the supply voltage is higher than required. The controller responds by gradually reducing the voltage until the target error rate is achieved.

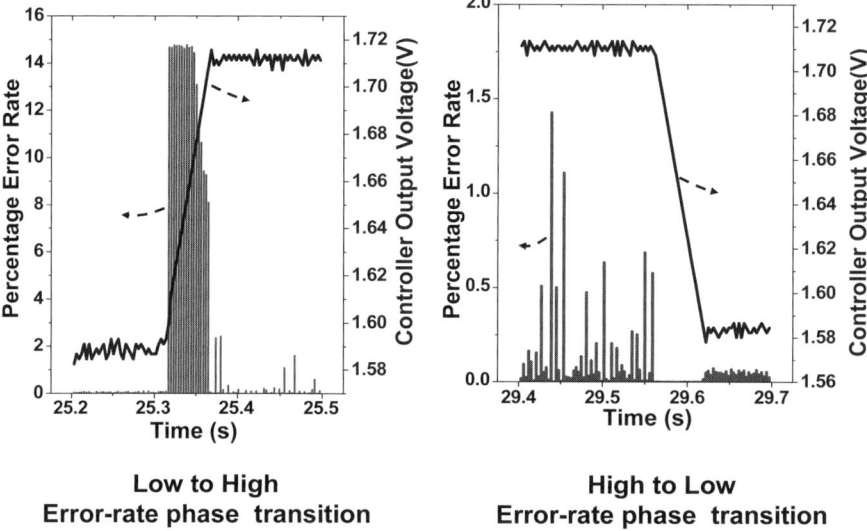

Low to High
Error-rate phase transition

High to Low
Error-rate phase transition

Figure 8.17 Voltage controller phase transition response. (a) Low to high transition. (b) High to low transition. (© IEEE 2005)

8.6 Ongoing Razor Research

Currently, research efforts on Razor are underway in ARM Ltd, UK. A deeper analysis of Razor as explained in the previous sections reveals several key issues that need to be addressed, before Razor can be deployed as mainstream technology.

The primary concern is the issue of Razor energy overhead. Since industrial strength designs are typically balanced, it is likely that significantly larger percentage of flip-flops will require Razor protection. Consequently, a greater number of delay buffers will be required to satisfy the short-path constraints. Increasing intra-die process variability, especially on the short paths, further aggravates this issue.

Another important concern is ensuring reliable state recovery in the presence of timing errors. The current scheme imposes a massive fan-out load on the pipeline *restore* signal. In addition, the current scheme cannot recover from timing errors in critical control signals which can cause undetectable state corruption in the shadow latch. Metastability on the *restore* signal further complicates state recovery. Though such an event is flagged by the *fail* signal, it makes validation and verification of a "Razor"-ized processor extremely problematic in current ASIC design methodologies.

An attempt is made to address these concerns by developing an alternative scheme for Razor, henceforth referred to as Razor II. The key idea in Razor II is to use the Razor flip-flop only for error detection. State recovery after a timing error occurs by a conventional replay mechanism from a check-pointed state. Figure 8.18 shows the pipeline modifications required to support such a recovery mechanism. The architectural state of the processor is check-pointed when an instruction has been validated by Razor and is ready to be committed to storage. The check-pointed state is buffered from the timing critical pipeline stages by several stages of stabilization which reduce the probability of metastability by effectively double-latching the pipeline output. Upon detection of a Razor error, the pipeline is flushed and system recovers by reverting back to the check-pointed architectural state and normal execution is resumed. Replaying from the

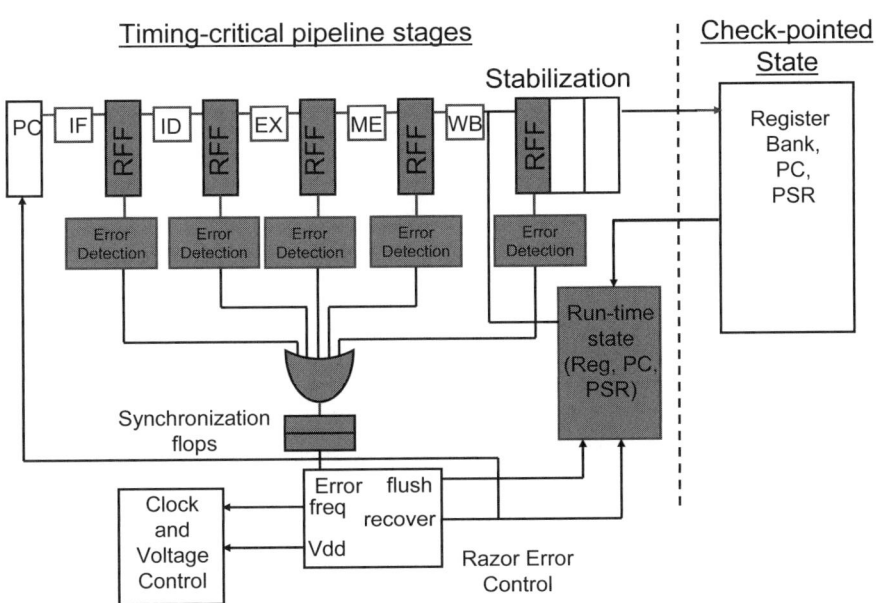

Figure 8.18 Pipeline modifications required for Razor II.

check-pointed state implies that a single instruction can fail in successive roll-back cycles, thereby leading to a deadlock. Forward progress in such a system is guaranteed by detecting a repeatedly failing instruction and executing the system at half the nominal frequency during recovery.

Error detection in Razor II is based on detecting spurious transitions in the D-input of the Razor flip-flop, as conceptually illustrated in Figure 8.19. The duration where the input to the RFF is monitored for errors is called the detection window. The detection window covers the entire positive phase of the clock cycle. In addition, it also includes the setup window in front of the positive edge of the clock. Thus, any transition in the setup window is suitably detected and flagged. In order to reliably flag potentially metastable events, safety margin is required to be added to the onset of the detection window. This ensures that the detection window covers the setup window under all process, voltage and temperature conditions. In a recent work, the authors have applied the above concept to detect and correct transient single event upset failures [33].

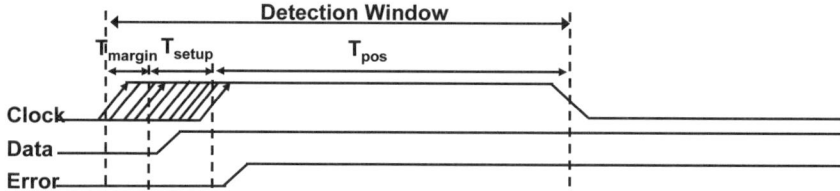

Figure 8.19 Transition detection-based error detection.

8.7 Conclusion

In this chapter, we presented a survey of different adaptive techniques reported in literature. We analyzed the concept of design margining in the presence of process variations and looked at how different adaptive techniques help eliminate some of the margins. We categorized these techniques as "always-correct" and "error detection and correction" techniques. We presented Razor as a special case study of the latter category and showed silicon measurement results on a chip using Razor for supply voltage control.

As process variations increase with each technology generation, adaptive techniques assume even greater relevance. However, deploying such techniques in the field is hindered either by their complexity as in the case of Razor or by the lack of substantial gains as in the case of canary circuits. Future research in this field needs to focus on combining effectiveness of Razor in eliminating design margins with the relative simplicity of the "always-correct" techniques. As uncertainties worsen, adaptive techniques provide a solution toward achieving computational correctness and faster design closure.

References

[1]   S.T. Ma, A. Keshavarzi, V. De, J.R. Brews, "A statistical model for extracting geometric sources of transistor performance variation," IEEE Transactions on Electron Devices, Volume 51, Issue 1, pp. 36–41, January 2004.

[2]   R. Gonzalez, B. Gordon, and M. Horowitz, "Supply and threshold voltage scaling for low power CMOS," IEEE Journal of Solid-State Circuits, Volume 32, Issue 8, August 1997.

[3]   S. Yokogawa, H. Takizawa, "Electromigration induced incubation, drift and threshold in single-damascene copper interconnects," IEEE 2002 International Interconnect Technology Conference, 2002, pp. 127–129, 3–5 June 2002.

[4]   W. Jie and E. Rosenbaum, "Gate oxide reliability under ESD-like pulse stress," IEEE Transactions on Electron Devices, Volume 51, Issue 7, July 2004.

[5]   International Technology Roadmap for Semiconductors, 2005 edition, http://www.itrs.net/ Links/2005ITRS/Home2005.htm.

[6]   M. Hashimoto, H. Onodera, "Increase in delay uncertainty by performance optimization," IEEE International Symposium on Circuits and Systems, 2001, Volume 5, pp. 379–382, 5, 6–9 May 2001.

[7]   S. Rangan, N. Mielke and E. Yeh, "Universal recovery behavior of negative bias temperature instability," IEEE Intl. Electron Devices Mtg., p. 341, December 2003.

[8]   G. Wolrich, E. McLellan, L. Harada, J. Montanaro, and R. Yodlowski, "A high performance floating point coprocessor," IEEE Journal of Solid-State Circuits, Volume 19, Issue 5, October 1984.

[9]   Trasmeta Corporation, "LongRun Power Management," http://www.transmeta.com/tech/longrun2.html

[10] Intel Corporation, "Intel Speedstep Technology," http://www.intel.com/support/processors/mobile/pentiumiii/ss.htm

[11] ARM Limited, http://www.arm.com/products/esd/iem_home.html

[12] T. Burd, T. Pering, A. Stratakos, and R. Brodersen, "A dynamic voltage scaled microprocessor system," International Solid-State Circuits Conference, February 2000.

[13] A.K. Uht, "Going beyond worst-case specs with TEATime," IEEE Micro Top Picks, pp. 51–56, 2004

[14] K.J. Nowka, G.D. Carpenter, E.W. MacDonald, H.C. Ngo, B.C Brock, K.I. Ishii, T.Y. Nguyen and J.L. Burns, "A 32-bit powerPC system-on-a-chip with support for dynamic voltage scaling and dynamic frequency scaling," IEEE Journal of Solid-State Circuits, Volume 37, Issue 11, pp. 1441–1447, November 2002

[15] T.D. Burd, T.A. Pering, A.J. Stratakos and R.W. Brodersen, "A dynamic voltage scaled microprocessor system," IEEE Journal of Solid-State Circuits, Volume 35, Issue 11, pp. 1571–1580, November 2000

[16] Berkeley Wireless Research Center, http://bwrc.eecs.berkeley.edu/

[17] M. Nakai, S. Akui, K. Seno, T. Meguro, T. Seki, T. Kondo, A. Hashiguchi, H. Kawahara, K. Kumano and M. Shimura, "Dynamic voltage and frequency management for a low power embedded microprocessor," IEEE Journal of Solid-State Circuits, Volume 40, Issue 1, pp. 28–35, January. 2005

[18] A. Drake, R. Senger, H. Deogun, G. Carpenter, S. Ghiasi, T. Ngyugen, N. James and M. Floyd, "A distributed critical-path timing monitor for a 65nm high-performance microprocessor," International Solid-State Circuits Conference, pp. 398–399, 2007.

[19] T. Kehl, "Hardware self-tuning and circuit performance monitoring," 1993 Int'l Conference on Computer Design (ICCD-93), October 1993.

[20] S. Lu, "Speeding up processing with approximation circuits," IEEE Micro Top Picks, pp. 67–73, 2004

[21] T. Austin, V. Bertacco, D. Blaauw and T. Mudge, "Opportunities and challenges in better than worst-case design," Proceedings of the ASP-DAC 2005, Volume 1, pp. 18–21, 2005.

[22] C. Kim, D. Burger and S.W. Keckler, IEEE Micro, Volume 23, Issue 6, pp. 99–107, November–December 2003.

[23] Z. Chishti, M.D. Powell, T. N. Vijaykumar, "Distance associativity for high-performance energy-efficient non-uniform cache architectures," Proceedings of the International Symposium on Microarchitecture, 2003, MICRO-36

[24] F. Worm, P. Ienne and P. Thiran, "A robust self-calibrating transmission scheme for on-chip networks," IEEE Transactions on Very Large Scale Integration, Volume 13, Issue 1, January 2005.

[25] R. Hegde and N. R. Shanbhag, "A voltage overscaled low-power digital filter IC," IEEE Journal of Solid-State Circuits, Volume39, Issue 2, February 2004.

[26] D. Roberts, T. Austin, D. Blaauw, T. Mudge and K. Flautner, "Error analysis for the support of robust voltage scaling," International Symposium on Quality Electronic Design (ISQED), 2005.

[27] L. Anghel and M. Nicolaidis, "Cost reduction and evaluation of a temporary faults detecting technique," Proceedings of Design, Automation and Test in Europe Conference and Exhibition 2000, 27–30 March 2000 pp. 591–598

[28] S. Das, D. Roberts, S. Lee, S. Pant, D. Blaauw, T. Austin, T. Mudge, K. Flautner, "A self-tuning DVS processor using delay-error detection and correction," IEEE Journal of Solid-State Circuits, pp. 792–804, April 2006.

[29] R. Sproull, I. Sutherland, and C. Molnar, "Counterflow pipeline processor architecture," Sun Microsystems Laboratories Inc. Technical Report SMLI-TR-94-25, April 1994.

[30] W. Dally, J. Poulton, Digital System Engineering, Cambridge University Press, 1998

[31] www.mosis.org

[32] www.xilinx.com

[33] D. Blaauw, S.Kalaiselvam, K. Lai, W.Ma, S. Pant, C. Tokunaga, S. Das and D.Bull "RazorII: In-situ error detection and correction for PVT and SER tolerance," International Solid-State Circuits Conference, 2008

[34] D. Ernst, N. S. Kim, S. Das, S. Pant, T. Pham, R. Rao, C. Ziesler, D. Blaauw, T. Austin, T. Mudge, K. Flautner, "Razor: A low-power pipeline based on circuit-level timing speculation," Proceedings of the 36th Annual IEEE/ACM International Symposium on Microarchitecture, pp. 7–18, December 2003.

[35] A. Asenov, S. Kaya, A.R. Brown, "Intrinsic parameter fluctuations in decananometer MOSFETs introduced by gate line edge roughness," IEEE Transactions on Electron Devices, Volume 50, Issue 5, pp. 1254–1260, May 2003.

[36] K. Ogata, "Modern control engineering," 4th edition, Prentice Hall, New Jersey, 2002.

Chapter 9 Variability-Aware Frequency Scaling in Multi-Clock Processors

Sebastian Herbert, Diana Marculescu

Carnegie Mellon University

9.1 Introduction

Variability is becoming a key concern for microarchitects as technology scaling continues and more and more increasingly ill-defined transistors are placed on each die. Process variations during fabrication result in a nonuniformity of transistor delays across a single die, which is then compounded by dynamic thermally dependent delay variation at runtime.

The delay of every critical path in a synchronously timed block must be less than the proposed cycle time for the block as a whole to meet that timing constraint. Thus, as both the amount of variation (due to ever-shrinking feature sizes as well as greater temperature gradients) and the number of critical paths (due to increasing design complexity and levels of integration) grow, the reduction in clock speed necessary to reduce the probability of a timing violation to an acceptably small level increases. However, the worst-case delay is very rarely exercised, and as a result, the overdesign that is necessary to deal with variability sacrifices large amounts of performance in the common case. Bowman et al. found that designs for the 50 nm technology node could lose an entire generation's worth of performance due to systematic within-die process variability alone [2].

A *variability-aware* microarchitecture is able to recover some of this lost performance. One such microarchitecture partitions a processor into multiple independently clocked frequency islands (FIs) [10, 14] and then uses this partitioning to address variations at the clock domain granularity. This chapter is an extension of the analysis of this microarchitecture performed by Herbert et al. [7].

A. Wang, S. Naffziger (eds.), *Adaptive Techniques for Dynamic Processor Optimization*,
DOI: 10.1007/978-0-387-76472-6_9, © Springer Science+Business Media, LLC 2008

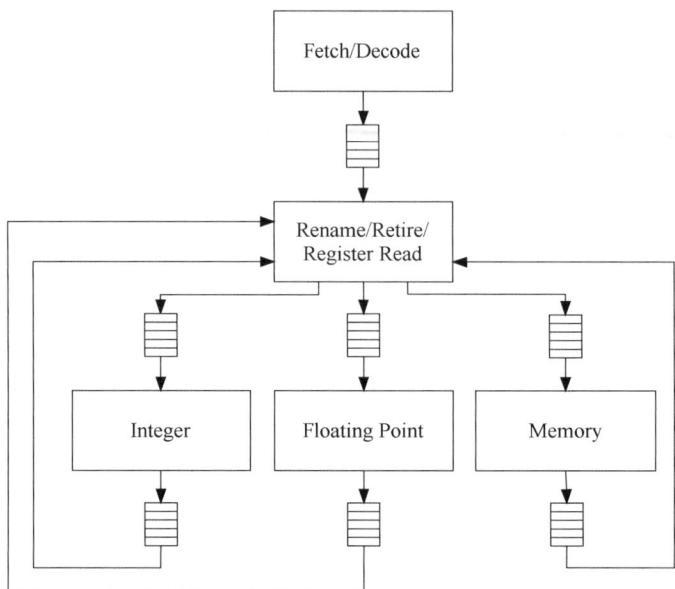

Figure 9.1 A microprocessor design using frequency islands.

Multi-clock designs using frequency islands provide increased flexibility over globally clocked designs. Each frequency island operates synchronously using its own local clock signal. However, arbitrary clock ratios are allowed between any pair of frequency islands, necessitating the use of asynchronous interfacing circuitry for inter-domain communication. For this reason, designs using frequency islands are often referred to as globally asynchronous, locally synchronous (GALS) designs.

An example of a frequency island design is shown in Figure 9.1. The processor core is divided into five clock domains. One contains the front-end fetch and decode logic, a second contains the register file, reorder buffer, and register renaming logic, and the execution units are split into integer, floating point, and memory domains. All communication between the domains must be synchronized by passing through a dual-clock FIFO.

Performing variability-aware frequency scaling using the FI partitioning addresses two sources of variability. First, it reduces the impact of random within-die process variability. As noted above, the probability of meeting a given timing constraint t_{max} decreases with both the amount of variability and the number of critical paths. While the amount of process variation cannot be addressed at the microarchitecture level, microarchitects can exercise some control over how often and where critical paths will be found.

Second, it addresses dynamic thermal variability that manifests itself as hotspots across the surface of the microprocessor die. At typical operating temperatures, transistor delay increases with temperature as a result of the effect of temperature on carrier mobility. Once again, an entire synchronously timed block must be clocked such that the delay through its hottest part meets its timing constraint, even though cooler parts could be run faster without creating *local* timing violations. If a microarchitecture has no thermal awareness, it is limited to always running at the frequency that results in correct operation at the maximum specified operating temperature.

Variability-aware frequency scaling (VAFS) sets the frequency of each clock domain as high as possible given that domain's worst *local* variations, rather than slowing down the entire processor to compensate for the worst *global* variations. Each clock domain in the FI processor has fewer critical paths than the processor as a whole, which shifts the mean of the maximum frequency distribution for each domain higher. Thus, the domains in the FI version can, on average, be clocked faster than the synchronous baseline to some degree, recovering some of the performance lost to process variation. This is a result of the fact that in the FI case, each clock domain's frequency is limited by its slowest local critical path rather than by the global slowest critical path, as in the fully synchronous case.

Thermal variability is addressed in a similar manner. In the synchronous case, the entire core must be slowed down to accommodate the temperature-induced increase in delay through its hottest block. For the FI case, the same is only true at the clock domain granularity. Thus, the impact of a hotspot on timing is isolated to the domain it is located in and does not require a global reduction in clock frequency.

9.2 Addressing Process Variability

9.2.1 Approach

The impact of parameter variations has been extensively studied at the circuit and device levels. However, with the increasing impact of variability on design yield, it has become essential to consider higher level models for parameter variation. Bowman et al. introduced the FMAX model with the aim of quantifying the impact of die-to-die and within-die variations on overall timing yield [2, 3]. They showed that the impact of variability on combinational circuits can be captured using two parameters: the logic depth of the circuit n_{cp} and the number of independent critical paths in the circuit N_{cp}. They observed that within-die (WID) variations tend to determine the mean of the worst-case delay distribution of a circuit, while die-to-die (D2D) variability determines its variance. Their model was validated against microprocessors from 0.25 μm to 0.13 μm technology nodes and was shown to accurately predict the mean, variance, and shape of the maximum frequency distribution. The FMAX model has subsequently been used in many studies on the effects of process variations at the microarchitecture level [9, 11, 12].

Typical microprocessor designs attempt to balance the logic depth across stages, so the number of critical paths N_{cp} is the dominant factor in determining the differences in how process variability affects each microarchitectural block. The delays of the N_{cp} independent critical paths are modeled as independent, identically distributed normal $\left(T_{cp,nom}, \sigma_{WID}^2 \right)$ random variables with probability density function (PDF) $f_{WID}(t)$ and cumulative distribution function (CDF) $F_{WID}(t)$. The effect of random within-die variability on a circuit block's delay is modeled as a random offset added to its nominal delay:

$$T_{cp,max} = T_{cp,nom} + \Delta T_{WID} \tag{9.1}$$

ΔT_{WID} is obtained by performing a max operation across N_{cp} critical paths, so the PDF for this random variable is given by

$$f_{\Delta T_{WID}}\left(\Delta t \right) = N_{cp} \times f_{WID}\left(T_{cp,nom} + \Delta t \right) \times \left(F_{WID}\left(T_{cp,nom} + \Delta t \right) \right)^{Ncp-1} \tag{9.2}$$

This equation has an intuitive interpretation. $f_{WID}\left(T_{cp,nom} + \Delta t \right)$ describes the probability of a particular single path having its delay increased by exactly Δt from nominal, while $\left(F_{WID}\left(T_{cp,nom} + \Delta t \right) \right)^{N_{cp}-1}$ gives the probability that every other path's delay is offset by an amount less than or equal to Δt (making the path that is offset by exactly Δt the slowest). The leading N_{cp} factor comes from the fact that any of the N_{cp} critical paths could be the one with the longest delay.

Figure 9.2 plots the worst-case delay distributions for $N_{cp} = (1, 2, 10)$ in terms of the path delay standard deviation. As N_{cp} increases, the standard deviation of the worst-case delay distribution decreases while its mean increases. Each of the clock domains in the FI partitioning has fewer critical paths than the microprocessor as a whole (since each clock domain is some smaller part of the entire processor). As a result, the mean of the FMAX distribution for each clock domain occurs at a higher frequency than the mean of the baseline FMAX distribution.

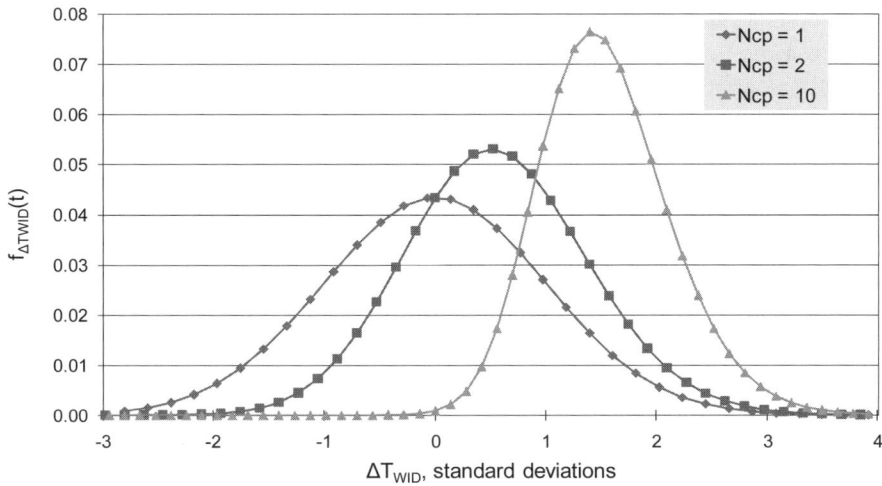

Figure 9.2 Delay distributions for $N_{cp} = (1, 2, 10)$.

Unfortunately, determining the number of *independent* critical paths in a given circuit in order to quantify this effect is not trivial. Correlations between critical path delays occur due to inherent spatial correlations in parameter variations and the overlap of critical paths that pass through one or more of the same gates. To overcome this problem, N_{cp} is redefined to be the *effective* number of independent critical paths that, when inserted into Equation (9.2), will yield a worst-case delay distribution that matches the statistics of the actual worst-case delay distribution of the circuit.

The proposed methodology estimates the effective number of independent critical paths for the two kinds of circuits that occur most frequently in processor microarchitectures: combinational logic and array structures. This corresponds roughly to the categorization of functional blocks as being either logic or SRAM dominated by Humenay et al. [9].

This methodology improves on the assumptions about the distribution of critical paths that have been made in previous studies. For example, Marculescu and Talpes assumed 100 total independent critical paths in a microprocessor and distributed them among blocks proportionally to device count [12], while Humenay et al. assumed that logic stages have only a single critical path and that an array structure has a number of critical paths equal to the product of the number of wordlines and number of bitlines [9]. Liang and Brooks make a similar assumption for register file SRAMs [11]. The proposed model also has the advantage of capturing the effects of "almost-critical" paths which would not be critical under nominal conditions, but are sufficiently close that they could become a block's slowest path in the face of variations. The model results presented here assume a 3σ of 20% for channel length [2] and wire segment resistance/capacitance.

9.2.2 Combinational Logic Variability Modeling

Determining the effective number of critical paths for combinational logic is fairly straightforward. Following the generic critical path model [2], the SIS environment is used to map uncommitted logic to a technology library of two-input NAND gates with a maximum fan-out of three. Gate delays are assumed to be independent normal random variables with mean equal to the nominal delay of the gate d_{nom} and standard deviation $\sigma_L / \mu_L \times d_{nom}$. Monte Carlo sampling is used to obtain the worst-case delay distribution for a given circuit, and then moment matching determines the value of N_{cp} that will cause the mean of analytical distribution from Equation (9.2) to equal that obtained via Monte Carlo.

This methodology was evaluated over a range of circuits in the ISCAS'85 benchmark suite and the obtained effective critical path numbers yielded distributions that were reasonably close to the actual worst-case delay distributions, as seen in Table 9.1. Note that the difference in the means of the two distributions will always be zero since they are explicitly matched. The error in the standard deviation can be as high as 25%, which is in line with the errors observed by Bowman et al. [3]. However, it is much lower when considering the combined effect of WID and D2D variations. Bowman et al. note that the variance in delay due to within-die variations is unimportant since it decreases with increasing N_{cp} and is dominated by the variance in delay due to die-to-die variations, which is independent of N_{cp} [2]. The error in standard deviation in the face of both WID and D2D variations is shown in the rightmost column of the table, illustrating this effect. Moreover, analysis of these results and others shows that most of the critical paths in a microprocessor lie in array structures due to their large size and regularity [9]. Thus, the error in the standard deviation for combinational logic circuits is inconsequential.

Such N_{cp} results can be used to assign critical path numbers to the functional units. Pipelining typically causes the number of critical paths in a circuit to be multiplied by the number of pipeline stages, as each critical path in the original implementation will now be critical in each of the stages. Thus, the impact of pipelining can be estimated by multiplying the functional unit critical path counts by their respective pipeline depths.

Table 9.1 Effective number of critical paths for ISCAS'85 circuits.

Circuit	Effective critical paths	% error in standard deviation	
		WID only	WID and D2D
C432	4.0	25.3	7.3
C499	11.0	19.7	4.5
C880	4.0	23.6	6.7
C2670	5.0	22.4	6.1
C6288	1.2	6.1	1.9

9.2.3 Array Structure Variability Modeling

Array structures are incompatible with the generic critical path model because they cannot be represented using two-input NAND gates with a maximum fan-out of three. As they constitute a large percentage of die area, it is essential to model the effect of WID variability on their access times accurately. One solution would be to simulate the impact of WID variability in a SPICE-level model of an SRAM array, but this would be prohibitively time consuming. An alternative is to enhance an existing high-level cache access time simulator, such as CACTI 4.1. CACTI has been shown to accurately estimate access times to within 6% of HSPICE values.

To model the access time of an array, CACTI replaces its transistors and wires with an equivalent RC network. Since the on-resistance of a transistor is directly proportional to its effective gate length L_{eff}, which is modeled as normally distributed with mean μ_L and standard deviation σ_L, R is normally distributed with mean R_{nom} and standard deviation $\sigma_L / \mu_L \times R_{nom}$.

To determine the delay, CACTI uses the first-order time constant of the network t_f, which can be written as $t_f = R \times C_L$, and the Horowitz model:

$$delay = t_f \sqrt{\alpha + \frac{\beta}{t_f}} \tag{9.3}$$

Here α and β are functions of the threshold voltage, supply voltage, and input rise time, which are assumed constant. The delay is a weakly nonlinear (and therefore strongly linear) function of t_f, which in turn is a linear function of R. Each stage delay in the RC network can therefore be modeled as a normal random variable. This holds true for all stages except the comparator and bitline stages, for which CACTI uses a second-order RC model. However, under the assumption that the input rise time is fast, these stage delays can be approximated as normal random variables as well.

Because the wire delay contribution to overall delay is increasing as technology scales, it is important to model random variations in wire dimensions as well as those in transistor gate length. CACTI lumps the entire resistance and capacitance of a wire of length L into a single resistance $L \times R_{wire}$ and a single capacitance $L \times C_{wire}$, where R_{wire} and C_{wire} represent the resistance and capacitance of a wire of unit length. Variations in the wire dimensions translate into variations in the wire resistance and capacitance.

R_{wire} and C_{wire} are assumed to be independent normal random variables with standard deviation σ_{wire}. This assumption is reasonable because the only physical parameter that affects both R_{wire} and C_{wire} is wire width, which has the least impact on wire delay variability [13]. Variability is modeled both along a single wire and between wires by decomposing a wire of length L into N segments, each with its own R_{wire} and C_{wire}. The standard deviation of the lumped resistance and capacitance of a wire of length L is thus $\sigma_{wire}/\sqrt{N}$. The length of each segment is assumed to be the feature size of the technology in which the array is implemented.

These variability models provide the delay distributions of each stage along the array access and the overall path delay distribution for the array. Monte Carlo sampling was used to obtain the worst-case delay distribution from the observed stage delay distributions, and the effective number of independent critical paths was then computed through moment matching.

This is highly accurate – in most cases, the estimated and actual worst-case delay distributions are nearly indistinguishable, as seen in Figure 9.3. Table 9.2 shows some effective independent critical path counts obtained with this model. Due to their regular structure, caches typically have more critical paths than the combinational circuits evaluated previously. Humenay et al. reached the same conclusion when comparing datapaths with memory arrays [9]. They assumed that the number of critical paths in an array was equal to the number of bitlines times the number of wordlines. The enhanced model presented here accounts for all sources of variability, including the wordlines, bitlines, decoders, and output drivers.

Table 9.2 Effective number of critical paths for array structures.

Array size	Wordlines	Bitlines	Effective critical paths
256 B	32	64	105
512 B	64	64	195
1024 B	128	64	415
2048 B	256	64	730

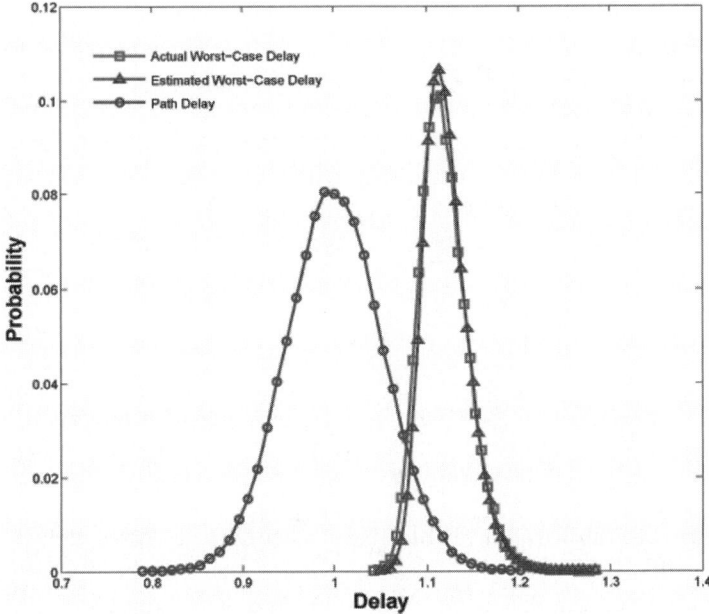

Figure 9.3 Estimated versus actual worst-case delay distribution for a 1 KB direct-mapped cache with 32 B blocks.

9.2.4 Application to the Frequency Island Processor

These critical path estimation methods were applied to an Alpha-like microprocessor, which was assumed to have balanced logic depth n_{cp} across stages. The processor is divided into five clock domains – fetch/decode, rename/retire/register read, integer, floating point, and memory. Table 9.3 details the effective number of independent critical paths in each domain. Using these values of N_{cp} in Equation (9.2) yields the probability density functions and cumulative distribution functions for the impact of variation on maximum frequency plotted in Figure 9.4.

The fully synchronous baseline incurs a 19.7% higher mean delay as a result of having 15,878 critical paths rather than only one. On the other hand, the frequency island domains are penalized by a best case of 13.0% and worst case of 18.7%. The resulting mean speedups for the clock domains relative to the synchronous baseline are calculated as:

$$speedup_{cp} = \frac{T_{cp,nom} + \mu_{\Delta t_{WID,synchronous}}}{T_{cp,nom} + \mu_{\Delta t_{WID,domain}}} \qquad (9.4)$$

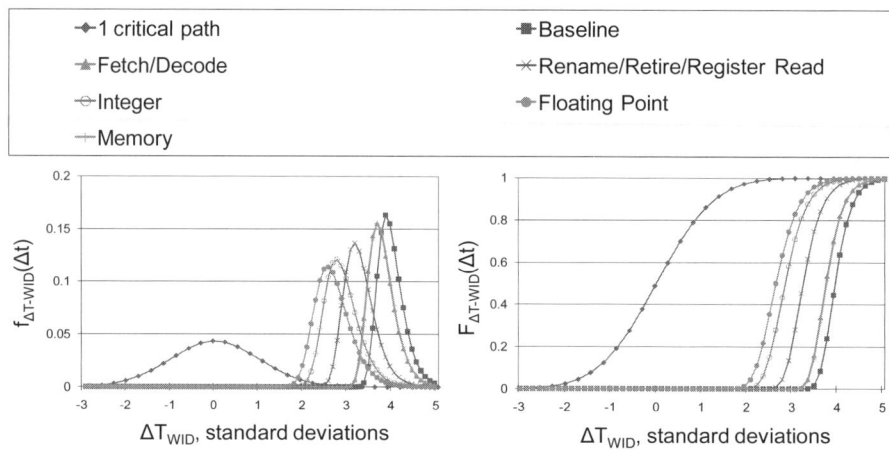

Figure 9.4 PDFs and CDFs for ΔT_{WID}.

Results are shown in Table 9.3, assuming a path delay standard deviation of 5%. This is between the values that can be extracted for the "half of channel length variation is WID" and "all channel length variation is WID" cases for a 50 nm design with totally random within-die process variations in Bowman et al.'s figure 11 [2].

These speedups represent the mean per-domain speedups that would be observed when comparing an FI design using VAFS to run each clock domain as fast as possible versus the fully synchronous baseline over a large number of fabricated chips. These results were verified with Monte Carlo analysis over one million vectors of 15,878 critical path delays. The mean speedups from this Monte Carlo simulation agreed with those in Table 9.3.

The exact speedups in Table 9.3 would *not* be seen on any single chip, as the slowest critical path (which limits the frequency of the fully synchronous processor) is also found in one of the five clock domains, yielding no speedup in that particular domain for that particular chip.

Table 9.3 Critical path model results.

Domain	Effective critical paths	$T_{cp,nom} + \mu_{\Delta t_{WID}}$	Speedup
Baseline	15,878	1.197	1.000
Fetch/Decode	6,930	1.187	1.008
Rename/Retire/Read	1,094	1.160	1.032
Integer	294	1.140	1.050
Floating Point	160	1.130	1.059
Memory	7,400	1.187	1.008

9.3 Addressing Thermal Variability

At runtime, there is dynamic variation in temperature across the die, which results in a further nonuniformity of transistor delays. Some units, such as caches, tend to be cool while others, such as register files and ALUs, may run much hotter. The two most significant temperature dependencies of delay are those on carrier mobility and that on threshold voltage.

Delay is inversely proportional to carrier mobility, μ. The BSIM4 model is used to account for the impact of temperature on mobility, with model cards generated for the 45 nm node by the Predictive Technology Model Nano-CMOS tool [17]. Values from the 2005 International Technology Roadmap for Semiconductors were used for supply and threshold voltage.

Temperature also affects delay indirectly through its effect on threshold voltage. Delay, supply voltage, and threshold voltage are related by the well-known alpha power law:

$$d \propto \frac{V_{DD}}{\left(V_{DD} - V_{TH}\right)^{\alpha}} \tag{9.5}$$

A reasonable value for α, the velocity saturation index, is 1.3 [7]. The threshold voltage itself is dependent on temperature, and this dependence is once again captured using the BSIM4 model.

Combining the effects on carrier mobility and threshold voltage

$$d \propto \frac{V_{DD}}{\mu_{eff}\left(T\right)\left(V_{DD} - V_{TH}\left(T\right)\right)^{\alpha}} \tag{9.6}$$

Maximum frequency is inversely proportional to delay, so with the introduction of a proportionality constant C, frequency is expressed as

$$f = C \frac{\mu_{eff}\left(T\right)\left(V_{DD} - V_{TH}\left(T\right)\right)^{\alpha}}{V_{DD}} \tag{9.7}$$

C is chosen such that the baseline processor runs at 4.0 GHz with $V_{DD} = 1.0$ V and $V_{TH} = 0.151$ V at a temperature of 145°C. The voltage parameters come from ITRS, while the baseline temperature was chosen based on observing that the 45 nm device breaks down at temperatures exceeding 150°C [7] and then adding some amount of slack. Thus, the baseline processor comes from the manufacturer clocked at 4.0 GHz with a specified maximum operating temperature of 145°C. Above this temperature, the transistors will become slow enough that timing constraints may not be met. However, normal operating temperatures will often be below this ceiling. VAFS exploits this thermal slack by speeding up cooler domains.

9.4 Experimental Setup

9.4.1 Baseline Simulator

The proposed schemes were evaluated using a modified version of the SimpleScalar simulator with the Wattch power estimation extensions [4] and HotSpot thermal simulation package [15]. The microarchitecture resembles an Alpha microprocessor, with separate instruction and data TLBs and the backend divided into integer, floating point, and memory clusters, each with their own instruction windows and issue logic. Such a clustered microarchitecture lends itself well to being partitioned into multiple clock domains. The HotSpot floorplan is adapted from one used by Skadron et al. [15] , and models an Alpha 21364-like core shrunken to 45 nm technology. The processor parameters are summarized in Table 9.4.

The simulator's static power model is based on that proposed by Butts and Sohi [5] and complements Wattch's dynamic power model. The model uses estimates of the number of transistors (scaled by design-dependent factors) in each structure tracked by Wattch. The effect of temperature on leakage power is modeled through both the exponential dependence of leakage current on temperature and the exponential dependence of leakage current on threshold voltage, which is itself a function of temperature. Thus, the equation for scaling subthreshold leakage current I_{leak} is

$$I_{leak}\left(T\right) = I_{leak}\left(T_0\right)e^{\frac{V_{TH}(T)}{T}} \qquad (9.8)$$

Table 9.4 Processor parameters.

Parameter	Value
Frequency	4.0 GHz
Technology	45 nm node, V_{DD} = 1.0 V, V_{TH} = 0.151 V
L1-I/D caches	32 KB, 64 B blocks, 2-way SA, 2-cycle hit time, LRU
L2 cache	2 MB, 64 B blocks, 8-way SA, 25-cycle hit time, LRU
Pipeline parameters	16 stages deep, 4 instructions wide
Window sizes	32 integer, 16 floating point, 16 memory
Main memory	100 ns random access, 2.5 ns burst access
Branch predictor	gshare, 12 bits of history, 4K entry table

A baseline leakage current at 25°C is taken from ITRS and then scaled according to temperature. HotSpot updates chip temperatures every 5 μs, at which point the simulator computes a leakage scaling factor for each block (at the same granularity used by Wattch) and uses it to scale the leakage power computed every cycle until the next temperature update.

9.4.2 Frequency Island Simulator

This synchronous baseline was the starting point for an FI simulator. It is split into five clock domains: fetch/decode, rename/retire/register read, integer, floating point, and memory. Each domain has a power model for its clock signal that is based on the number of pipeline registers within the domain. Inter-domain communication is accomplished through the use of asynchronous FIFO queues [6], which offer improved throughput over many other synchronization schemes under nominal FIFO operation.

Several versions of the FI simulator were used in the evaluation. The first is the baseline version (FI-B), which splits the core into multiple clock domains but runs each one at the same 4.0 GHz clock speed as the synchronous baseline (SYNCH). This baseline FI processor does not implement any variability-aware frequency scaling; all of the others do.

The second FI microarchitecture speeds up each domain as a result of the individual domains having fewer critical paths than the microprocessor as a whole. The speedups are taken from Table 9.3, and this version is called FI-CP. In the interests of reducing simulation time, only the mean speedups were simulated. These represent the average benefit that an FI processor would display over an equivalent synchronous processor on a per-domain basis over the fabrication of a large number of dies.

The third version, FI-T, assigns each domain a baseline frequency that is equal to the synchronous baseline's frequency, but then scales each domain's frequency for its temperature according to Equation (9.7) after every chip temperature update (every 20,000 ticks of a 4.0 GHz reference clock).

A final version, FI-CP-T, uses the speeds from FI-CP as the baseline domain speeds and then applies thermally aware frequency scaling. Both FI-T and FI-CP-T perform dynamic frequency scaling using an aggressive Intel XScale-style DFS system as in [16].

9.4.3 Benchmarks Simulated

In order to accurately account for the effects of temperature on leakage power and power on temperature, simulations are iterated for each workload and configuration, feeding the output steady-state temperatures of one run back in as the initial temperatures of the next in search of a consistent operating point. This iteration continues until temperature and power values converge, rather than performing a set number of iterations. With this methodology, the initial temperatures of the first run do not affect the final results, but only the number of iterations required.

The large number of runs required per benchmark prevented simulation of the entire suite of SPEC2000 benchmarks due to time constraints. Simulations were completed for seven of the benchmarks: the *164.gzip*, *175.vpr*, *197.parser*, and *256.bzip2* integer benchmarks and the *177.mesa*, *183.equake*, and *188.ammp* floating point benchmarks.

The simulation methodology addresses time variability by simulating three points within each benchmark, starting at 500, 750, and 1,000 million instructions and gathering statistics for 100 million more. The one exception was *188.ammp*, which finished too early. Instead, it was fast-forwarded 200 million instructions and then run to completion. Because the FI microprocessor is globally asynchronous, space variability is also an issue (e.g., the exact order in which clock domains tick could have a significant effect on branch prediction performance as the arrival time of prediction feedback will be altered). The simulator randomly assigns phases to the domain clocks, which introduces slight perturbations into the ordering of events and so averages out possible extreme cases over three runs per simulation point per benchmark. Both types of variability were thus addressed using the approaches suggested by Alameldeen and Wood [1].

9.5 Results

The FI configurations are compared on execution time, average power, total energy, and energy delay^2 in Figure 9.5.

9.5.1 Frequency Island Baseline

Moving from a fully synchronous design to a frequency island, one (FI-B) incurs an average 7.5% penalty in execution time. There is a fair amount of variation between benchmarks in the significance of the performance degradation. Both *164.gzip* and *197.parser* run about 11% slower, while *177.mesa* and *183.equake* only suffer a slowdown of around 2%. Broadly, floating point applications are impacted less than integer ones since many of their operations inherently have longer latencies, reducing the significance of the latency added due to crossing FI boundaries (*188.ammp* seems to be an exception). Workloads which exhibit large numbers of stalls due to waiting on memory or other resources are those which observe the smallest performance penalties, since the extra latency due to FI is almost completely hidden behind these stalls.

Due to the use of small local clock networks and the stretching of execution time, the FI processor draws 10.7% less power per cycle, resulting in a consumption of 4.0% less energy than the synchronous baseline over the execution of the same instructions. Energy-delay^2 is increased by 11.3% in making the move to the baseline FI architecture, making it uncompetitive for all but the most power-limited applications (in which case the 10% reduction in power draw might not be large enough to be significant).

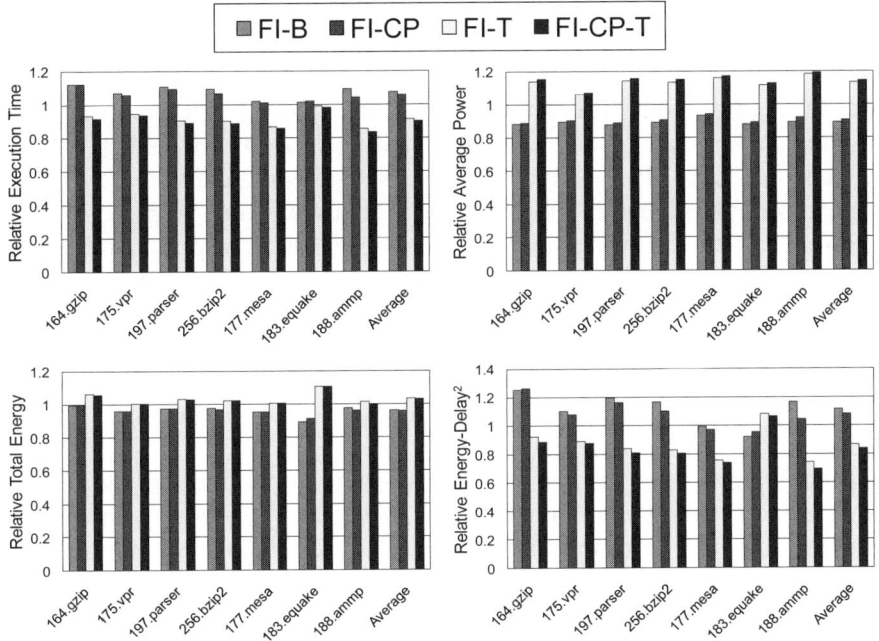

Figure 9.5 Simulation results relative to the synchronous baseline.

9.5.2 Frequency Island with Critical Path Information

FI-CP adds the speedups calculated from the critical path information in Section 9.2.4 to the FI baseline. Despite the average per-domain speedup in FI-CP being 3.1%, execution time decreases by only 1.4% because of the mismatch between speedups. The fetch and memory domains are barely sped up at all as a result of the large number of critical paths in the first-level caches, in keeping with the findings of Humenay et al. that the L1 caches are limiters of clock frequency in modern microprocessors [9]. This decreases the average number of executed instructions per clock tick for each back-end domain because of two factors. First, instructions are entering their window at a relatively reduced rate due to the low instruction cache speedup. Second, load-dependent instructions must wait relatively longer for operands due to the low data cache speedup. Thus, although the computation domains cycle more often, there is a much smaller increase in the amount of work that can actually be done in a fixed amount of time.

Benchmarks which are computation limited see the largest improvements, while those that are memory limited gain little. *183.equake* actually appears to suffer an increase in execution time, which is likely due to simulation variability. As a result of the faster clocking of domains, the average power drawn per cycle increases very slightly when enabling these speedups (by about 1.4%). However, the faster execution leads to essentially no change in energy usage and an overall energy-delay^2 reduction of 2.9%. These small improvements alone do not create a design which is competitive with the fully synchronous baseline.

FI-CP suffers from the domain partitioning used, which is performed based on the actual functionality of blocks without taking into account the number of critical paths that they contain. A better partitioning might use some metric that relates the number of critical paths in a block to its criticality to performance. However, "criticality to performance" can be difficult to quantify, since the critical path through the core may be different for different applications.

Moreover, there is overhead associated with every domain and domain boundary crossing. Combining domains can reduce the required number of domain boundary crossings as well as design complexity, but will also reduce the power savings introduced by the FI clocking scheme (since it merges some small clock networks to create a single larger one). Furthermore, it reduces the flexibility of the FI microarchitecture and might impact opportunities for VAFS or dynamic voltage/frequency scaling. On the other hand, splitting a clock domain into multiple smaller domains requires the opposite set of trade-offs to be evaluated.

9.5.3 Frequency Island with Thermally Aware Frequency Scaling

FI-T applies thermally aware frequency scaling to the FI baseline, running each domain as fast as possible given its current temperature rather than always assuming the worst-case temperature. FI-T offers significantly better performance than FI-B or FI-CP. In fact, accounting for dynamic thermal variation results in an average execution time reduction of 8.7% when compared to the fully synchronous baseline. As expected, the performance improvement enabled by thermally aware frequency scaling is highly dependent on the behavior (both thermal and otherwise) of the workload under consideration. *188.ammp* runs cool and so sees a large performance boost, finishing in 14.4% less time on the FI processor with thermally aware frequency scaling than on the synchronous baseline. However, many other benchmarks see similar frequency increases, but do not display as large an execution time reduction. For example, *183.equake* is memory-bound and so gains relatively little (only a 1% speedup relative to the synchronous baseline), despite the large increases in the clock domain frequencies of the clock domains.

Two things are required to see a significant performance gain from thermally aware frequency scaling: sufficient thermal headroom and application behavior which can take advantage of the resulting frequency increases in the core. This translates into a large amount of variation in the change in energy-efficiency brought about by enabling thermally aware frequency scaling.

Since FI-T runs the domain clocks somewhat faster than the baseline speed, a significant average power penalty of 13.3% relative to the synchronous baseline is observed. This corresponds to a 3.4% increase in the amount of energy used to perform the same amount of computation. However, the larger reduction in the amount of time required to perform the computation leads to an average energy-delay^2 13.4% lower than the synchronous baseline's energy.

FI-T suffers somewhat from naïvely speeding up domains regardless of whether this improves performance or not. The most egregious example is the speeding up of the floating point domain in the integer benchmarks. This may even adversely affect performance because each clock tick dissipates some power, regardless of whether there are any instructions in the domain or not. This results in higher local temperatures, which may spill over into a neighboring domain which is critical to the performance and causes it to be clocked at a lower speed.

One solution is to use some control scheme similar to those used for DVFS to decide whether a domain should actually be sped up and by how much. Equation (9.7), which describes the scaling of frequency with temperature, also includes the dependence of frequency on supply voltage, so DVFS could possibly be integrated with VAFS. An integrated control system would be required to prevent the two schemes from pulling clock frequency in opposite directions. This area requires further research.

Like FI-CP, FI-T could also benefit from a more intelligent domain partitioning. Since each domain's speed is limited by its hottest block, it might make sense to group blocks into domains based on whether they tend to run cool, hot, or in between. However, while there are some functional blocks which can be identified as generally being hotspots (e.g., the integer register file and scheduling logic), the temperature at which other blocks run is highly workload-dependent (e.g., the entire floating point unit).

9.5.4 Frequency Island with Critical Path Information and Thermally Aware Frequency Scaling

The results for FI-CP-T, which applies both variability-aware frequency schemes, show that the two are largely additive. A 10.0% reduction in execution time is achieved at the cost of 14.6% higher average power; the total energy penalty is 3.0%. This actually represents a reduction in the amount of energy consumed relative to FI-T. The reduction in execution time from FI-T to FI-CP-T is 1.4%, the same as that observed in moving from FI-B to FI-CP. An initial fear when combining FI-CP and FI-T was that the higher baseline speeds as a result of the -CP speedups would result in a sufficient increase in temperature to reduce the -T speedups by an equal amount, resulting in a scheme that offered no better performance than FI-T and was more complex. However, these results show that the speedups applied by FI-CP and FI-T are largely independent. The final energy-delay^2 reduction offered by full VAFS is 16.1%.

The synergy between the two schemes is due to the fact that the caches tend to run cool. As a result, thermally aware frequency scaling speeds up the clock domains containing the L1 caches slightly more than the others, which helps to mitigate the lack of speedup for the caches in FI-CP. Thus, the speedups of the computation domains due to considering critical path information can be better taken advantage of.

9.6 Conclusion

Variability is one of the major concerns that microprocessor designers will have to face as technology scaling continues. It is potentially easier for a frequency island design to address variability as a result of the processor being partitioned into multiple clock domains, which allows the negative effects of variability on maximum frequency to be localized to the domain they occur in. This variability-aware frequency scaling can be used to address both process and thermal variabilities.

The effects of random within-die process variability will be difficult to mitigate using a simple FI partitioning of the core. The large number of critical paths in a modern processor means that even decoupling groups of functional blocks with relatively low critical path counts from those with higher ones does not yield a large improvement in their mean frequencies.

On the other hand, exploiting the thermal slack between current operating temperatures and the maximum operating temperature by speeding up cooler clock domains proves to have significant performance and energy-efficiency benefits. An FI processor with such thermally aware frequency scaling is capable of overcoming the performance disadvantages inherent to the FI design style to achieve better performance than a similarly organized fully synchronous microprocessor.

As technology continues to scale, the magnitude of process variations will increase due to the need to print ever-smaller features, while thermal variation also worsens due to greater transistor density causing a higher difference in power densities across the chip. It will soon be the case that such variations can no longer be handled below the microarchitecture level and abstracted away, and the benefits from creating a variability-tolerant or variability-aware microarchitecture will outweigh the increased work and design complexity involved.

Acknowledgments

The authors thank Siddharth Garg for his assistance with generating the critical path model results.

References

[1]   A. Alameldeen and D. Wood, "Variability in Architectural Simulations of Multi-threaded Workloads", HPCA'03: Proceedings of the 9th International Symposium on High-Performance Computer Architecture, 2003, pp. 7–18

[2]   K. Bowman, S. Duvall and J. Meindl, "Impact of Die-to-die and Within-die Parameter Fluctuations on the Maximum Clock Frequency Distribution for Gigascale Integration", IEEE Journal of Solid-State Circuits, February 2002, Vol. 37, No. 2, pp. 183–190

[3]   K. Bowman, S. Samaan and N. Hakim, "Maximum Clock Frequency Distribution with Practical VLSI Design Considerations", ICICDT'04: Proceedings of the International Conference on Integrated Circuit Design and Technology, 2004, pp. 183–191

[4]   D. Brooks, V. Tiwari and M. Martonosi, "Wattch: A Framework for Architectural-level Power Analysis and Optimizations", ISCA'00: Proceedings of the 27th International Symposium on Computer Architecture, 2000, pp. 83–94

[5]   J. Butts and G. Sohi, "A Static Power Model for Architects", MICRO 33: Proceedings of the 33rd Annual ACM/IEEE International Symposium on Microarchitecture, 2000, pp. 191–201

[6]   T. Chelcea and S. Nowick, "Robust Interfaces for Mixed Systems with Application to Latency-insensitive Protocols", DAC'01: Proceedings of the 38th annual Design Automation Conference, 2001, pp. 21–26

[7]   S. Herbert, S. Garg and D. Marculescu, "Reclaiming Performance and Energy Efficiency from Variability", PAC2'06: Proceedings of the 3rd Watson Conference on Interaction Between Architecture, Circuits, and Compilers, 2006

[8]   H. Hua, C. Mineo, K. Schoenfliess, A. Sule, S. Melamed and W. Davis, "Performance Trend in Three-dimensional Integrated Circuits", IITC'06: Proceedings of the 2006 International Interconnect Technology Conference, 2006, pp. 45–47

[9]   E. Humenay, D. Tarjan and K. Skadron, "Impact of Parameter Variations on Multi-core Chips", ASGI'06: Proceedings of the 2006 Workshop on Architectural Support for Gigascale Integration, 2006

[10] A. Iyer and D. Marculescu, "Power and Performance Evaluation of Globally Asynchronous Locally Synchronous Processors", ISCA'02: Proceedings of the 29th International Symposium on Computer Architecture, 2002, pp. 158–168

[11] X. Liang and D. Brooks, "Mitigating the Impact of Process Variations on Processor Register Files and Execution Units", MICRO 39: Proceedings of the 39th Annual ACM/IEEE International Symposium on Microarchitecture, 2006, pp. 504–514

[12] D. Marculescu and E. Talpes, "Variability and Energy Awareness: A Microarchitecture-level Perspective", DAC'05: Proceedings of the 42nd annual Design Automation Conference, 2005, pp. 11–16

[13] M. Orshansky, C. Spanos and C. Hu, "Circuit Performance Variability Decomposition", IWSM'99: Proceedings of the 4th International Workshop on Statistical Metrology, 1999, pp. 10–13

[14] G. Semeraro, G. Magklis, R. Balasubramonian, D. Albonesi, S. Dwarkadas and M. Scott, "Energy-efficient Processor Design Using Multiple Clock Domains with Dynamic Voltage and Frequency Scaling", HPCA'02: Proceedings of the 8th International Symposium on High-Performance Computer Architecture, 2002, pp. 29–42

[15] K. Skadron, M. Stan, W. Huang, S. Velusamy, K. Sankaranarayanan and D. Tarjan, "Temperature-aware Microarchitecture", ISCA'03: Proceedings of the 30th International Symposium on Computer Architecture, 2003, pp. 2–13

[16] Q. Wu, P. Juang, M. Martonosi and W. Clark, "Formal Online Methods for Voltage/Frequency Control in Multiple Clock Domain Microprocessors", ASPLOS-XI: Proceedings of the 11th International Conference on

Architectural Support for Programming Languages and Operating Systems, 2004, pp. 248–259

[17] W. Zhao and Y. Cao, "New Generation of Predictive Technology Model for Sub-45nm Design Exploration", ISQED'06: Proceedings of the 7th International Symposium on Quality Electronic Design, 2006, pp. 585–590

Chapter 10 Temporal Adaptation – Asynchronicity in Processor Design

Steve Furber, Jim Garside

The University of Manchester, UK

10.1 Introduction

Throughout most of the history of the microprocessor, designers have employed an approach based on the use of a central clock to control functional units within the processor. While there are situations – such as the musicians in a symphony orchestra or the crew of a rowing boat – where global synchrony is a vital aspect of the overall functionality, a microprocessor is not such a system. Here the clock is merely a design convenience, a constraint on how the system's components operate that simplifies some design issues and allows the use of a well-developed VLSI design flow where the designer can analyse the entire system state at any instant and use this to influence the transition to the next state. The clock has become so dominant in modern processor design that few designers ever stop to consider dispensing with it; however, it is not *necessary* – synchronisation may be restricted to places where it is essential to function.

Although a tremendous aid in simplifying a complex design task, the globally clocked model does have its drawbacks. In engineering terms, perhaps, the greatest problem is the difficulty of sustaining the fundamental assumption of the model, which is that the clock arrives simultaneously at every latch in the system. This not only is a considerable headache in its own right but also results directly in undesirable side effects such as power wastage and high levels of electromagnetic emission. However, here the primary concern is adaptivity and, in this too, the synchronous model is an obstacle.

A. Wang, S. Naffziger (eds.), *Adaptive Techniques for Dynamic Processor Optimization*, DOI: 10.1007/978-0-387-76472-6_10, © Springer Science+Business Media, LLC 2008

The benefits of designing a microprocessor that operates without a clock have been explored by various groups around the world [1–5]. These include the ability to operate in, and adapt to, an environment with highly variable memory access characteristics and highly variable power requirements [6], and offer potential performance gains by allowing different parts of a system to run at their own 'natural' speeds rather than throttling them all to the rate of the slowest.

This chapter explores some of the issues that arise in the design of a clockless microprocessor and summarises the developments that have led recently to the release of asynchronous microprocessor designs into the commercial marketplace [7, 8].

10.2 Asynchronous Design Styles

Discarding the clock is not a step to be taken lightly, but can bring benefits. Before examining these in detail, it is useful to define what is meant by the alternative to clocked design, which is commonly termed 'asynchronous design'.

An asynchronous design is one where different subsystems operate independently. Some synchronisation of disparate parts *is* necessary, but this is generally only when two subsystems need to communicate. In overview, this means that different parts of the system are operating at different rates, sometimes pausing to rendezvous with a neighbour before proceeding. Without a central timing reference, there needs to be some other means to determine when each operation is complete: this is done locally, which is why asynchronous circuits are often referred to as 'self-timed'.

There are two principal models of how self-timing may be implemented. The first is to introduce a timing model into each unit which acts as a 'clock' for that subsystem; note, however, that the 'period' of this clock may vary from cycle to cycle and it will be stopped when there is no work to do. This model is referred to as a 'bundled-data' model because, like a synchronous circuit, a set of data lines with approximately equal timing properties are 'bundled' together [9]. This method has a small control overhead – comparable in cost to a clock network – and relies on assumptions about the timing behaviour of the data buses – again, like a synchronous circuit. Output synchronisation is achieved by presenting the data output, asserting an output 'request' and waiting for a handshake 'acknowledge' signal to be returned; this prevents subsequent operations overrunning a unit which may be temporarily stalled.

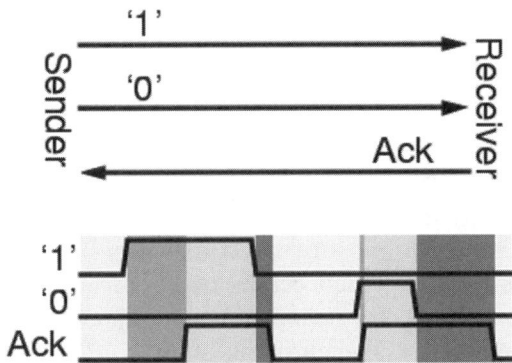

Figure 10.1 Dual-rail asynchronous communication.

An alternative model encodes the timing with the data signals so that arbitrary delays may be introduced without affecting the functionality of the circuit. The simplest, but by no means only, means to do this is 'two-wire' (commonly termed 'dual-rail') encoding (Figure 10.1) [10] where each data bit is carried by two binary signals that together indicate one of three conditions: a zero has arrived, a one has arrived, or nothing has arrived yet (so that data bit is currently invalid). Circuits constructed in this manner will function regardless of any delays and will deliver its best speed for any operation; however, the penalty is that they are approximately twice the size of 'conventional' circuits.

This second model is intrinsically robust with respect to variations in operating conditions. If an element speeds up or slows down – for example, as a result of a change in temperature or supply voltage – the circuit will continue to operate correctly, unless some external real-time condition is violated. However, in the bundled-data model, it is also possible for a delay element to track environmental conditions; the delay can be engineered as a close analogue of the circuit it is modelling, using the same gates, close by on the silicon so that both manufacturing and environmental conditions are similar. While not quite as robust as the dual-rail model, this has been proven reliable in practice – for example, in the Amulet processors [11]. Indeed, in some circumstances, this is the only practical way to manage a delay model: a good example would be a memory where a single extra column, hard-wired to a known value, provides a good timing reference for the whole array. Similarly in a processor's datapath, an extra bit which evaluates to a prescribed value can be added to provide an accurate timing model. This represents about a 3% overhead for a 32-bit processor rather than something around 100% overhead in a dual-rail system.

This adaptivity to environmental conditions means that voltage scaling on self-timed circuits is trivially easy to manage. All that is needed is to vary the voltage; the operating speed and power will adapt automatically. Similarly, the circuits will slow down if they become hot, but they will still function correctly. This has been demonstrated repeatedly with experimental asynchronous designs.

A great deal is said about voltage scaling elsewhere in this book, so it is sufficient here to note that most of the complexity of voltage scaling is in the clock control system, which ceases to be an issue when there is no clock to control! Instead, this chapter concentrates on other techniques which are facilitated by the asynchronous style.

10.3 Asynchronous Adaptation to Workload

Power – or, rather, *energy* – efficiency is important in many processing applications. As described elsewhere, one way of reducing the power consumption of a processor is reducing the clock (or instruction) frequency, and energy efficiency may then also be improved by lowering the supply voltage. Of course, if the processor is doing nothing useful, the energy efficiency is very poor, and in this circumstance, it is best to run as few instructions as possible. In the limit, the clock is stopped and the processor 'sleeps', pending a wake-up event such as an interrupt. Synchronous processors sometimes have different sleep modes, including gating the clock off but keeping the PLL running, shutting down the PLL, and turning the power off. The first of these still consumes noticeable power but allows rapid restart; the second is more economical but takes considerable time to restart as the PLL must be allowed to stabilise before the clock is used. This is undesirable if, for example, all that is required is the servicing of interrupts in a real-time system. It is a software decision as to which of these modes to adopt; needless to say this software also imposes an energy overhead.

An asynchronous processor has fewer modes. If the processor is powered it is either running as fast as it can under the prevailing environmental conditions or stalled waiting for some input or output. Because there is no external clock, if one subsystem is caused to stall any stage, waiting for its outputs will stall soon afterwards, as will stages trying to send input to it. In this way, a single gate anywhere in the system can rapidly bring the whole system to a halt. For example, Figure 10.2 shows an asynchronous processor pipeline filling from the prefetch unit; here the system is halted by a 'HALT' operation reaching the execution stage, at which point the preceding pipeline fills up and stalls while the subsequent stages stall because they are starved of input. When the halt is rescinded, the system will resume where it left off and come to full speed almost instantaneously. Thus, power management is extremely easy to implement and requires almost no software control.

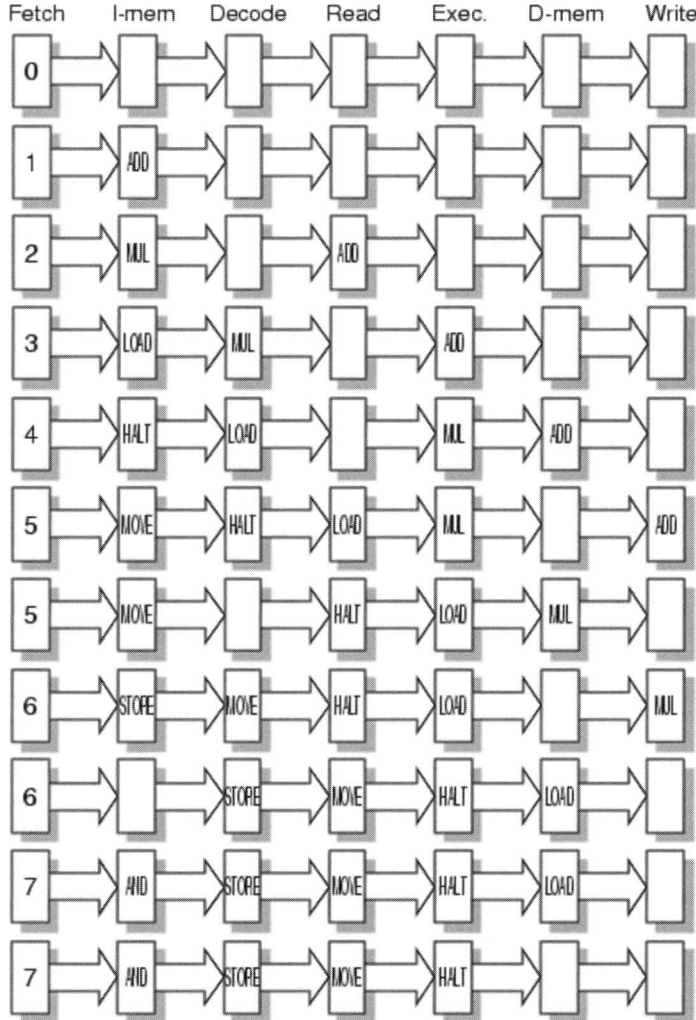

Figure 10.2 Processor pipeline halting in execution stage.

In the Amulet processors, a halt instruction was retrofitted to the ARM instruction set [12] by detecting an instruction which branches to itself. This is a common way to implement an idle task on the ARM and causes the processor to 'spin' until it is interrupted, burning power to no effect. In Amulet2 and Amulet3, this instruction causes a local stall which rapidly propagates throughout the system, reducing the dynamic power to zero. An interrupt simply releases the stall condition, causing the processor to resume and recognise the interrupt. This halt implementation is transparent – as the effect of stopping is not distinguishable from the effect of repeating an instruction which does not alter any data – except in the power consumption.

Perhaps the most useful consequence of asynchronous systems only processing data on demand is that this results in power savings throughout the system. If a multiplier (for example) is not in use, it is not 'clocked' and therefore dissipates no dynamic power. This can be true of any subsystem, but it is particularly important in infrequently used blocks.

Of course, it is possible to gate clocks to mimic this effect, but clock gating can easily introduce timing compatibility problems and is certainly something which needs careful attention by the designer. Asynchronous design delivers an optimal 'clock gating' system without any additional effort on the part of the designer.

10.4 Data-Dependent Timing

A well-engineered synchronous pipeline will usually be 'balanced' so that the critical path in each stage is approximately the same length. This allows the circuit to be clocked at its maximum frequency, without performance being wasted as a result of potentially faster stages being slowed to the common clock rate. Good engineering is not easy, and considerable effort may need to be expended to achieve this.

The same principle holds in a self-timed system although the design constraints are different. A self-timed pipeline will find its own operating speed in a similar fashion to traffic in a road system; a queue will form upstream of a choke point and be sparser downstream. In a simulation, this makes it clear where further design attention is required; this is usually – but not always – the slowest stage. One reason why a particularly slow stage may not slow the whole system is that it is on a 'back road' with very little traffic. There is no requirement to process each operation in a fixed period, so the system may adapt to its operating conditions. Here are some examples:

- In a memory system, some parts may go faster than others; cache memories rely on this property and can be exploited even in synchronous systems as a cache miss will stall for multiple clock cycles waiting for an external response. This is the 'natural' behaviour of an asynchronous memory where response is a single 'cycle' but the length of the cycle is varied according to need. An advantage in the asynchronous system is that it is easier to vary more parameters, and these can be altered in more 'subtle' ways than simply in discrete multiples of clock cycles.

- Not all operations take the same evaluation time: some operation evaluation is data dependent. A simple example is a processor's ALU operation which typically may include options to MOVE, AND, ADD or MULTIPLY operands. A MOVE is a fast operation and an AND, being a bitwise operation, is a similar speed. ADDs, however, are hampered by the need to propagate carries across the datapath and therefore are considerably slower. Multiplication, comprising repeated addition, is of course slower still. A typical synchronous ALU will probably set its critical path to the ADD operation and accept the inefficiency in the MOVE. Multiplication may then require multiple clock cycles, with a consequent pipeline stall, or be moved to a separate subsystem. An asynchronous ALU can accommodate all of these operations in a single cycle by varying the length of the cycle. This simplifies the higher-level design – any stalls are implicit – and allows faster operations to complete faster. It is sometimes said that self-timed systems can thus deliver 'average case performance'; in practice, this is not true because it is likely that the operation subsequent to a fast operation like MOVE will not reach the unit immediately it is free, or the fast operation could be stalled waiting for a previous operation to complete. Having a 50:50 mixture of 60mph cars and 20mph tractors does not mean the traffic flows at 40mph! However, if the slow operations are quite rare – such as multiplication in much code – then the traffic can flow at close to full speed most of the time while the overall model remains simple.

- It is possible to exploit data dependency at a finer level. Additions are slow because of carry propagation. To speed them up requires considerable effort, and hence hardware, and hence energy is typically expended in fast carry logic of some form. This ensures that the critical path – propagating a carry from the least to the most significant bit position – is as short as possible. However, operations which require long carry propagation distances are comparatively rare; the effort, hardware, and power are expended on something which is rarely used. Given random operands, the longest carry chain in an N-bit adder is $O(N)$, but the *average* length is $O(\log_2(N))$; for a 32-bit adder the longest is about 6× the average. If a variable-length cycle is possible, then a simple, energy-efficient, ripple-carry adder can produce a correct result in a time comparable to a much larger (more expensive, power-consuming) adder.

Unfortunately, this is not the whole story because there is an overhead in detecting the carry completion and, in any case, 'real' additions do not use purely random operands [13]. Nevertheless, a much cheaper unit can supply respectable performance by adapting its timing to the operands on each cycle. In particular, an incrementer, such as is used for the programme counter, can be built very efficiently using this principle.

- At a higher level, it is possible to run different subsystems deliberately at different rates. As a final example, the top level of the memory system for Amulet3 is – as on many modern processors – split across separate instruction and data buses to allow parallelism of access [14]. Here these buses run to a unified local memory which is internally partitioned into interleaved blocks. Provided two accesses do not 'collide', these buses run independently at their own rates, and the bandwidth of the more heavily loaded instruction bus – which is simpler because it can only perform read operations – is somewhat higher than that of the read/write, multi-master data bus. In the event that two accesses collide in a single block, the later-arriving bus cycle is simply stretched to accommodate the extra latency. Adaptability here gives the designer freedom: slowing the instruction bus down to match the two speeds would result in lower performance, as would slowing the data bus to exactly half the instruction bus speed.

The flexibility of asynchronous systems allows a considerable degree of modularity in those systems' development. Provided interfaces are compatible, it is possible to assemble systems and be confident that they will not suffer from timing-closure problems – a fact which has been known for some time [15]. It would be nice to say that such systems would always work correctly! Unfortunately, this is not the case as, as in any complex asynchronous system, it is possible to engineer in deadlocks; it is only timing incompatibilities which are eliminated. Where this *is* exploitable is in altering or upgrading systems where a module – such as a multiplier – can be replaced with a compatible unit with different properties (e.g. higher speed or smaller area) with confidence that the system will not need extensive resimulation and recharacterisation.

Perhaps the most important area to emerge from this is at a higher level, i.e. in Systems-on-Chip (SoCs) using a GALS (Globally Asynchronous, Locally Synchronous) approach [16]. Here conventional clocked IP blocks are connected via an asynchronous fabric, effectively eliminating the timing-closure problems at the chip level – at least from a functional viewpoint. This can represent a considerable time-saving for the ASIC designer.

10.5 Architectural Variation in Asynchronous Systems

A pipelined architecture requires a succession of state-holding elements to capture the output from one stage and hold it for the next. In a synchronous architecture, these pipeline registers may be edge triggered (i.e. D-type flip-flops) for simplicity of design; if this is too expensive then transparent latches may be used, typically using a two-phase, non-overlapping clock with alternating stages on opposite phases. The use of transparent latches has largely been driven out in recent times by the need to accommodate the limitations of synthesis and static timing analysis tools in high-productivity design flows, so the more expensive and power-hungry edge-triggered registers have come to dominate current design practice.

10.5.1 Adapting the Latch Style

In some self-timed designs (e.g. dual-rail), the latches may be closely associated with the control circuits; however, a bundled-data datapath closely resembles its synchronous counterpart. Because data is not transferred 'simultaneously' in all parts of the system, the simplicity (cheapness) of transparent latches is usually the preferred option. Here the 'downstream' latch closes and then allows the 'upstream' latch to open at any subsequent time. This operation can be seen in Figure 10.3 where transparent latches are unshaded and closed latches shaded.

Here there is a design trade-off between speed and power. Figure 10.3 depicts an asynchronous pipeline in which the latches are 'normally open' – i.e. when the pipeline is empty all its latches are transparent; at the start the system thus looks like a block of combinatorial logic. As data flows through, the latches close behind it to hold it stable (or, put another way, to delay subsequent changes) and then open again when the next stage has captured its inputs. In the figure this is seen as a wave of activity as downstream latches close and, subsequently, the preceding latch opens again. When the pipeline is empty (clear road ahead!), this model allows data to flow at a higher speed than is possible in a synchronous pipeline because the pipeline latency is the sum of the critical paths in the stages rather than the multiple of the worst-case critical path and the pipeline depth.

The price of this approach is a potential increase in power consumption. The data 'wave front' will tend to skew as it flows through the logic, which can cause the input of a gate to change more times than it would if the wave front were re-aligned at every stage. This introduces glitches into the data which result in wasted energy due to the spurious transitions which can propagate considerable distances.

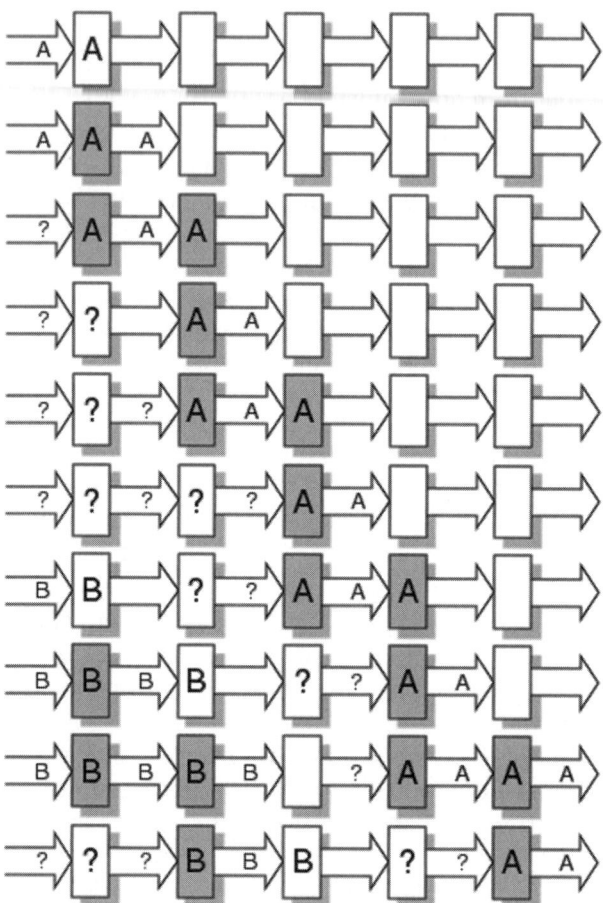

Figure 10.3 Pipeline with 'normally open' latches. Open latches are unshaded; closed latches are shaded.

To prevent glitch propagation, the pipeline can adopt a 'normally closed' architecture (Figure 10.4). In this approach, the latches in an empty pipeline remain closed until the data signalls its arrival, at which point they open briefly to 'snap up' the inputs. The wave of activity is therefore visible as a succession of briefly transparent latches (unshaded in the figure).

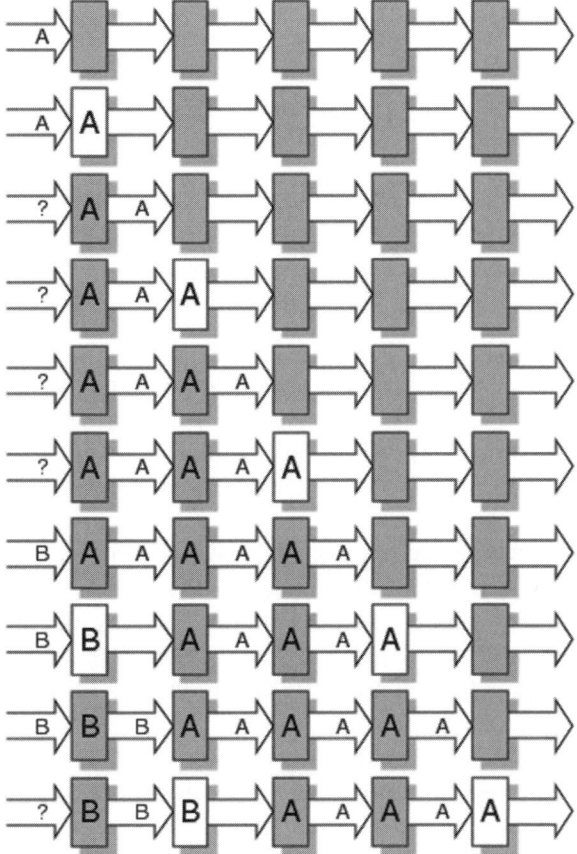

Figure 10.4 Pipeline with 'normally closed' latches. Open latches are unshaded; closed latches are shaded.

Their outputs therefore change nearly simultaneously, re-aligning the data wave front and reducing the chance of glitching in the subsequent stage. The disadvantage of this approach is that data propagation is slowed waiting for latches, which are not retaining anything useful, to open.

These styles of latch control can be mixed freely. The designer has the option of increased speed or reduced power. If the pipeline is filled to its maximum capacity, the decision is immaterial because the two behaviours can be shown to converge. However, in other circumstances a choice has to be made. This allows some adaptivity to the application at design time, but the principle can be extended so that this choice can be made dynamically according to the system's loading.

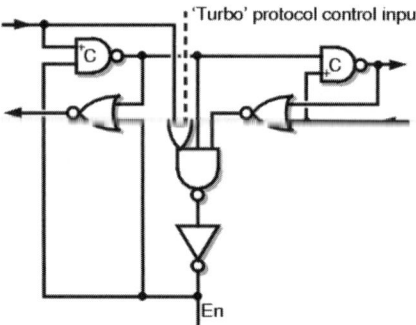

Figure 10.5 Configurable asynchronous latch controller.

The two latch controllers can be very similar in design – so much so that a single additional input (two or four additional transistors, depending on starting point) can be used to convert one to the other (Figure 10.5). Furthermore, provided the change is made at a 'safe' time in the cycle, this input can be switched dynamically. Thus, an asynchronous pipeline can be equipped with both 'sport' and 'economy' modes of operation using 'Turbo latches' [17].

The effectiveness of using normally closed latches for energy conservation has been investigated in a bundled-data environment; the result depends strongly on both the pipeline occupancy and, as might be expected, the variation in the values of the bits flowing down the datapath.

The least favourable case is when the pipeline is fully occupied, when even a normally open latch will typically not open until about the time that new data is arriving; in this case, there is no energy wastage due to the propagation of earlier values. In the 'best' case, with uncorrelated input data and low pipeline occupancy, an energy saving of ~20% can be achieved at a price of ~10% performance, or vice versa.

10.5.2 Controlling the Pipeline Occupancy

In the foregoing, it has tacitly been assumed that processing is handled in pipelines. Some applications, particularly those processing streaming data, naturally map onto deep pipelines. Others, such as processors, are more problematic because a branch instruction may force a pipeline flush and any speculatively fetched instructions will then be discarded, wasting energy. However, it is generally not possible to achieve high performance without employing pipelining.

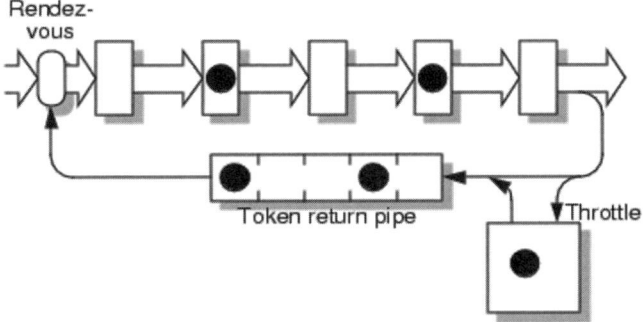

Figure 10.6 Occupancy throttling using token return mechanism.

In a synchronous processor, the speculation depth is effectively set by the microarchitecture. It is possible to leave stages 'empty', but there is no great benefit in doing so as the registers are still clocked. In an asynchronous processor, latches with nothing to do are not 'clocked', so it is sensibly possible to throttle the input to leave gaps between instruction packets and thus reduce speculation, albeit at a significant performance cost. This can be done, for example, when it is known that a low processing load is required or, alternatively, if it is known that the available energy supply is limited. Various mechanisms are possible: a simple throttle can be implemented by requiring instruction packets to carry a 'token' through the pipeline, collecting it at fetch time and recycling it when they are retired (Figure 10.6). For full-speed operation, there must be at least as many tokens as there are pipeline stages so that no instruction has to wait for a token and flow is limited purely by the speed of the processing circuits. However, to limit flow, some of the tokens (in the return pipeline) can be removed, thus imposing an upper limit on pipeline occupancy. This limit can be controlled dynamically, reducing speculation and thereby cutting power as the environment demands.

An added bonus to this scheme is that if speculation is sufficiently limited, other power-hungry circuits such as branch prediction can be disabled without further performance penalty.

10.5.3 Reconfiguring the Microarchitecture

Turbo latches can alter the behaviour of an asynchronous pipeline, but they are still latches and still divide the pipeline up into stages which are fixed in the architecture. However, in an asynchronous system adaptability can be extended further; even the stage sizes can be altered dynamically!

A 'normally open' asynchronous stage works in this manner:

1. Wait for the stage to be ready and the arrival of data at the input latch;
2. Close the input latch;
3. Process the data;
4. Close the output latch;
5. Signal acknowledgement;
6. Open the input latch.

Such latching stages operate in sequence, with the whole task being partitioned in an arbitrary manner.

If another latch was present halfway through data processing (step 3, above), this would subdivide the stage and produce the acknowledgement earlier than otherwise. The second half of the processing could then continue in parallel with the recovery of the earlier part of the stage, which would then be able to accept new data sooner. The intermediate latch would reopen again when the downstream acknowledgement (step 5, above) reached it, ready to accept the next packet. This process has subdivided what was one pipeline stage into two, potentially providing a near doubling in throughput at the cost of some extra energy in opening and closing the intermediate latch.

In an asynchronous pipeline, interactions are always local and it is possible to alter the pipeline depth *during* operation knowing that the rest of the system will accommodate the change. It is possible to tag each data packet with information to control the latch behaviour. When a packet reaches a latch, it is forced into local synchronisation with that stage. Instead of closing and acknowledging the packet the controller can simply pass it through by keeping the latch transparent and forwarding the control signal. No acknowledgement is generated; this will be passed back when it appears from the subsequent stage. In this manner, a pipeline latch can be removed from the system, altering the microarchitecture in a fundamental way. In Figure 10.7, packet 'B' does not close – and therefore 'eliminates' – the central latch; this and subsequent operations are slower but save on switching the high-capacitance latch enable.

Of course, this change is reversible; a latch which has been deactivated can spot a reactivation command flowing through and close, reinstating the 'missing' stage in the pipeline. In Figure 10.8, packet 'D' restores the central latch allowing the next packet to begin processing despite the fact that (in this case) packet 'C' appears to have stalled.

Why might this be useful? The technique has been analysed in a processor model using a range of benchmarks [18–20]. As might be expected, collapsing latches and combining pipeline stages – in what was, initially, a reasonably balanced pipeline – reduces overall throughput by, typically,

50–100%. Energy savings are more variable: streaming data applications that contain few branches show no great benefit; more 'typical' microprocessor applications with more branches exhibit ~10% energy savings and, as might be expected, the performance penalty is at the lower end of the range. If this technique is to prove useful, it is certainly one which needs to be used carefully and applied dynamically, possibly under software control; however, it can provide benefits and is another tool available to the designer.

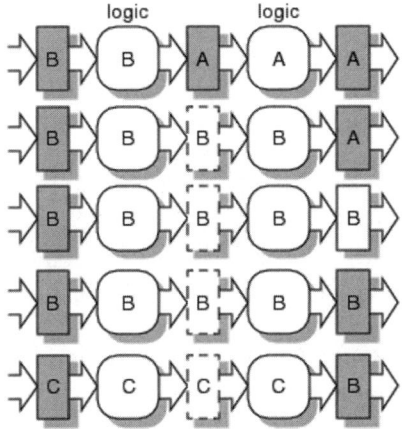

Figure 10.7 Pipeline collapsing and losing latch stage.

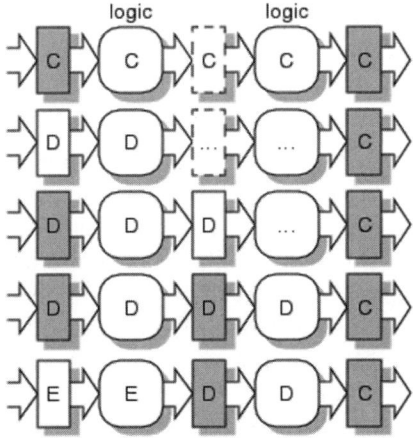

Figure 10.8 Pipeline expanding and reinstating latch stage.

10.6 Benefits of Asynchronous Design

Asynchronous operation brings diverse benefits to microprocessors, but these are in general hard to quantify. Unequivocal comparisons with clocked processors are few and far between. Part of the difficulty lies in the fact that there are many ways to build microprocessors without clocks, each offering its own trade-offs in terms of performance, power efficiency, adaptability, and so on. Exploration of asynchronous territory has been far less extensive than that of the clocked domain, so we can at this stage only point to specific exemplars to see how asynchronous design can work out in practice.

The Amulet processor series demonstrated the feasibility, technical merit, and commercial viability of asynchronous processors. These full-custom designs showed that asynchronous processor cores can be competitive with clocked processors in terms of area and performance, with dramatically reduced electromagnetic emissions. They also demonstrated modest power savings under heavy processing loads, with greatly simplified power management and greater power savings under variable event-driven workloads.

The Philips asynchronous 80C51 [7] has enjoyed considerable commercial success, demonstrating good power efficiency and very low electromagnetic emissions. It is a synthesised processor, showing that asynchronous synthesis is a viable route to an effective microprocessor, at least at lower performance levels.

The ARM996HS [8], developed in collaboration between ARM Ltd and Handshake Solutions, is a synthesised asynchronous ARM9 core available as a licensable IP core with better power efficiency (albeit at lower performance) than the clocked ARM9 cores. It demonstrated low current peaks and very low electromagnetic emissions and is robust against current, voltage, and temperature variations due to the intrinsic ability of the asynchronous technology to adapt to changing environmental conditions.

All of the above designs employ conventional instruction set architectures and have implemented these in an asynchronous framework while maintaining a high degree of compatibility with their clocked predecessors. This compatibility makes comparison relatively straightforward, but may constrain the asynchronous design in ways that limit its potential. More radical asynchronous designs have been conceived that owe less to the heritage of clocked processors, such as the Sun FLEET architecture [21], but there is still a long way to go before the comparative merits of these can be assessed quantitatively.

10.7 Conclusion

Although almost all current microprocessor designs are based on the use of a central clock, this is not the only viable approach. Asynchronous design, which dispenses with global timing control in favour of local synchronisation as and when required, introduces several potential degrees of adaptation that are not readily available to the clocked system. Asynchronous circuits intrinsically adapt to variations in supply voltage (making dynamic voltage scaling very straightforward), temperature, process variability, crosstalk, and so on. They can adapt to varying processing requirements, in particular enabling highly efficient event-driven, real-time systems. They can adapt to varying data workloads, allowing hardware resources to be optimised for typical rather than very rare operand values, and they can adapt very flexibly (and continuously, rather than in discrete steps) to variable memory response times. In addition, asynchronous processor microarchitectures can adapt to operating conditions by varying their fundamental pipeline behaviour and effective pipeline depth.

The flexibility and adaptability of asynchronous microprocessors make them highly suited to a future that holds the promise of increasing device variability. There remain issues relating to design tool support for asynchronous design, and a limited resource of engineers skilled in the art, but the option of global synchronisation faces increasing difficulties, at least some of which can be ameliorated through the use of asynchronous design techniques. We live in interesting times for the asynchronous microprocessor; only time will tell how the balance of forces will ultimately resolve.

References

[1] A.J. Martin, S.M. Burns, T.K. Lee, D. Borkovic and P.J. Hazewindus, "The Design of an Asynchronous Microprocessor", ARVLSI: Decennial Caltech Conference on VLSI, ed. C.L. Seitz, MIT Press, 1989, pp. 351–373.

[2] S.B. Furber, P. Day, J.D. Garside, N.C. Paver and J.V. Woods, "AMULET1: A Micropipelined ARM", Proceedings of CompCon'94, IEEE Computer Society Press, San Francisco, March 1994, pp.476–485.

[3] A. Takamura, M. Kuwako, M. Imai, T. Fujii, M. Ozawa, I. Fukasaku, Y. Ueno and T. Nanya, "TITAC-2: A 32-Bit Asynchronous Microprocessor Based on Scalable-Delay-Insensitive Model", Proceedings of ICCD'97, October 1997, pp. 288–294.

[4] M. Renaudin, P. Vivet and F. Robin, "ASPRO-216: A Standard-Cell Q.D.I. 16-Bit RISC Asynchronous Microprocessor", Proceedings of Async'98, IEEE Computer Society, 1998, pp. 22–31. ISBN:0-8186-8392-9.

[5]   S.B. Furber, J.D. Garside and D.A. Gilbert, "AMULET3: A High-Performance Self-Timed ARM Microprocessor", Proceedings of ICCD'98, Austin, TX, 5–7 October 1998, pp. 247–252. ISBN 0-8186-9099-2.

[6]   S.B. Furber, A. Efthymiou, J.D. Garside, M.J.G. Lewis, D.W. Lloyd and S. Temple, "Power Management in the AMULET Microprocessors", IEEE Design and Test of Computers, ed. E. Macii, March–April 2001, Vol. 18, No. 2, pp. 42–52. ISSN: 0740-7475.

[7]   H. van Gageldonk, K. van Berkel, A. Peeters, D. Baumann, D. Gloor and G. Stegmann, "An Asynchronous Low-Power 80C51 Microcontroller", Proceedings of Async'98, IEEE Computer Society, 1998, pp. 96–107. ISBN:0-8186-8392-9.

[8]   A. Bink and R. York, "ARM996HS: The First Licensable, Clockless 32-Bit Processor Core", IEEE Micro, March 2007, Vol. 27, No. 2, pp. 58–68. ISSN: 0272-1732.

[9]   I. Sutherland, "Micropipelines", Communications of the ACM, June 1989, Vol. 32, No. 6, pp.720–738. ISSN: 0001-0782.

[10] J. Sparsø and S. Furber (eds.), "Principles of Asynchronous Circuit Design – A Systems Perspective", Kluwer Academic Publishers, 2002. ISBN-10: 0792376137 ISBN-13: 978-0792376132.

[11] S.B. Furber, D.A. Edwards and J.D. Garside, "AMULET3: A 100 MIPS Asynchronous Embedded Processor", Proceedings of ICCD'00, 17–20 September 2000.

[12] D. Seal (ed.), "ARM Architecture Reference Manual (Second Edition)", Addison-Wesley, 2000. ISBN-10: 0201737191 ISBN-13: 978-0201737196.

[13] J.D. Garside, "A CMOS VLSI Implementation of an Asynchronous ALU","Asynchronous Design Methodologies", eds. S.B. Furber and M. Edwards, Elsevier 1993, IFIP Trans. A-28, pp. 181–207.

[14] D. Hormdee and J.D. Garside, "AMULET3i Cache Architecture", Proceedings of Async'01, IEEE Computer Society Press, March 2001, pp. 152–161. ISSN 1522-8681 ISBN 0-7695-1034-4.

[15] W.A. Clark, "Macromodular Computer Systems", Proceedings of the Spring Joint Conference, AFIPS, April 1967.

[16] D.M. Chapiro, "Globally-Asynchronous Locally-Synchronous Systems", Ph.D. thesis, Stanford University, USA, October 1984.

[17] M. Lewis, J.D. Garside and L.E.M. Brackenbury, "Reconfigurable Latch Controllers for Low Power Asynchronous Circuits", Proceedings of Async'99, IEEE Computer Society Press, April 1999, pp. 27–35.

[18] A. Efthymiou, "Asynchronous Techniques for Power-Adaptive Processing", Ph.D. thesis, Department of Computer Science, University of Manchester, UK, 2002.

[19] A. Efthymiou and J.D. Garside, "Adaptive Pipeline Depth Control for Processor Power-Management", Proceedings of ICCD'02, Freiburg, September 2002, pp. 454–457. ISBN 0-7695 1700-5 ISSN 1063-6404.

[20] A. Efthymiou and J.D. Garside, "Adaptive Pipeline Structures for Speculation Control", Proceedings of Async'03, Vancouver, May 2003, pp. 46–55. ISBN 0-7695-1898-2 ISSN 1522-8681.

[21] W.S. Coates, J.K. Lexau, I.W. Jones, S.M. Fairbanks and I.E. Sutherland, "FLEETzero: An Asynchronous Switching Experiment", Proceedings of Async'01, IEEE Computer Society, 2001, pp. 173–182. ISBN:0-7695-1034-5.

Chapter 11 Dynamic and Adaptive Techniques in SRAM Design

John J. Wuu

Advanced Micro Devices, Inc.

11.1 Introduction

The International Technology Roadmap for Semiconductors (ITRS) predicted in 2001 that by 2013, over 90% of SOC die area will be occupied by memory [7]. Such level of integration poses many challenges, such as power, reliability, and yield. In addition, as transistor dimensions continue to shrink, transistor threshold voltage (V_T) variation, which is inversely proportional to the square root of the transistor area, continues to increase. This V_T variation, along with other factors contributing to overall variation, is creating difficulties in designing stable SRAM cells that meet product density and voltage requirements.

This chapter examines various dynamic and adaptive techniques for mitigating some of these common challenges in SRAM design. The chapter first introduces innovations at the bitslice level, which includes SRAM cells and immediate peripheral circuitry. These innovations seek to improve bitcell stability and increase the read and write margins, while reducing power. Next, the power reduction techniques at the array level, which generally involve cache sleeping and methods for regulating the sleep voltage, as well as schemes for taking the cache into and out of sleep are discussed. Finally, the chapter examines the yield and reliability, which are issues that engineers and designers cannot overlook, especially as caches continue to increase in size. To improve reliability, one must account for test escapes, latent defects, and soft errors; thus the chapter concludes with a discussion of error correction and dynamic cache line disable or reconfiguration options.

A. Wang, S. Naffziger (eds.), *Adaptive Techniques for Dynamic Processor Optimization*,
DOI: 10.1007/978-0-387-76472-6_11, © Springer Science+Business Media, LLC 2008

11.2 Read and Write Margins

Figure 11.1 illustrates the basic 6-Transistor (6T) SRAM cell, with back-to-back inverters holding the storage node values and access transistors allowing access to the storage nodes. In Figure 11.1, the transistors in the inverter are labeled as M_P and M_N, while the access transistor is labeled as M_A. M_P, M_N, and M_A are used when referring to transistors in the SRAM cell.

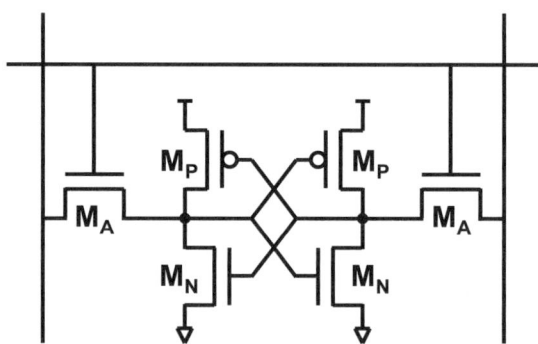

Figure 11.1 Basic SRAM cell.

As basic requirements, a SRAM cell must maintain its state during a read access and be capable of changing state during a write operation. In other words, a cell must have positive read and write margins.

While there are many different methods to quantify a SRAM cell's read margin, graphically deriving the Static Noise Margin (SNM) through the butterfly curve, as introduced in [17] and described in Chapter 6, remains a common approach. In addition to its widespread use, the butterfly curve can conveniently offer intuitive insight into a cell's sensitivity to various parameters. The butterfly curve is used to facilitate the following discussion.

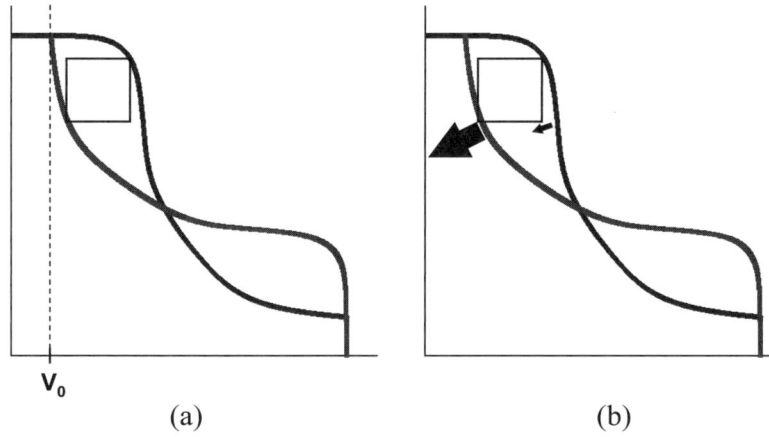

V_0

(a)   (b)

Figure 11.2 Butterfly curves.

Figure 11.2a illustrates the butterfly curve of a typical SRAM cell. As introduced in Chapter 6, SNM is defined by the largest square that can fit between the two curves. Studying the butterfly curve indicates that to enlarge the size of the SNM square, designers must lower the value of V_0. Since V_0 is determined by the inverse of M_N/M_A, the M_N/M_A ratio must be high. The large arrow in Figure 11.2b illustrates the effect of increasing the M_N/M_A ratio. However, increasing M_N has the side effect of decreasing the M_P/M_N ratio (highlighted by the small arrow), which would slightly decrease SNM. Following this logic concludes that increasing M_P to achieve higher M_P/M_N ratio could also improve a cell's SNM.

On the other hand, to achieve good write margin, a "0" on the bitline must be able to overcome M_P holding the storage node at "1" through M_A. Therefore, decreasing M_A and increasing M_P to improve SNM would negatively impact the write margin. In recent process nodes, with voltage scaling and increased device variation, it is becoming difficult to satisfy both read and write margins.

The following sections will survey the range of techniques that seek to dynamically or adaptively improve SRAM cells' read and write margins.

11.2.1 Voltage Optimization Techniques

Because a SRAM cell's stability is highly dependent on the supply voltage, voltage manipulation can impact a cell's read and write margins.

Voltage manipulation techniques can be roughly broken down into row and column categories, based on the direction of the voltage manipulated cells.

11.2.1.1 Column Voltage Optimization

To achieve high read and write margins, a SRAM cell must be stable during a read operation and unstable during a write operation. One way to accomplish this is by providing the SRAM cell with a high VDD during read operations and a low VDD during write operations. One example [22] is the implementation of a dual power supply scheme. As shown in Figure 11.3, multiplexers allow either the high or the low supply to power the cells on a column-by-column basis. During standby operation, the low supply is provided to all the cells to decrease leakage power. Cell stability can be maintained with this lower voltage because the cells are not being accessed; thus, the bitlines do not disturb the storage nodes through M_A. When the cells are accessed in a read operation, the row of cells with the active wordline (WL) experiences a read-disturb, which reduces the stability of the SRAM cells; therefore, the high supply is switched to all the columns to improve the accessed cells' stability. During a write operation, the columns that are being written to remain on the low supply, allowing easy overwriting of the cells. Assuming a column-multiplexed implementation, the columns not being written to are provided with the higher supply, just like in a read operation, to prevent data corruption from the disturb.

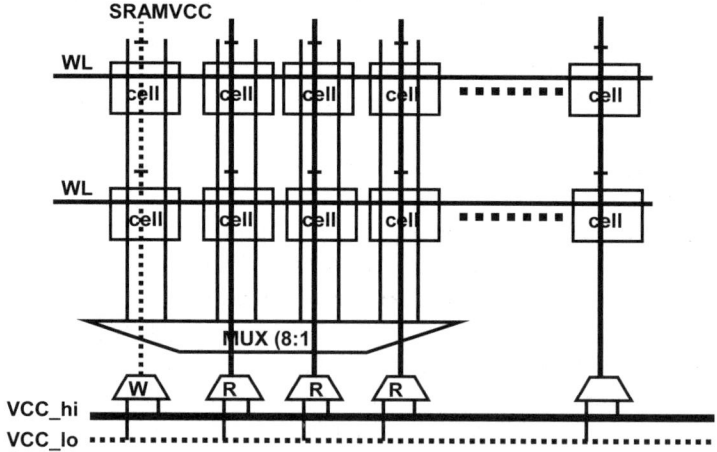

Figure 11.3 Dual supply column-based voltage optimization [22]. (© 2006 IEEE)

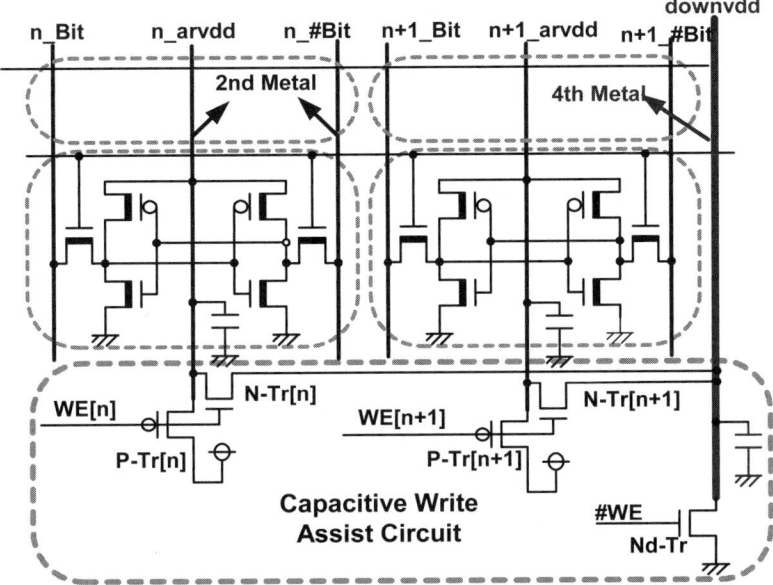

Figure 11.4 Charge sharing for supply reduction [14]. (© 2007 IEEE)

Since extra supplies are not always available in product design, another example [14] uses charge sharing to lower the supply to the columns being written to. As shown in Figure 11.4, "downvdd" is precharged to VSS. For a write operation, supplies to the selected columns are disconnected from VDD, and shorted to "downvdd". The charge sharing lowers the supply's voltage to a level determined by the ratio of the capacitances, allowing writes to occur easily.

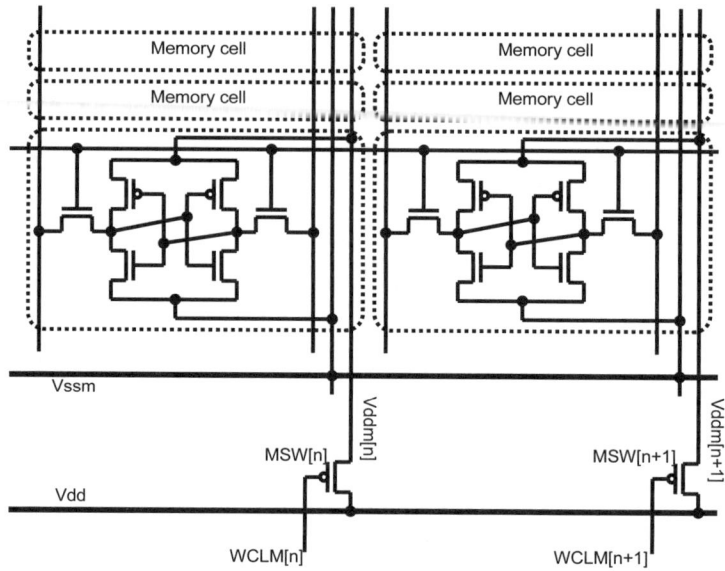

Figure 11.5 Write column supply switch off [21]. (© IEEE 2006)

Yet another example [21] uses a power-line-floating write technique to assist write operations. Instead of switching in a separate supply or charge sharing the supply, as in previous examples, the supply to the write columns is simply switched off, floating the column supply lines at VDD (Figure 11.5). As the cells are written to, the floating supply line (Vddm) discharges through the "0" bitline, as shown in Figure 11.6a. The decreased supply voltage allows easy writing to the cells. As soon as the cell flips to its intended state, the floating supply line's discharge path is cut off, preventing the floating supply line from fully discharging (Figure 11.6b).

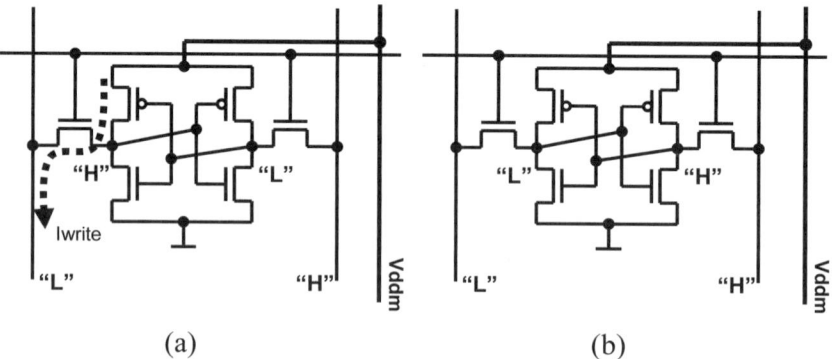

Figure 11.6 Power-line-floating write [21]. (© IEEE 2006)

In all column voltage manipulation schemes, nonselected cells must retain state with the lowered supply.

11.2.1.2 Row Voltage Optimization

Similar to the previous section, designers can apply voltage manipulation in the row direction as well. However, unlike column-based voltage optimization, row-based voltage optimization generally cannot simultaneously optimize for both read and write margins in the same operation, as needed in a column-multiplexed design. Therefore, row-based voltage manipulation tends to be more suitable for non-column-multiplexed designs where all the columns are written to in a write operation.

The most obvious method to apply row-based voltage optimization is to raise the supply for the row of accessed cells in a read operation, or to lower the supply for the row of cells being written to. In addition, the following are some other examples of row-based voltage optimization.

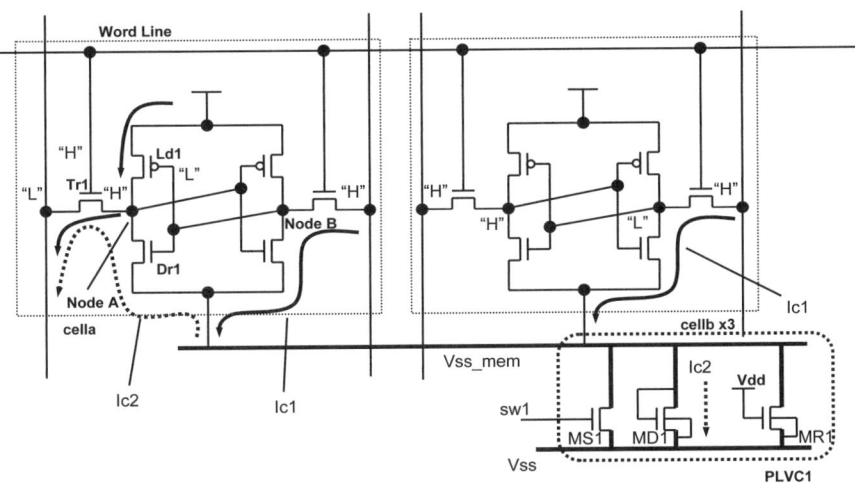

Figure 11.7 Raised source line write [20]. (© IEEE 2004)

In [20], the SRAM cells' source line (SL) (i.e., source terminals of M_Ns in Figure 11.1) is disconnected from VSS during write operations. The SL is allowed to float until it is clamped by an NFET diode (Figure 11.7). The raised SL (Vss_mem in Figure 11.7) decreases the drive of the PFETs, which allows easy overwriting of the cell. (In this specific example, the floating SL is shared among all the cells in the array, not just the cells in a row. However, designers can apply the same technique on a row-by-row basis at the cost of area overhead.) A variation of this technique would disconnect the SL during both write and standby operations to achieve power savings, and connect the SL to VSS only during read operations when the extra stability margin is needed. The drawback to this variation is the additional delay needed to restore SL to VSS before a read operation can begin.

A similar example [13] also floats SL during write operations. In addition, the SL is driven to a negative voltage during read operations. This allows for faster bitline development, as well as more stable cells during read operations.

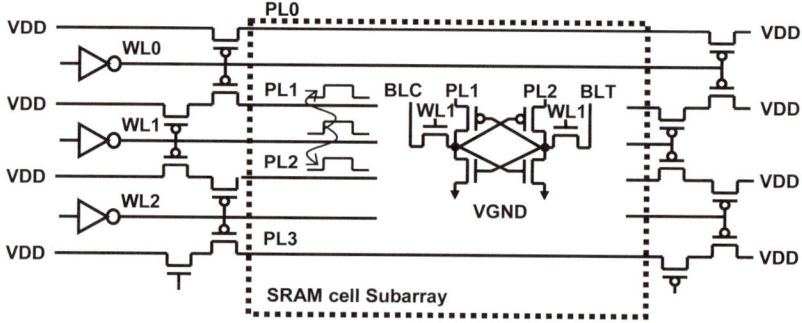

Figure 11.8 Supply line coupling [3]. (© IEEE 2004)

If a separate supply is not available, another way to boost the internal supply of SRAM cells during a read access to achieve higher stability is through coupling. In [3], wordline wires are routed next to the row's supply lines. As seen in Figure 11.8, as the wordline rises, it disconnects the supply lines from VDD, and couples the voltages of the supply lines higher than VDD. Assuming insignificant current is sourced from the supply line during a read access, the bootstrapped supply increases the drive on M_{NS} and improves the cell's stability. However, for cell designs with low M_N/M_A ratios, the "0" storage node may rise higher than M_N's threshold voltage, causing the floating supply lines to discharge.

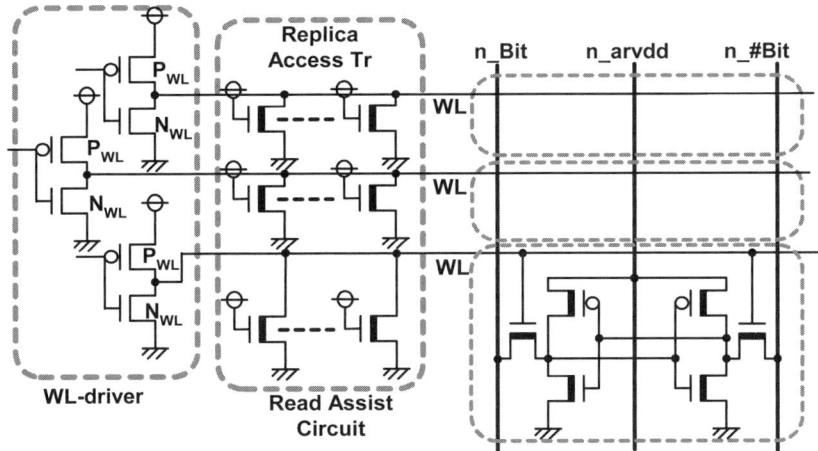

Figure 11.9 Wordline driver using RATs [14]. (© IEEE 2007)

In [14], instead of increasing the SRAM cell's supply to improve stability, the WL voltage is reduced slightly. Reduced wordline voltage degrades the drive of M_A, which essentially improves the M_N/M_A ratio. This implementation makes additional efforts to account for global threshold voltage variations. Figure 11.9 illustrates the scheme, using "replica access transistors" (RATs) that have almost the same physical topology as M_A to lower the WL voltage. In general, lower V_{TN} causes SRAM cells to be less stable. Therefore, the RATs lower WL more when V_{TN} is low, and less when V_{TN} is high, to achieve balance between read margin and read speed.

11.2.2 Timing Control

Aside from voltage manipulation, designers can also improve cell stability by decreasing the amount of time the cell is under stress during a read operation. For example, in a design that uses differential sensing, a small bitline voltage drop could be sufficient for sensing the bitcell value. Leaving on the wordline longer than necessary would allow the bitlines to continue to disturb the "0" storage node, leading marginal SRAM cells to flip their values.

In typical designs, the wordline shutoff is triggered on phase or cycle boundaries. If the optimal wordline shutoff time does not align with phase or cycle boundaries, or if the designer prefers to have the wordline high time independent of the frequency, then the designer could employ a pulsed wordline scheme, such as the one used in [11]. The challenge is to design the appropriate pulse width that is just long enough for reads to complete successfully across different process corners and operating conditions.

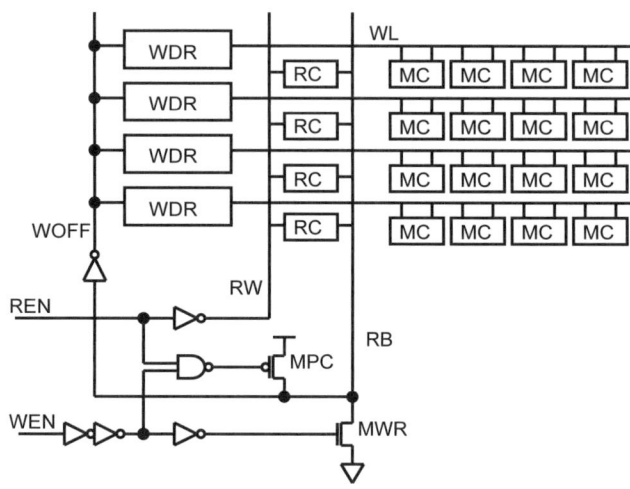

Figure 11.10 Read and write replica circuits [21]. (© IEEE 2006)

In [15], a read replica path, which uses 12 dummy SRAM cells, was used for generating the shutoff edge for wordlines. The dummy SRAM cells, which resemble real SRAM cells but have internal values hardwired, help the replica path to track the variation in normal read paths. In addition to the read replica circuits [21], a write replica circuit was also added. In general, read operations take more time to complete than write operations. Therefore, it is advantageous to shut off the wordline during a write operation as soon as the write is completed successfully, which will prevent unselected columns in a column-multiplexed design from continuing to discharge the bitlines, resulting in wasted power. Figure 11.10 is an example illustrating the read and write replica paths together. The replica bitline (RB) is precharged to VDD through MPC before read or write operations begin. For a read operation, REN activates to "0", causing the read-replica wordline (RW) to turn on the read dummy cells' (RC) wordline. The RC's discharge RB, which turns off the wordlines through the WOFF signal. In a write operation, RB is discharged through MWR, which also triggers WOFF. In general, higher V_{TN} requires the write time to be longer. Therefore, dies with higher V_{TN} would have a slower discharge through MWR, providing the write operation more time to complete.

The above illustration is just one example of designs using replica circuits. The danger of replica circuits, of course, is no replica can perfectly track real paths through all process and operating corners. For example, the write replica circuit above does not track PFET variations, which also impact write margin. However, tracking some variation can usually yield more optimal designs than no tracking at all.

11.3 Array Power Reduction

With power-per-performance becoming an important parameter, engineers pay increasing attention to reducing the power of embedded SRAM arrays, which often occupy a large percentage of the total die area. Since activity factor is generally low for large caches, leakage power represents a significant, if not the dominant, portion of the overall cache power. Devices in a SRAM cell typically have channel lengths much greater than the process minimum for variation control; thus, subthreshold leakage has traditionally been limited. However, subthreshold leakage has worsened with recent technology nodes and more importantly, gate leakage (and in some cases, junction leakage) is getting significantly worse with oxide scaling. As a result, SRAM leakage power now requires careful attention. Because leakage power has a strong dependence on voltage, many have experimented with or implemented with "sleeping" the cache's supply.

11.3.1 Sleep Types

In general, cache "sleep" involves providing inactive SRAM cells, which do not experience read-disturb, with a lowered supply to achieve power savings. The lowered supply must be high enough to allow the inactive cells to maintain their data. Then, before the cells are accessed, they are "woken up" by providing a higher supply that can fulfill both read-disturb and access speed requirements.

The most straightforward implementation of cache sleep involves providing the cache with two separate, external supplies. However, a second supply is an expensive solution, so realistic implementations often choose to generate and regulate the second supply locally. In general, these implementations fall into two categories – active and passive. "Active sleep" schemes try to actively maintain the reduced voltage at a certain level, while "passive sleep" schemes rely on voltage division or threshold voltage to determine the reduced voltage.

11.3.1.1 Active Sleep

Khellah et al. [10] used an op-amp to help control the reduced supply; Figure 11.11 illustrates its general concept. When the arrays are active, "wake" causes SramVSS to be connected to VSS through the strong NFET. During idle mode, the strong NFET is turned off, allowing SramVSS to float. SramVSS will rise due to array leakage, but the op-amp will prevent SramVSS from rising above VREF. Of course, VDD – VREF must be greater than the SRAM cells' standby VccMin, which is the minimum voltage at which cells are stable, to maintain cell data. In this implementation, VREF is externally supplied for ease of controllability. Also, an "early wake" signal is provide ahead of "wake", to reduce the ground-bounce noise due to sudden discharge of SramVSS.

Jumel et al. [8] used a similar concept as the previous example, but took it a step further. As shown in Figure 11.12, an on-chip bandgap reference generates a reference voltage that is stable across PVT. In addition, the voltage regulator is designed to track VDD, so a higher VDD would also allow SramVSS to rise, maintaining VDD – SramVSS close to VccMin. Finally, the output of this regulator is trimmed on a die-by-die basis at wafer probe to account for process variations.

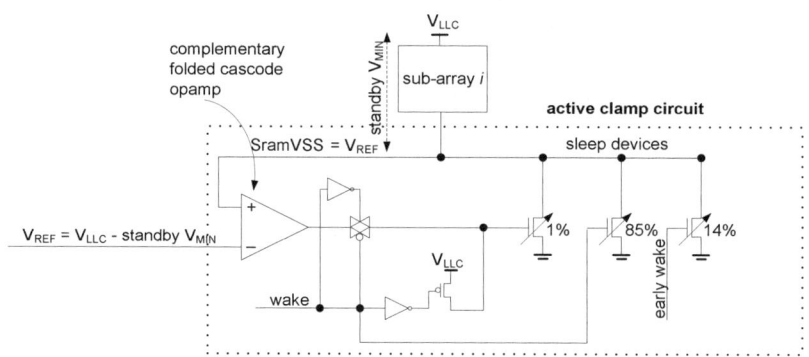

Figure 11.11 Active sleep control [10]. (© IEEE 2006)

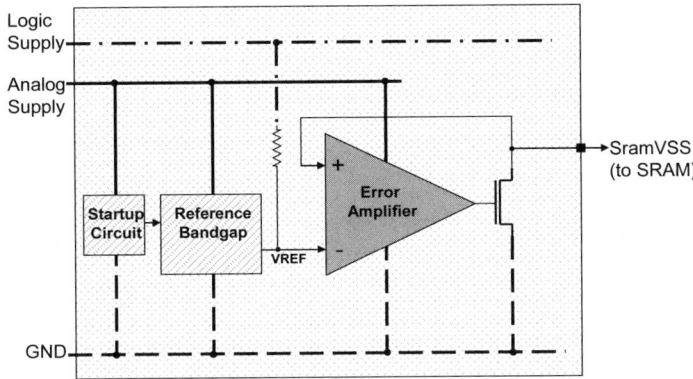

Figure 11.12 Active sleep control with bandgap reference and VDD tracking [8].
(© IEEE 2006) Courtesy of Philippe Royannez: Texas Instruments, Inc.

11.3.1.2 Passive Sleep

One straightforward way to generate a reduced supply is to use a diode,
such as in [1] and illustrated in Figure 11.13. When SramVSS rises to the
diode's threshold voltage, the diode would clamp SramVSS. The
downside to this scheme is its inflexibility, as the clamping voltage is
determined primarily by just the threshold voltage, and cannot be
optimized for different supply voltages.

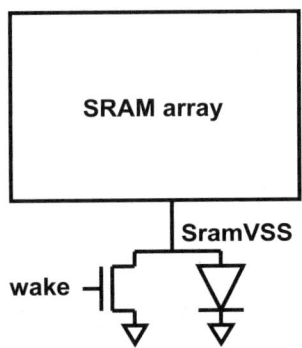

Figure 11.13 Diode clamping sleep voltage.

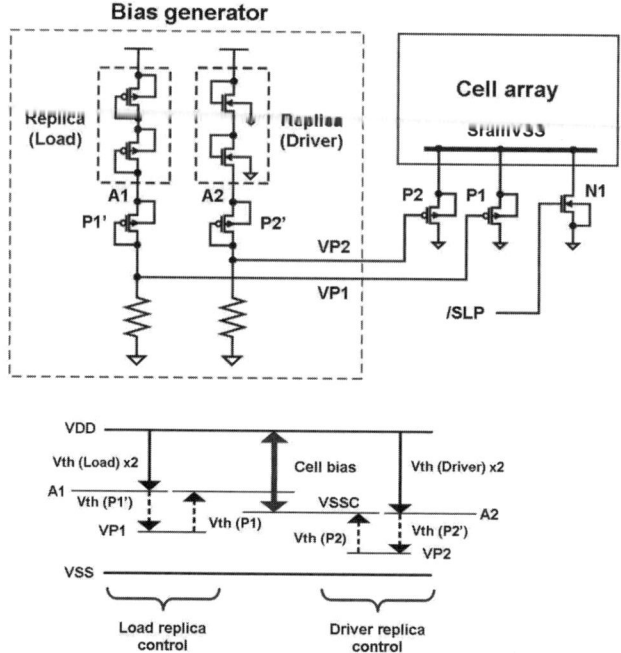

Figure 11.14 Bias generator with replica transistors [18]. (© IEEE 2006)

The example shown in Figure 11.14 aims to remove the SRAM supply's dependency on VDD [18]. Rather than setting the array supply to VDD – V_T, which can vary depending on VDD, the array supply depends only on transistor threshold voltages, as specified in Equation (11.1).

$$\text{Array supply} = 2 * \text{Max}(V_T(M_N), V_T(M_P)) \tag{11.1}$$

In this implementation, the array supply voltage specified in Equation (11.1) is assumed to be sufficient for satisfying VccMin requirements. To adapt to different PVT conditions, the bias generator is built using replica transistors. The two replica load PFETs drop A1's voltage to

$$A1 = VDD - 2 * V_T(M_P) \tag{11.2}$$

Similarly, the two replica driver NFETs drop A2's voltage to

$$A2 = VDD - 2 * V_T(M_N) \tag{11.3}$$

Finally, the matching P1 and P1' FETs clamp SramVSS at A1, while the matching P2 and P2' FETs clamp SramVSS at A2. The resulting SramVSS is the lower of A1 and A2, producing Equation (11.1).

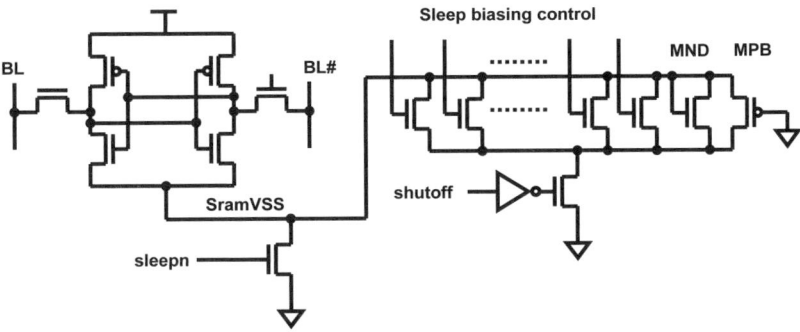

Figure 11.15 Passive sleep with parallel pull-down transistors [4]. (© IEEE 2007)

In yet another example of passive cache sleep [4], a group of NFETs of different sizes were built in parallel between SramVSS and VSS, as shown in Figure 11.15. In this implementation, VSS is gated by a shut-off FET to support cache power-down. During silicon characterization, the optimal combination of these NFETs is determined to maximize leakage power savings while maintaining cell stability. To provide better immunity from temperature variation, MND and MPB were added to the bias generator. In high temperature regions, the increased cell leakage would cause SramVSS to rise, reducing the supply to the memory cells and compromising stability. In such regions, the reduced V_Ts for MND and MPS due to the high temperature would strengthen the pull-down, and reduce the amount that SramVSS rises.

11.3.2 P Versus N Sleep

All the examples shown above use N-sleep, which provides the SRAM cells with true VDD and regulates SramVSS. Before accessing the SRAM cells, NFETs are used to restore SramVSS to VSS. Of course, designers can also implement the complementary P-sleep. In P-sleep designs, SRAM cells are provided with true VSS and a regulated SramVDD. Before accessing the P-sleep SRAM cells, PFETs are used to restore SramVDD to VDD.

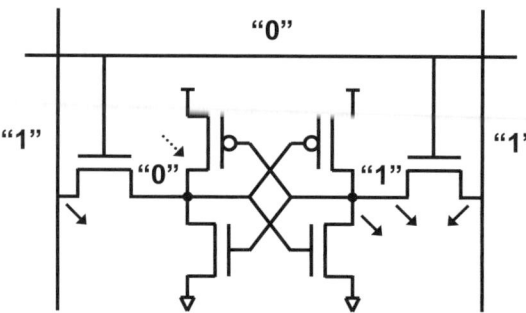

Figure 11.16 Junction leakage paths in SRAM cell.

At first glance, N-sleep seems the obvious favorite because the superior current driving capability of NFETs allows for smaller wake-up transistors, thus producing more efficient designs. However, designers must consider additional factors to make the appropriate choice. For example, the VSS net in a SRAM array often has more capacitance than the VDD net, so the larger SramVSS capacitance that must be discharged may negate the increase in the NFET's drive strength per. Also, P-sleep could provide additional power savings to processes that have non-negligible junction leakage. Figure 11.16 shows the junction leakage components in a typical SRAM cell, which includes 4 N-diffusion to body paths (solid arrows) and 1 P-diffusion to N-well path (dotted line arrow). Because of the greater number of N-diffusion to body paths, and because junction leakage from N-diffusion is usually worse than the junction leakage from P-diffusion, lowering VDD reduces the junction leakage more than raising VSS would. This is especially important for designs that leverage the bias circuitry to help shut off portions of the cache, such as in [4]. Shutting off VSS would cause SramVSS to rise, but the rise would eventually be halted by the increase in N-diffusion junction leakage as more N-diffusions are no longer at VSS. Shutting off VDD, on the other hand, could allow SramVDD to drop more significantly as P-diffusion leakage is less severe than N-diffusion leakage. Therefore, the proper choice between P and N sleep should be evaluated based on the specific process and SRAM cell design.

11.3.3 Entering and Exiting Sleep

The goal for sleep mode is to reduce power consumption. However, each time the cache enters or exits sleep mode, some active power is dissipated.

For example, the "wake" transistors must be turned on and off as the cache exits and enters sleep. If the cache enters and exits sleep modes too often, the active power dissipated for entering and exiting sleep may negate any power savings achieved through supply reduction.

To address this problem, one implementation [4] relies on the locality nature of cache accesses. A counter counts a set number of cycles after an array is accessed before allowing the array to return to sleep. If another access arrives before the counter finishes, the counter resets and begins again after the most recent access. Designer could optimize the number of wait cycles to balance leakage power reduction and active power dissipation for entering and exiting sleep.

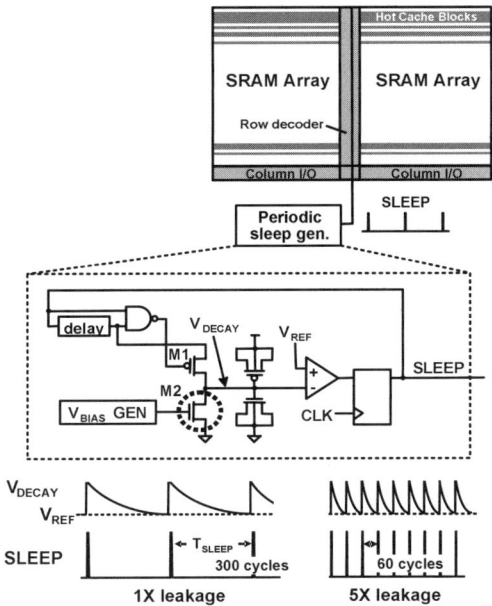

Figure 11.17 PVT aware sleep frequency scheme [12]. (© IEEE 2006)

Another implementation tries to take into account PVT variation to determine the frequency of entering sleep [12]. It assumes that, for corners with high leakage, the cache must be more aggressively placed into sleep mode; for corners with less leakage, the cache can be placed into sleep less often to reduce active power dissipation. Therefore, as shown in Figure 11.17, it uses a self-decay circuit to periodically generate the sleep pulse. For high leakage situations, the circuit would generate sleep pulses much more frequently, as the leakage power savings outweigh the active power dissipation. Also, rather than having the sleep signal generated periodically, this concept can also be modified to allow cache accesses to trigger the start of the decay circuit. In such a scheme, the cache would be placed under sleep if no access to the cache were made after a certain amount of time, as determined by the self-decay circuit.

11.3.4 Dynamic Cache Power Down

Due to the temporal locality of caches, a line that has not been used for a long time is not likely to be used again. Therefore, one could power down portions of a cache that have not been accessed recently. Because those SRAM cells do not need to retain state, VccMin does not need to be maintained, potentially realizing additional power savings. In one conceptual implementation [9], a small binary counter is provided for each cache line. The counter is clocked by a very slow clock and is reset every time the cache line is accessed. When the counter saturates, the supply to the corresponding line is shut off, powering down that line. Such an implementation requires significant area overhead. In addition, designers must take care to ensure that the power introduced by the additional circuitry (e.g., counter, control logic, slow clock) is not greater than the power savings.

In a different example, a scheme dubbed "cache-by-demand" in a microprocessor product [16] allows the microarchitecture to dynamically identify low usage of the cache, then powers down a portion of the cache to effectively reduce the size of the cache.

11.3.5 Data Bus Encoding

For large caches occupying large die area with wide data words, the active power dissipated by the long data bus routes could be significant. In such situations, designers can use data bus encoding for power reduction. Many encoding algorithms and techniques exist, but this chapterdescribes only the Bus-Invert Code [5], a simple example, to illustrate the use of data encoding.

Bus-Invert Code uses one extra bit to store data inversion information. The design keeps track of the previous data sent over the long wires, and compares it with the new data. If more than half of the bits are changing values, then the new data is inverted before being sent to the long wires. In addition, the inversion bit is set to keep track of the data's polarity. Such a scheme reduces the worst case number of transitions to half of the total number of bits, thus saving worst case power and improving worst case di/dt. However, savings for typical, random data would be lower, as the percentage of signals changing between two lines of data is typically much less than 100%.

11.4 Reliability

With caches becoming larger, transistor dimensions getting smaller, and process variation getting worse, designers must pay careful attention to reliability, which falls roughly into two categories. The first involves soft errors, which are errors caused by alpha particle or cosmic ray strikes, and are transitional in nature. The second type involves hard errors, which could include latent defects and test escapes.

11.4.1 Soft Errors

Techniques for mitigating soft errors have been widely known for many years, so this section offers only a brief description. To account for soft errors, arrays usually require either parity or Error Correcting Code (ECC). Parity allows for detection of single-bit errors (or, more exactly, an odd number of errors), but offers no way to correct an error. In other words, parity turns potential Silent Data Corruption (SDC) into Detected, Uncorrected Error (DUE). Although this may be sufficient for small arrays, large arrays generally require ECC. Hamming Code is widely used in SRAM arrays, which can correct single-bit errors and detect double-bit errors. Since column multiplexing usually provides physical separation between neighboring bits in the same ECC word, it becomes very unlikely for multiple bits in the same ECC word to flip from a single soft error strike. Therefore, Single-Error-Correct, Double-Error-Detect (SECDED) codes are generally sufficient for SRAM arrays under typical use.

11.4.2 Hard Errors

Hard errors such as latent defects and test escapes are not detected during silicon testing, but can surface in the field. Rather than accepting these failures as true defects, dynamic techniques exist for arrays to tolerate such failures.

11.4.2.1 Cache Line Disable

One dynamic technique dynamically disables cache lines by setting the MESI bits in the tag into a special "never valid" state when failures are detected [19]. Once the line enters the "never-valid" state, it will no longer be used by the processor. Because only a small number of lines are expected to contain failures, disabling a very small percentage of the overall cache has negligible impact on performance.

This technique can use different algorithms for determining lines to disable. For example, when an ECC error is detected, it could rewrite the corrected line back to the cache and attempt to read the line again. If ECC error is detected again, then the line is determined to be defective [19]. The drawback to this method, however, is that latent defects often go through a period of intermittent failures before becoming permanent defects. During this period, the defect may not remanifest itself while the corrected data is reread, causing defective lines not to be disabled as early as possible. Another implementation uses a table to keep track of ECC failures [4]. When a line causes an ECC failure for the first time, it is simply corrected, and logged into this table. Since soft errors are not likely to occur twice in the same line, if ECC error is detected in a line that already exists in the table, then the failure is determined to be caused by a defect, and the line is disabled.

11.4.2.2 Cache Line Remap

Another approach to disabling defective lines is by remapping them to different locations. In [2], a defective line is remapped to a different column in the same wordline. Since multiple lines now share the same physical location, the column address is added to the tag bits to differentiate the lines. During a tag lookup, both the tag and the column index are compared against the stored tag and column index bits, and a "hit" is only signaled if both match. This implementation essentially reduces the size of the cache without disabling any specific addresses.

11.4.2.3 Defect Correction

The cache line disable and remapping techniques discussed above have the drawback of reducing the actual cache size. Although the impact to performance could be negligible, these techniques can face other issues. For example, in applications where determinism is important, these techniques may pose additional challenges.

By correcting defects rather than disabling defective lines, designers can avoid such issues. Although ECC is typically used to address soft errors, it could also be used as a powerful tool to correct hard defects. Assuming the same level of protection against soft errors is required as caches that use SECDED ECC, such an implementation would require double-error-correct, triple-error-detect (DECTED) ECC. Out of the two correctable-error budget, designers can allocate one bit for hard defect correction, while reserving the other for soft error protection. Under this usage, designers must always enable correction for at least one error bit to avoid frequent exceptions for error handling. In applications where soft-error resiliency is not important, designers could use traditional SECDED ECC with error correction always enabled.

Such implementations would be able to tolerate one hard defect per ECC word. In other words, these techniques can correct a very large number of random, single-bit defects. Since random dopant fluctuation is expected to be the main limiter to SRAM VccMin as technology continues to scale, defects at voltages just below VccMin tend to be random, single-bit failures. Therefore, these error correction features can allow memory arrays to operate below normal VccMin and help SRAM continue to scale into future technology nodes.

11.5 Conclusion

This chapter surveyed dynamic and adaptive techniques in the area of SRAM design that seek to improve read and write margins, reduce power, and improve reliability.

Dynamic voltage optimization, especially column-based techniques that can independently improve both read and write margins in a column-multiplexed design, can be very effective. Silicon results from [22] demonstrated 10x reduction in random single-bit failures, when applying a 100mV offset from the wordline supply. However, such voltage optimization techniques can be expensive in product designs. For example, the area overhead to implement these techniques can be significant, and supply switching can increase the overall delay. More importantly, the additional supply can pose many difficulties. A dedicated second supply to the cache is the most robust solution, but it is costly and not always available to designers. Other solutions, such as charge sharing, supply floating, and supply coupling, are risky to design and can have limitations such as unreliable low frequency operations.

Similar "second-supply" challenges exist for cache sleep techniques as well. Unless a separate sleep supply is available, the sleep voltage must be regulated internally via reference voltages or voltage division schemes.

To account for noise and variation, sufficient margin must be added at various stages to ensure robust operation across PVT corners. Therefore, in designs where VccMin is significantly less than VDD, appreciable power savings can be achieved with cache sleep; however, in designs where VccMin is close to VDD (e.g., 200mV), realistic power savings may be limited after taking into account the various voltage margins.

Finally, the ability to tolerate a large number of random bit failures, either by correction or by other means, can become increasingly important as SRAM continues to scale. In addition to providing protection against test escapes and latent defects, which could help lower test time, test complexity, defect rate in the field, and cost, such features can also lead to lower VccMin and/or smaller SRAM cells. These features are attractive because they do not rely on risky circuit design, nor do they require a separate supply. Rather they are effective architectural features that designers can implement using known design techniques.

References

[1]   Agarwal A, Roy, K (2003) A Noise Tolerant Cache Design to Reduce Gate and Sub-threshold Leakage in the Nanometer Regime. Proc. ISLPED, pp 18–21

[2]   Agarwal A, Paul B, Roy K (2004) A Novel Fault Tolerant Cache to Improve Yield in Nanometer Technologies. Proc. IOLTS, pp 149–154

[3]   Bhavnagarwala A, Kosonocky S, Kowalczyk S, Joshi R, Chan Y, Srinivasan U, Wadhwa J (2004) A Transregional CMOS SRAM with Single, Logic VDD and Dynamic Power Rails. Symp. VLSI Circuits Dig. Tech. Papers, pp 292–293

[4]   Chang J, Huang M, Shoemaker J, Benoit J, Chen SL, Chen W, Chiu S, Ganesan R, Leong G, Lukka V, Rusu S, Srivastava D (2007) The 65-nm 16-MB Shared On-Die L3 Cache for the Dual-Core Intel Xeon Processor 7100 Series. IEEE J. Solid-State Circuits vol 42 no 4, pp 846–852

[5]   Cheng W, Pedram M (2001) Memory Bus Encoding for Low Power: A Tutorial. Proc. ISQED, pp 26–28

[6]   Dorsey J, Searles S, Ciraula M, Johnson S, Bujanos N, Wu D, Braganza M, Meyers S, Fang E, Kumar R (2007) An Integrated Quad-Core Opteron Processor. ISSCC Dig. Tech. Papers, pp 102–103

[7]   International Technology Roadmap for Semiconductors (2001)

[8]   Jumel F, Royannez P, Mair H, Scott D, Er Rachidi A, Lagerquist R, Chau M, Gururajarao S, Thiruvengadam S, Clinton M, Menezes V, Hollingsworth R, Vaccani J, Piacibello F, Culp N, Rosal J, Ball M, Ben-Amar F, Bouetel L, Domerego O, Lachese JL, Fournet-Fayard C, Ciroux J, Raibaut C, Ko U (2006) A Leakage Management System Based on Clock Gating Infrastructure for a 65-nm Digital Base-Band Modem Chip. Symp. VLSI Circuits Dig. Tech. Papers, pp 214–215

[9] Kaxiras S, Hu Z (2001) Cache Decay: Exploiting Generational Behavior to Reduce Cache Leakage Power. Int. Symp. Comput. Architect., pp 240–25

[10] Khellah M, Kim SN, Howard J, Ruhl G, Sunna M, Ye Y, Tschanz J, Somasekhar D, Borkar N, Hamzaoglu F, Pandya G, Farhang A, Zhang K, De V (2006) A 4.2GHz 0.3mm2 256kb Dual-Vcc SRAM Building Block in 65nm CMOS. ISSCC Dig. Tech. Papers, pp 2572–2573

[11] Khellah M, Ye Y, Kim NS, Somasekhar D, Pandya G, Farhang A, Zhang K, Webb C, De V (2006) Wordline & Bitline Pulsing Schemes for Improving SRAM Cell Stability in Low-Vcc 65nm CMOS Designs.   Symp. VLSI Circuits Dig. Tech. Papers, pp 9–10

[12] Kim C, Kim JJ, Chang IJ, Roy K (2006) PVT-Aware Leakage Reduction for On-Die Caches With Improved Read Stability. IEEE J. Solid-State Circuits vol 41 no 1, pp 170–178

[13] Mizuno H, NaganoT (1995) Driving Source-Line (DSL) Cell Architecture for Sub-1-V High-Speed Low-Power Application. Symp. VLSI Circuits Dig. Tech. Papers, pp 25–26

[14] Ohbayashi S, Yabuuchi M, Nii K, Tsukamoto Y, Imaoka S, Oda Y, Yoshihara T, Igarashi M, Takeuchi M, Kawashima H, Yamaguchi Y, Tsukamoto K, Inuishi M, Makino H, Ishibashi K, Shinohara H (2007) A 65-nm SoC Embedded 6T-SRAM Designed for Manufacturability With Read and Write Operation Stabilizing Circuits. IEEE J. Solid-State Circuits vol 42 no 4, pp 820–829

[15] Osada K, Shin JL, Khan M, Liou Y, Wang K, Shoji K, Kuroda K, Ikeda S, Ishibashi K (2001) Universal-Vdd 0.65-2.0-V 32-kB Cache Using a Voltage-Adapted Timing-Generation Scheme and a Lithographically Symmetrical Cell. IEEE J. Solid-State Circuits vol 36 no 11, pp 1738–1744

[16] Sakran N, Yuffe M, Mehalel M, Doweck J, Knoll E, Kovacs A (2007) The Implementation of the 65nm Dual-Core 64b Merom Processor. ISSCC Dig. Tech. Papers, pp 106–107

[17] Seevinck E, List FJ, Lohstroh J (1987) Static-Noise Margin Analysis of MOS SRAM Cells. IEEE J. Solid-State Circuits vol 22 no 5, pp 748–754

[18] Takeyama Y, Otake H, Hirabayashi O, Kushida K, Otsuka N (2006) A Low Leakage SRAM Macro With Replica Cell Biasing Scheme. IEEE J. Solid-State Circuits vol 41 no 4, pp 815–822

[19] Wuu J, Weiss D, Morganti C, Dreesen M (2005) The Asynchronous 24MB On-chip Level-3 Cache for a Dual-core Itanium Family Processor. ISSCC Dig. Tech. Papers, pp 488–489

[20] Yamaoka M, Shinozaki Y, Maeda N, Shimazaki Y, Kato K, Shimada S, Yanagisawa K, Osada K (2004) A 300MHz 25uA/Mb Leakage On-Chip SRAM Module Featuring Process-Variation Immunity and Low-Leakage-Active Mode for Mobile-Phone-Application Processor. ISSCC Dig. Tech. Papers, pp 494–495

[21] Yamaoka M, Maeda N, Shinozaki Y, Shimazaki Y, Nii K, Shimada S, Yanagisawa K, Kawahara T (2006) 90-nm Process-Variation Adaptive Embedded SRAM Modules With Power-Line-Floating Write Technique. IEEE J. Solid-State Circuits vol 41 no 3, pp 705–711

[22] Zhang K, Bhattacharya U, Chen Z, Hamzaoglu F, Murray D, Vallepalli N, Wang Y, Zheng B, Bohr M (2006) A 3-GHz 70-Mb SRAM in 65-nm CMOS Technology With Integrated Column-Based Dynamic Power Supply. IEEE J. Solid-State Circuits vol 41 no 1, pp 146–151

Chapter 12 The Challenges of Testing Adaptive Designs

Eric Fetzer, Jason Stinson, Brian Cherkauer, Steve Poehlman

Intel Corporation

In this chapter, we describe the adaptive techniques used in the Itanium® 2 9000 series microprocessor previously known as Montecito [1].

Montecito features two dual-threaded cores with over 26.5 MB of total on die cache in a 90nm process technology [2] with seven layers of copper interconnect. The die, shown in Figure 12.1, is 596 mm^2 in size, contains 1.72 billion transistors, and consumes 104 W at a maximum frequency of 1.6 GHz. To manufacture a product of such complexity, a sophisticated series of tests are performed on each part to ensure reliable operation throughout its service at a customer installation. Adaptive features often interfere with these tests. This chapter discusses three adaptive features on Montecito: active de-skew for reliable low skew clocks, Cache Safe Technology® for robust cache operation, and Foxton Technology® for power management. Traditional test methods are discussed, and the specific impacts of active de-skew and the power measurement system for Foxton are highlighted. Finally, we analyze different power management systems and consider their impacts on manufacturing.

12.1 The Adaptive Features of the Itanium 2 9000 Series

12.1.1 Active De-skew

The large die of the Montecito design results in major challenges in delivering a low skew global clock to all of the clocked elements on the die. Unintended clock skew directly impacts the frequency of the design by shortening the sample edge of the clock relative to the driving edge of a different clock. Random and systematic process variation in both the

A. Wang, S. Naffziger (eds.), *Adaptive Techniques for Dynamic Processor Optimization*, DOI: 10.1007/978-0-387-76472-6_12, © Springer Science+Business Media, LLC 2008

27.7 mm

21.5 mm

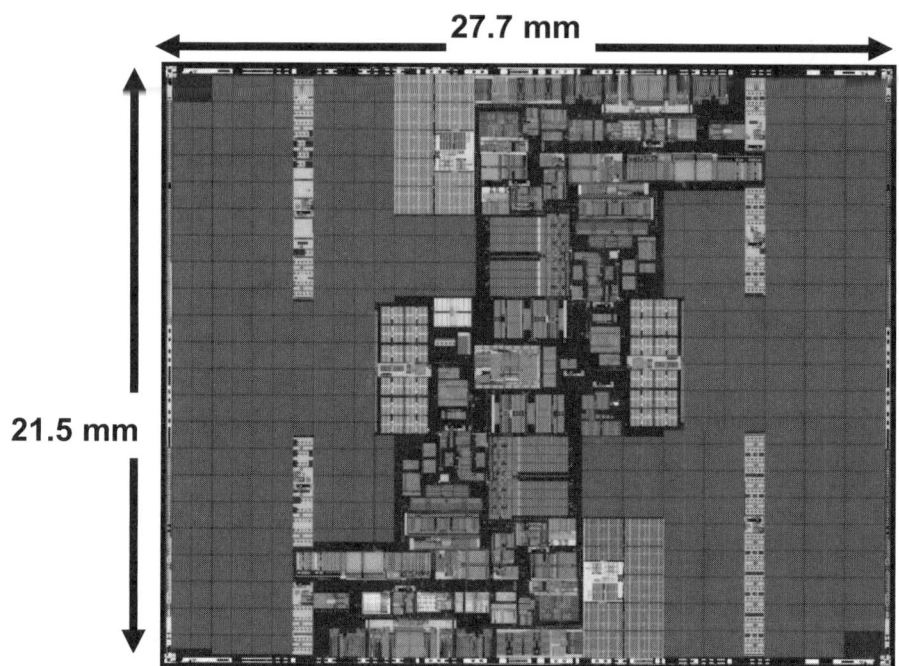

Figure 12.1 Montecito die micrograph.

transistor and metal layers makes it difficult to accurately design a static clock distribution network that will deliver a predictable clock edge placement throughout the die. Additionally, dynamic runtime effects such as local voltage droop, thermal gradients, and transistor aging further add to the complexity of delivering a low skew clock network. As a result of these challenges, the Montecito design implemented an adaptive de-skewing technique to significantly reduce the clock skew while keeping power consumption to a minimum.

Traditional methods of designing a static low skew network include both balanced Tree and Grid approaches (Figure 12.2). The traditional Tree network uses matching buffer stages and either layout-identical metal routing stages (each route has identical length/width/spacing) or delay-identical metal routing (routes have different length/width/spacing but same delay). A Grid network also uses matched buffer stages but creates a shorted "grid" for the metal routing, where all the outputs of a particular clock stage are shorted together.

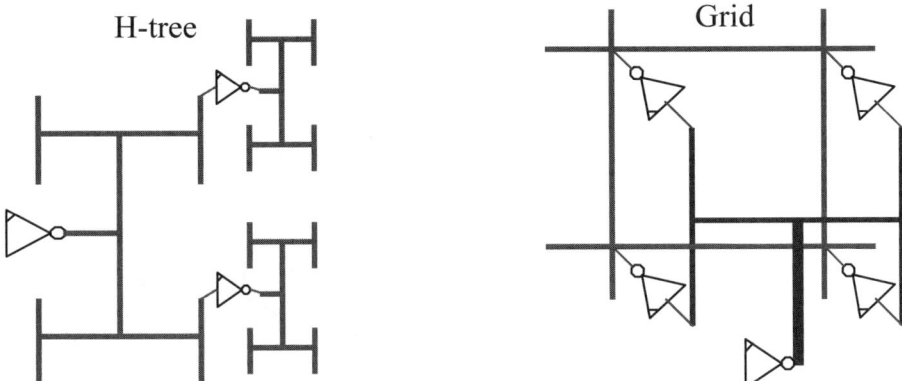

Figure 12.2 Example H-tree and grid distributions.

The benefit of a Tree approach is the relatively low capacitance of the network compared to the Grid approach. This results in significantly lower power dissipation. For a typical modern CPU design, the Grid approach consumes 5–10% of total power, as compared to the Tree approach, which can be as low as 1–2%. However, the Grid approach is both easier to design and more tolerant of in-die variation. A Tree network requires very balanced routes, which take significant time to fine-tune and optimally place among area-competing digital logic. The Grid network is much easier to design, as the grid is typically defined early in design and included as part of the power distribution metallization. The Grid network is also more tolerant of variation—since all buffers at a given stage are shorted in a Grid network, variation in devices and metals is effectively averaged out by neighboring devices/metals. While this results in very low skew, it also further increases power by creating temporary short circuits between neighboring skewed buffers. For a fully static network, the Grid approach is generally the lowest skew approach, but results in a significant power penalty.

The Montecito design could not afford the additional power consumption of a Grid approach. An adaptive de-skew system [3] was integrated with a Tree network to achieve low skew while simultaneously keeping power to a minimum. The de-skew system compares dozens of end points along the clock distribution network against their neighbors and then adjusts distribution buffer delays, using a delay line, to compensate for any skew. Ultimately, a single reference point (zone 53 in Figure 12.3) is used at the golden measure and all of the other zones (43, 44, etc.) align to it hierarchically.

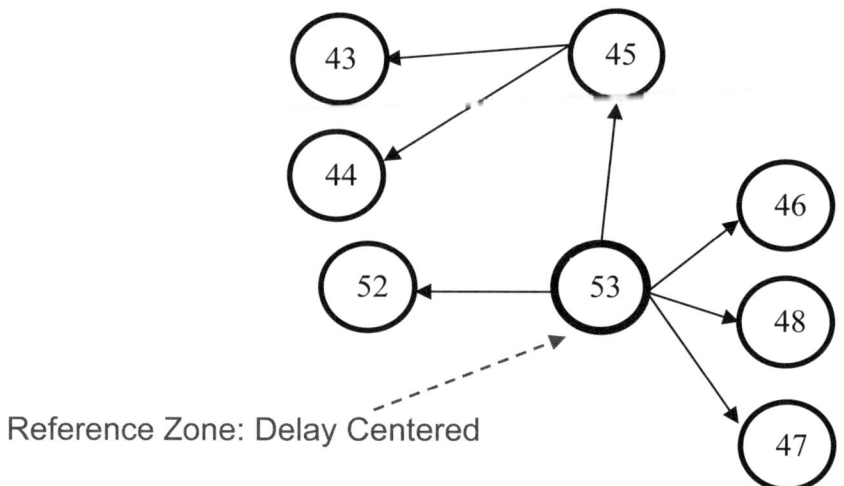

Figure 12.3 Partial comparator connectivity for active De-skew. (© IEEE 2005)

Similar de-skewing techniques have been used in past designs [4, 5]; however, these projects have de-skewed the network at startup (power-on) or determined a fixed setting at manufacturing test. The Montecito implementation keeps the de-skew correction active even during normal operation. This has the benefit of correcting for dynamic effects such as voltage droop and thermal gradient induced skew.

The de-skew comparison uses a circuit called a phase comparator. The phase comparator (Figure 12.4) takes two clock inputs from different regions of the die (ina and inb). In the presence of skew, either cvda or cvdb will rise before the other, which will in turn cause either up or down to assert. The output of the phase comparator is fed to a programmable delay buffer to mitigate the skew.

Empirically it has been shown that the adaptive de-skew method on Montecito decreases the clock skew by 65% when compared to uncompensated routes. Additionally, using different workloads, the de-skew network has been demonstrated to help mitigate the impact of voltage and temperature on the clock skew.

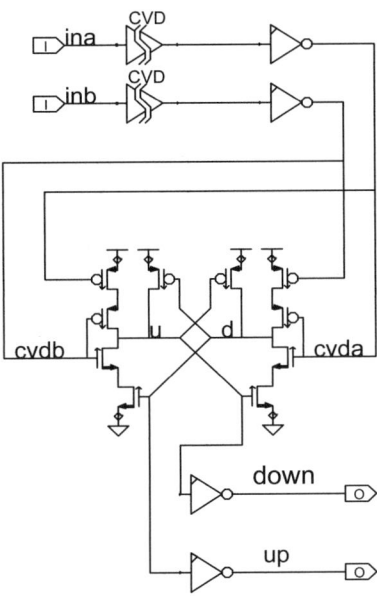

Figure 12.4 Montecito phase comparator circuit. (© IEEE 2005)

12.1.2 Cache Safe Technology

Montecito has a 24 MB last-level cache (LLC) on-die. As a result of its large size, the cache is susceptible to possible latent permanent or semi-permanent defects that could occur over the lifetime of the part. The commonly used technique of Error Correction Codes (ECC) was insufficient to maintain reliability in the presence of such defects which significantly add to the multi-bit failure rate. As a result, the design implements an adaptive technique called Cache Safe Technology (CST) to dynamically disable cache lines with latent defects during operation of the CPU.

Like most large memory designs, the Montecito LLC is protected with a technique called Error Correction Codes (ECC) [6]. For each cache line, additional bits of information are stored that make it possible to detect and reconstruct a corrected line of data in the presence of bad bits. "Temporary" bad cache bits typically arise from a class of phenomenon collectively called Soft Errors [7]. Soft Errors are the result of either alpha particles or cosmic rays and cause a charge to be induced into a circuit node. This induced charge can dynamically upset the state of memory elements in the design. Large caches are more susceptible simply because of their larger area. Soft Errors occur at a statistically predictable rate (called Soft Error Rate, or SER), so for any size cache the depth of protection needed from Soft Errors can be determined. In the case of Montecito, the LLC implements a single bit correction/double bit detection scheme to reduce the impact of transient Soft Errors to a negligible level.

The ECC scheme starts to break down in the presence of permanent cache defects. While the manufacturing flow screens out all initial permanent cache defects during testing, it is possible for latent defects to manifest themselves after the part has shipped to a customer. Latent defects include such mechanisms as Negative Bias Temperature Instability [8] (NBTI, which induces a shift in V_{th}), Hot Carrier or Erratic Bit [9] (gate oxide-related degradation), and electro-migration (shifts in metal atoms causing opens and shorts). Montecito implements CST to address these in-field permanent cache defects.

The CST monitors the ECC events in the cache and permanently disables cache lines that consistently show failures. At the onset of an in-field permanent cache defect on a bit, ECC will correct the line when it is read out. The CST handler will detect that an ECC event occurred and request a second read from the same cache line. If the bit is corrected on the second read, the handler will determine that the line has a latent defect. The data is moved to a separate area of the cache, and CST marks the line as invalid for use. The line remains invalid until the machine is restarted. In this manner, a large number of latent defects can be handled by CST while using ECC only to handle the temporary bit failures of Soft Errors.

12.1.3 Foxton Technology

Montecito features twice the number of cores of its predecessor and a large LLC, yet it reduces total power consumption to 104W compared to 130W for its predecessor. This puts the chip under very tight power constraints. By monitoring power directly, Montecito can adaptively adjust its power consumption to stay within a specified power envelope. It does this through a technique called Foxton Technology [10]. This prevents overdesign of the system components such as voltage regulators and cooling solutions, while reducing the guard-bands required to guarantee that a part stays within the specification. Foxton Technology implementation is divided into two pieces: power monitoring and reaction.

Power monitoring is accomplished through a mechanism that measures both voltage and resistance to back calculate the current. If the resistance of a section of the power delivery is known (R_{pkg}), and the voltage drop across that resistance is known ($V_{conn} - V_{die}$), then power can be calculated simply as:

$$Power = \frac{V_{die} * (V_{conn} - V_{die})}{R_{pkg}}$$

Power is delivered to the Montecito design by a voltage regulator, via an edge connector, through a substrate on which the die is mounted.

Edge Connector

Die
(Under heat spreader)

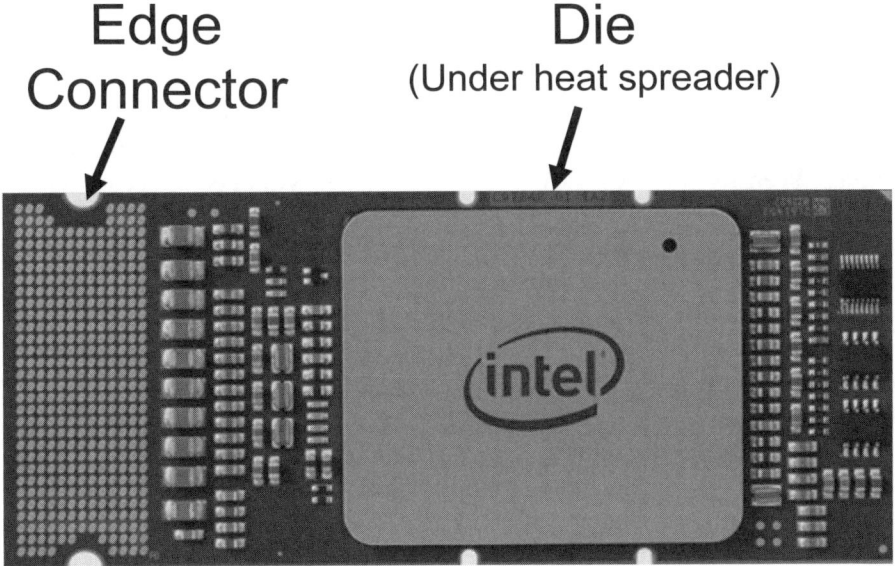

Figure 12.5 Montecito package.

The section of power delivery from the edge connector (Figure 12.5) to the on-die grid is used as the measurement point to calculate power. Montecito has four separate supplies (Vcore, Vcache, Vio, and V_{fixed}), which all need to be monitored or estimated in order to keep the total power below the specification.

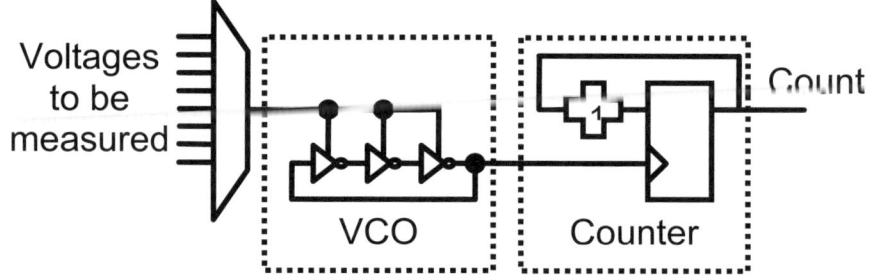

Figure 12.6 Measurement block diagram.

To calculate the voltage drop, the voltages at the edge connector and on-die grid need to be measured. A voltage-controlled ring oscillator (VCO) is used to provide this measurement (Figure 12.6). The higher the voltage, the faster the VCO will transition. By attaching a counter to the output of the VCO, a digital count can be generated that is representative of the voltage seen by the VCO. To convert counts to voltages, a set of on-die reference voltages are supplied to the VCO to create a voltage-to-count lookup table. Once the table is created, voltage can be interpolated between entries in the lookup table. Linearity is critical in this interpolation—the VCOs are designed to maintain strong linearity in the voltage range of interest. Dedicated low resistance trace lines route the two points (edge connector voltage and on-die voltage) to the VCOs on the microprocessor.

To calculate the resistance, R_{pkg}, a special calibration algorithm is used. Because package resistance varies both from package to package and with temperature, the resistance value is not constant. Using on-die current sources to supply known current values, the calibration runs periodically to compute package resistance. By applying a known current across the resistance, and measuring the voltage drop, the resistance can be calculated.

Once the power is known, the Montecito design has two different mechanisms to adjust power consumption to stay within its power specification. The first is an architectural method, which artificially throttles instruction execution to reduce power. By limiting the number of instructions that can be executed, the design will have less activity and hence lower power. Aggressive clock gating in the design (shutting off the clock to logic that is not being used) is particularly important in helping to reduce power when instruction execution is throttled.

The second method of power adjustment dynamically adjusts both the voltage and frequency of the design. If the power, voltage, and frequency of the current system are known, it is a simple matter to recalculate the new power when voltage and frequency are adjusted. A small state machine called a charge-rationing controller (QRC) is provided in the design to make these calculations and determine the optimal voltage and frequency to adhere to the power specification. The voltage regulator used with the Montecito design can be digitally controlled by the processor, enabling the voltage to be raised and lowered by the QRC. The on-die clock system also has the ability to dynamically adjust frequency in increments of 1/64th of a clock cycle. Using this method, the QRC can control both the frequency and voltage of the design in real time, enabling it to react to the power monitoring measurements. This second mechanism was used as a proof of concept on the Montecito design and is expected to be utilized in future designs.

12.2 The Path to Production

12.2.1 Fundamentals of Testing with Automated Test Equipment (ATE)

All test methods rely on two fundamental properties of the content and automatic test equipment environments: determinism and repeatability. Determinism is the ability to predict the outcome of a test by knowing the input stimulus. Determinism is required for defect-free devices to match the logic simulation used to generate test patterns. Repeatability is the ability to do the same thing over and over and achieve the same result. This is not same as determinism in that it does not guarantee that the result is known. In testing, given the same electrical environment (frequency, voltage, temp, etc.), the same results should be achievable each and every time a test runs passing or failing.

12.2.2 Manufacturing Test

Manufacturing production test is focused on screening for defects and the determination of the frequency, power, and voltage that the device operates at (a process known as "binning") (Figure 12.7). Production testing is typically done in three environments [11, 12].

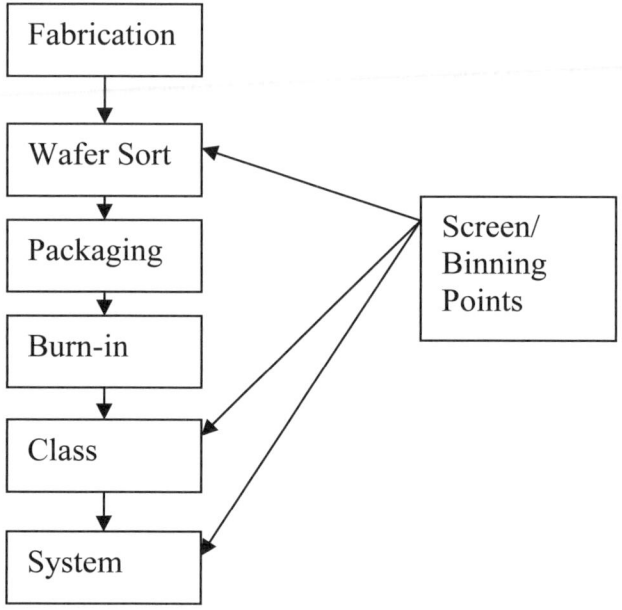

Figure 12.7 Test flow.

The first of these environments is wafer sort. Wafer sort is usually the least "capable" test environment as there are limitations in what testing can be performed through a probe card, as well as thermal limitations [13,12]. Power delivery and I/O signal counts for most modern VLSI designs exceed the capabilities of probe cards to deliver full functionality. With these limitations, wafer testing is usually limited to power supply shorts, shorts and opens on a limited number of I/O pads and basic functionality. Basic functionality testing is performed through special test modes and "backdoor" features that allow access to internal state in a limited pin count environment. This type of testing is referred to as "structural test" and is distinguished from "functional test", which uses normal pathways to test the device. Structural testing is typically focused on memory array structures and logic. The arrays are tested via test modes that change the access to the cache arrays to enable testing via BIST (built-in-self-test) or DAT (direct access testing, where the tester can directly access the address and data paths to an array). Logic is often tested using scan access to apply test patterns generated by ATPG (automated pattern generation) tools and/or BIST.

12.2.3 Class or Package Testing

After dice which pass wafer sort have been assembled, they are passed through burn-in, which operates the part at high voltages and temperatures revealing defects in manufacture while stabilizing device characteristics. The next test step is called class testing, or package test. The package enables fully featured voltage, frequency, and thermal testing of the processor—this is known as "functional testing". Class testers and device handlers are very complex and expensive pieces of equipment that handle all of the various testing requirements of a large microprocessor. Power delivery, high speed I/O, large diagnostic memory space, and thermal dissipation requirements significantly drive up cost and complexity of a functional tester.

The tester and test socket need to be able to meet all power and thermal delivery needs to fully test to the outer envelope of the design specifications, and thus meet or exceed any real customer system environments. This includes frequency performance testing to determine if the processor meets the "binning" frequency. A common practice for processor designs is to support multiple frequency "bins", which takes advantage of natural manufacturing variation by creating multiple variations of a product that are sold at different frequencies. The class socket is the primary testing socket used to determine the frequency bin to which a given part should belong. Frequency testing allows for screening of speed-sensitive defects as well as any electrical marginality that may exist as a result of the design or manufacturing process. In addition to frequency binning, binning based on power requirements is also becoming a common practice with today's high-performance processors due to customer demands for more power efficient products.

While wafer sort does not typically support full pin counts or full frequency testing, the class socket is fully featured for both. The class socket environment supports large pin counts and the high frequencies needed to test the processor at full input/output requirements, and provides the needed power connections to the power supplies in the tester. Support for the full complement of processor pins allows for normal functional testing (running code) in the class socket.

Traditional test pattern content is generated by simulating the case on a functional logic model [14]. The test case is simulated using the logic model for the processor, and the logic values at the pads are captured for each bus clock cycle during the simulation. The captured simulation data is then post-processed into a format which the tester can use to provide the stimulus and the expected results for testing of the processor. The operation of the processor in the tester socket must be deterministic and match the simulation environment exactly to enable the device under test to pass. As the expected data is stored in the tester and is compared at specific clock cycles relative to the stimulus data, this is called "stored response" testing (Figure 12.8). This is distinguished from "transaction" testing, which does not require cycle accurate deterministic behavior. Transaction testing usually requires an "intelligent" agent to effectively communicate with the device under test in its own functional protocol— this is usually accomplished through the use of another IC (e.g., a chipset) or a full system. Transaction testing at the class socket is still in early development phases within the industry—currently this type of testing is usually reserved for the final socket (system-based).

Drive Data Bus	Expect Addr Bus	Acquire Addr Bus
ZZZZ ZZZZ	FFFF FFB0	FFFF FFB0
ZZZZ ZZZZ	FFFF FFB0	FFFF FFC0
9090 9090	XXXX XXXX	FFFF FFC0
9090 9090	XXXX XXXX	FFFF FFC0
ZZZZ ZZZZ	FFFF FFC0	FFFF FFD0
ZZZZ ZZZZ	FFFF FFC0	FFFF FFD0
9090 9090	XXXX XXXX	FFFF FFD0
9090 90F4	XXXX XXXX	FFFF FFD0
7777 7777	FFFF FFD0	0000 0010

Figure 12.8 Stored response tester interface.

In recent years, cache resident testing has become more popular [15,16,13,17] for microprocessors. The test case is loaded into the processor's internal caches and then executed wholly from within the cache memory. The execution of the test case must use only the state internal to the processor to pass—there can't be any reliance on the external bus or state that wasn't preconditioned when the test case was loaded. Test cases are written in such a manner that they produce a passing signature, which is then stored in either a register or cache memory location. If the expected signature isn't read out correctly at the end of the test case, the processor has failed the test pattern and is deemed a "failure".

The expected "signature" can be developed in several ways. One approach is to preload the test case into the logic simulator and run it to completion. At the end of the simulation, the "signature" for the test case is recorded. As the case is simulated on the logic model, the "signature" can contain cycle-accurate timing for all signals. This enables the resulting test to not only check for architecturally correct behavior but also timing correct behavior—for instance, that the test completed within a specific number of core cycles. However, the cost of this level of timing accuracy is very slow simulation speeds through the logic model. Ultimately, simulation speed will limit the size and complexity of tests that take this approach.

A second approach to determine the "signature" is to code in the expected result. The test case only has to be valid assembly code that can be compiled into a format usable by the cache and initialization code. Without the use of logic simulation, care must be taken to prevent the inclusion of any clock cycle-dependent signals. This method of signature generation enables architectural validation, but not cycle-accurate timing validation. The benefit of this approach is that very large, complex diagnostics can be developed since they aren't dependent on very slow logic model simulations. Much faster architectural-level simulators may be used to compute the final "signature". Additionally, this approach enables non-deterministic architectural features to be tested. For instance, power management features that dynamically change the bus ratio based on power and/or thermal values are not deterministic because of the random nature of such values. This approach enables testing of such non-deterministic features.

12.2.4 System Testing

Platform or system testing provides flexibility not available in either the wafer sort or class tester environments. With the added flexibility comes higher cost and complexity, as the "tester" must be instrumented to determine when failures occur. The system test is transaction based—meaning that it intelligently communicates with the device under test rather than simply driving inputs and comparing outputs based on a predetermined set of vectors (stored response). This enables significantly longer tests compared to the class testers as only the transactions need to be stored in memory. In a stored response test, every single bus cycle of input and output data needs to be stored in memory. This makes the system test socket much more flexible in terms of the test content that can be used to test the device. However, with the flexibility comes less controllability and observability of the device under test. This makes it more difficult to determine pass/fail conditions as well as direct the test diagnostics at specific conditions or areas of the design.

There are two basic methods of checking pass/fail used for system test. The first is to use a "golden" device that subsequent devices under test are compared against. This is done by developing a custom system that allows for both the golden device and the device under test to receive the same stimulus from the system, and the golden device controls the system via its outputs that are compared to the device under test. If any differences occur, the device under test becomes a failed part and thus is no longer a candidate to be shipped to the customer. This approach has the limitation that all "good" devices must perform in exactly the same way as the golden device. With modern power management schemes and control, it is no longer true that all "good" devices perform in exactly the same way.

The second approach is to develop test content that is self-checking. This means that the content must test the desired functionality in such a way to produce a passing signature in memory or internal cache location. A system controller will examine the test pattern result and determine if it is passing or failing.

The test system controller must monitor power supply voltages, current, and temperature of the processor under test to determine if the device is performing within the required specifications. Thus, the system controller must have an additional level of automation and monitoring beyond just the test pattern content control. This adds to the overall cost and complexity of the system test step. Similar to the class test and wafer sort, the system test will follow a preprogrammed step of operations and decision tree to "bin" the processor. These steps will take longer than on the class tester because the test cases have less controllability and observability and must run longer to achieve similar design coverage. Also, the overhead of communicating with other devices external to the system adds significantly to the test time. However, the lower cost of a system test (relative to class test) combined with the ability to run very long diagnostic content actually enables this socket to provide the highest test coverage. Also, since system test does not require cycle-accurate determinism, it's the ideal location to test non-deterministic architectural features such as power and thermal management.

12.3 The Impact of Adaptive Techniques on Determinism and Repeatability

Adaptive circuits can impact both determinism and repeatability when testing and validating systems. Adaptive systems will often behave differently at different times. These differences negatively impact automated systems for observing chip behavior.

For instance, active clock de-skewing, when directly observed, will generate a unique and unpredictable result. At any moment in time small voltage fluctuations or thermal variations can cause the state of the system to have an apparently random value. When observed, these values will appear to be non-deterministic (unpredictable) and non-repeatable (different on each run). Without special consideration, direct observations of the active de-skew state would not be usable for manufacturing test.

As previously described, the Foxton power management system can dynamically adjust frequency to control power. Such changes in frequency can cause the chip to execute code over more cycles to reduce power. If this behavior is activated during traditional ATE testing, the tests will come back as failures since the ATE expects results to be cycle accurate.

The next few sections describe the validation and testing of these two features, active de-skew and Foxton technology, on the Itanium 2 processor and the techniques used to resolve the determinism and repeatability issues.

12.3.1 Validation of Active De-skew

The validation of the active de-skew system took many brute force methods. The first step in validating the de-skew system is measuring the behavior of the delay line, which is used to adjust the delay of the clock buffers. To do this, fixed delay line trim control values were scanned into the delay line to characterize delay line performance for a given setting. The impact to the delay was measured using an oscilloscope connected to a clock observability output pin on the processor package. The tick size for each of the 128 settings is about 1.5 ps. To achieve this resolution in measurement, tens of thousands of samples needed to be taken for each data point. Figure 12.9 shows the results from a typical part. Notice the inflection points that can be found at settings 32 and 96. These indices, along with setting 64, represent transition points inside the circuit. The inflections are caused by the slight additional capacitive load from routing in the layout. Setting 64, being the center point of the design, was better matched than 32 and 96 and as a result has a smaller deviation. This extra capacitance slows down the edge rate and increases the impact of particular trim settings.

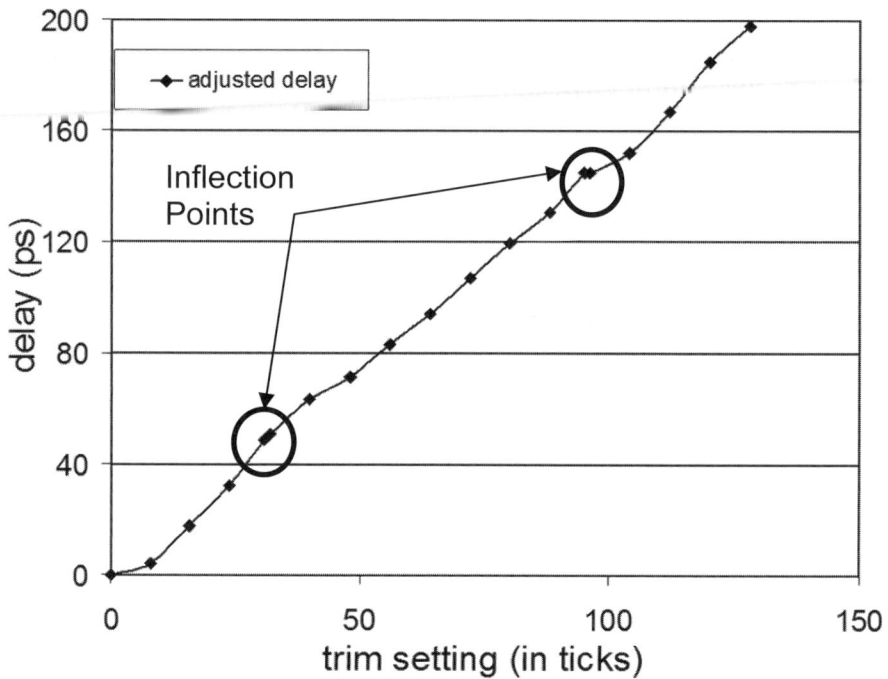

Figure 12.9 Active de-skew buffer linearity. (© IEEE 2005)

To check the comparator outputs (Figure 12.4) "up" and "down" that control the delay line in normal operation, the delay line can be scanned to each possible value. As the delay line value is increased, the comparator output changes (representing the changing relationship of the clocks). The comparator output can be viewed directly via scan and must be monotonic for the de-skew system to work.

Verification of the delay line and comparators in isolation doesn't prove that the active de-skew system actually removes skew. Measuring CPU frequency with the de-skew system on and off gives an indication of operation but is still not sufficient. It is possible, and probable, that a critical timing path (a circuit whose performance limits die frequency) can actually be improved by skew. In this case, active de-skew could actually slow down a part. To prevent this situation, the system has the ability to add fixed offsets that can be introduced via firmware to improve these paths.

To determine if active de-skew is removing skew appropriately, a more direct approach was necessary. Through the use of optical emission probing [18], it is possible to measure delays in the clock distribution accurately. Figure 12.10 shows a waveform of emitted light intensity for several different clock drivers across the die, which is indicative of the clock distribution delay to each of the drivers.   With de-skew off, a selected clock trace exhibits significant skew.   With de-skew on; the particular clock trace is in the middle of the pack.   While Figure 12.10 only shows a single outlier to represent the range of skews observed, a complete population of clocks would show a full spectrum of delays between the earliest and latest waveforms representing the total skew. The optical probing infrastructure inhibits heat removal, so absolute skew measurements using this method are not reliable representations of skew for a normal CPU.   Such probing is very time consuming and can only be performed on a few parts.   Probing did reveal that skew found with de-skew off in one core was not necessarily in the other core.   Since the cores are identical in layout and simply mirrored, the only possible differences between them are process, voltage, and temperature variations.

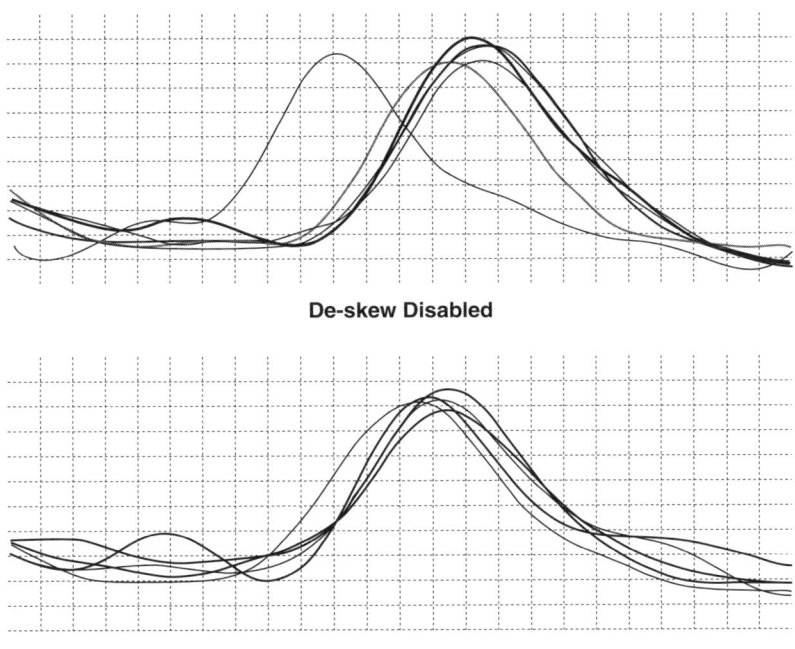

De-skew Disabled

De-skew Enabled

Figure 12.10 Light emission waveforms for clock drivers. (© IEEE 2005)

While these methods of verifying the behaviors of active de-skew are appropriate for small numbers of parts, they are not feasible for volume production testing. Oscilloscope measurements and optical probing can require several hours per part to perform the measurements. Furthermore optical probing requires part disassembly to observe the photons. As a result, different solutions were required for volume testing in production

12.3.2 Testing of Active De-skew

As discussed earlier, direct observation of values to be compared on a tester generally have to be repeatable and deterministic. To get around this limitation in ATE, codes can be used to create tester detectable patterns. This data is not checked as the test is running, but instead the data is collected by the tester for further analysis. In the case of active de-skew, thermometer codes were used to evaluate the correctness of the delay line trim settings. Typically, the delay line would store the trim setting in a binary value. For 128 possible settings, this would only require 7 bits to store. However, with data stored in this manner, if one of the bits is wrong, how would it be detected? Since there is no expected, known value to compare against, there is no way to determine whether the observed value reflects correct operation. By using a thermometer code, at the expense of 121 extra storage elements, the integrity of the code can be checked by the tester. In the example in Figure 12.11, a single error in the code shows up as a second 0 to 1 transition in the thermometer code. By intentionally skewing the inputs to comparators in the de-skew system, functionality can be checked for all possible clock relationships and delay line values.

Binary Encoded Value (7 bits Required)

Desired	Observed	Undetectable Error
0011001	0011011	

Thermometer Encoded Value   (128 bits Required)

Desired	Observed	Error Identified by multiple $0 \rightarrow 1$ transitions
...00011111..	...00010111...	
...00011111..	...01011111...	

Figure 12.11 Thermometer code example.

Testing the rest of the chip with clock de-skew operating also has challenges. One technique for testing circuits is to stop the clock, scan in a test vector, burst the clock for a few cycles, and scan out a result. This method of testing has many direct conflicts with active de-skew.

- Stopping the clock prevents the active de-skew system from updating properly. Without updates, the active de-skew system will not compensate for skew that results from changing environmental conditions during testing.
- If the clock skew, while in a system, is not the same as the skew during testing by ATE, the manufacturing test can end up being an inaccurate measurement of actual processor speed in a system.
- Stop clock conditions create voltage overshoot/droop events that do not exist while clocks are continuously operating.

As a result, special consideration must be taken when testing using this method. The de-skew system must be preloaded with fixed values during ATE testing that represent skew conditions in the system. The de-skew system must also be disabled while actually running the test to prevent a response to the "artificial" voltage event that results from this kind of test.

12.3.3 Testing of Power Measurement

Central to Montecito's Foxton power management system is the ability to measure power consumed at any given moment. The power measurement system utilizes a small microcontroller, an input selectable VCO (voltage controlled oscillator) and the natural parasitic resistance of the package, as described earlier. Testing power measurement requires ensuring that each individual component is within the required specifications.

The power microcontroller, unlike the other parts of the microprocessor, runs at its own fixed frequency. While the processor can dynamically change frequencies, the power controller needs a constant known frequency for its understanding of time. As a result, all communication between the microcontroller and the rest of the processor is asynchronous. In order to test the microcontroller directly, a BIST engine and custom test patterns are used. Due to its asynchronous nature when communicating with the core, and its non-deterministic outputs when measuring power, testing the microcontroller at speed is more involved than simply running code on the processor and checking its results. To test the microcontroller features, its firmware is replaced with special self-checking content. This content is run and stores its results which are then scanned out using cache resident structural test content as described previously.

In a normally functioning system, the VCO counts are calibrated to known voltages by using a band-gap reference on the package and a resistor ladder on the die (Figure 12.12). The microcontroller samples the VCO count for each voltage, V_{ladder}, available from the resistor ladder.

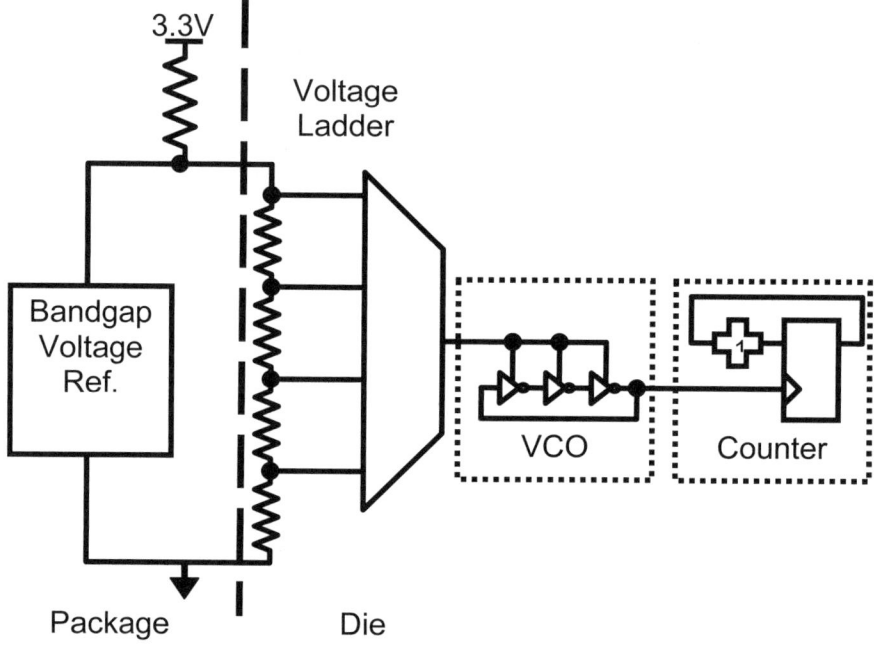

Figure 12.12 VCO calibration circuit diagram.

To perform a measurement, the firmware receives a count from the VCO and interpolates between the two nearest datapoints. Using Table 12.1, it can be seen how a count of 19350 would be translated into a voltage of 1.018V through interpolation.

Table 12.1 Voltage vs. VCO count example.

Voltage	Count (example)
1.000	19250
1.007	19291
1.015	19337
1.023	19375
1.030	19410

During wafer sort, where bare die is tested, the on-package band-gap reference is not available and the band-gap reference is replaced by a fixed voltage. Firmware is loaded into the microcontroller to evaluate the linearity and gain of the voltage/VCO count table. In Figure 12.13, this process is shown using both a good and a bad part.

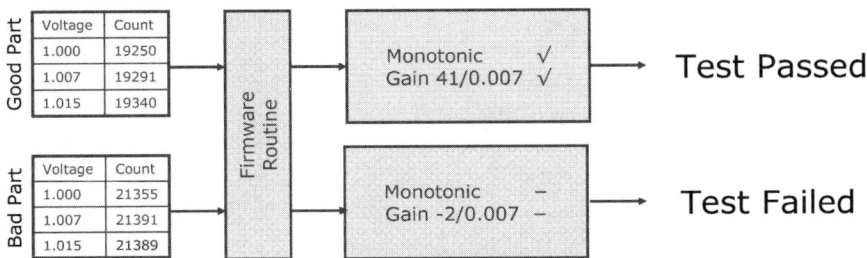

Figure 12.13 Process for evaluating VCO table.

For the bad part, an increase in voltage from 1.007 to 1.015 caused a decrease in VCO count from 21391 to 21389. This behavior would cause the count 21390 to make to both 1.011V and 1.006V making voltage measurement far too inaccurate to measure power accurately.

With the testing of the VCO complete, the on-package parasitic resistance can be measured. If the resistance is too low, not enough voltage delta will be generated under load to get an accurate power measurement. If the resistance is too high, significant power is wasted in the package itself.

By measuring the voltage drop across the connector (V_{c1}–V_{d1}) using the VCO while the chip is idle and consuming standby current I_0 and then measuring the voltage drop (V_{c2}–V_{d2}) while the chip is under a known additional current load, I_{Delta}, the package resistance can be computed using a simple formula (Figure 12.14). This formula, once again, is applied using special firmware in the microcontroller and the range is tested to be within acceptable limits (Figure 12.15).

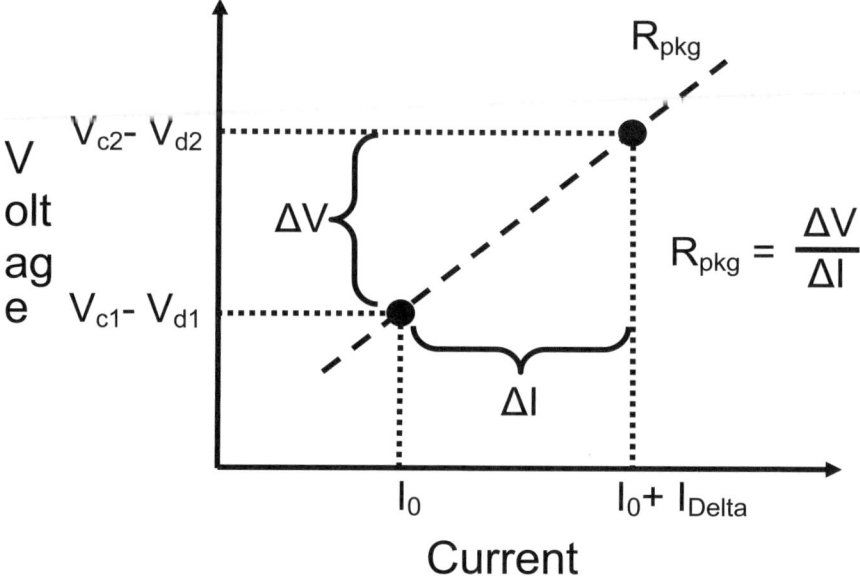

Figure 12.14 Graphical representation of Rpkg measurement. (© IEEE 2006)

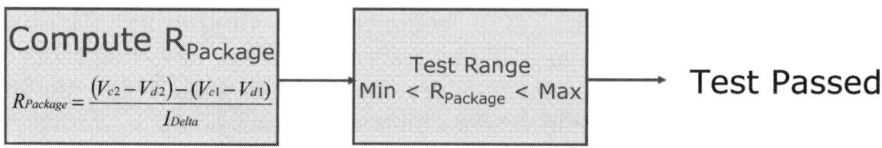

Figure 12.15 Computation and test of R_{pkg}.

12.3.4 Power Measurement Impacts on Other Testing

During operation the package resistance is not a constant. As package temperature increases so does the resistance of the package. The temperature of the processor is a function of processor activity and ambient temperature of the system. As a result, the resistance of the package must be recomputed every few microseconds to keep the power measurement accurate. To do this, the processor must be briefly interrupted so known currents (I_0 and I_0+I_{Delta}) can be passed through the connector. This interruption stalls any running code, which is a slight performance impact. Due to the asynchronous nature of the microcontroller interface, this stall is not deterministic and cannot be anticipated by the test infrastructure. In a standard ATE, this delay would be seen as a malfunction and the test would fail. As a result, power measurement, and all functionality that relies on it, must be disabled for the testing of standard content.

The asynchronous nature of the microcontroller interface is not the only limitation. Even if the design managed a repeatable and deterministic interface between power measurement and the processor, the system would still need to be disabled during testing. In order to guarantee robust functionality over the lifetime of operation, parts are tested well beyond their normal operation limits. This ensures that as silicon performance degrades with continued use, the part stays with specification. In testing beyond the normal limits, the part will exceed its maximum specified power. This would cause the power management system to measure a power that is "too high" and place the chip in a reduced performance mode. Figure 12.16 shows a typical shmoo of a part with a frequency limiting critical path. This path forces the frequency of the part to be reduced at low voltage.

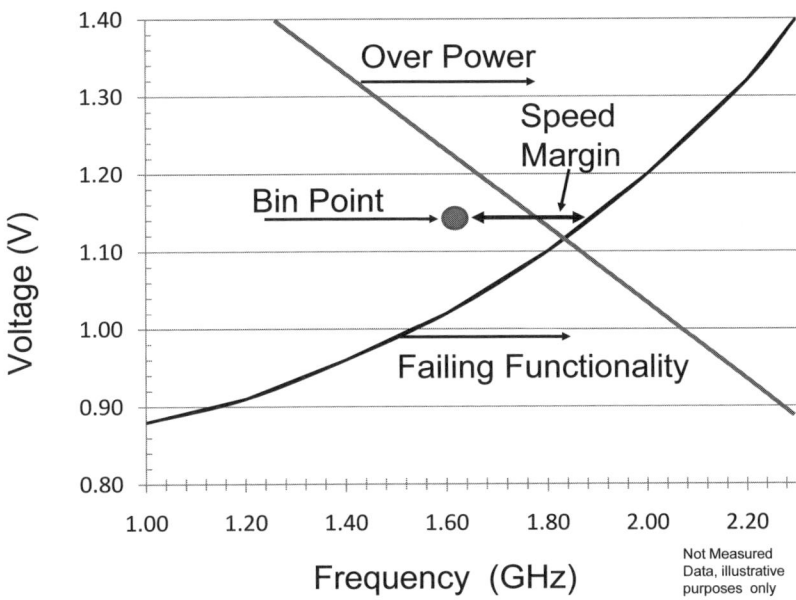

Figure 12.16 Speed-path shmoo with max power line.

The bin point (the point at which the part will operate when in use by a customer) requires a speed margin, or guard-band, be applied from the failing region. While the bin point is well below the maximum power line, the bin point combined with the necessary speed margin exceeds the maximum power. If power measurement were enabled the processor would observe this excess power during test and limit the instructions being executed to lower the power. This change in behavior would cause the test to fail and eliminates the ability to test with the margin required.

12.3.5 Test Limitations and Guard-Banding

In traditional testing, margin (also known as guard-band) is used to ensure reliable operation when the part is operating in less than ideal conditions. Guard-bands are required for many reasons including:

- Tester limitations: The accuracy of voltage supplied and thermal control on the tester is limited.
- Content limitations: System traces for large applications used to measure power often need to be approximated (reduced in size) when run on a tester.
- Transistor aging: As silicon is stressed over time, transistor performance degrades.

Adaptive circuit techniques are often used to enable reductions in these guard-bands. For example, a part that can measure its own power and can react to it can adjust its own consumption to stay within the required envelope. For a non-adaptive design, the worst-case power code must be tested and a guard-band applied to ensure no future code exceeds the test power consumption. An adaptive design can have less guard-band because if future code draws more power from the chip, the part will "do the right thing." However, it is not quite this simple. The adaptive part requires guard-bands for each of its measurement and adjustment systems. In the case of power measurement, there is error in the package resistance measurement due to thermal drift. There are also inaccuracies in the voltage measurement caused by power supply noise and VCO non-linearities. As a result, implementation details determine whether or not actual guard-banding is reduced. In the case of power measurement on the Itanium 2, the sum of guard-bands for power measurement circuitry is less than 5%, while the potential error in power code is significantly larger, making adaptation a win.

The Itanium 2 has a thermal management system very similar to power measurement. Using the same VCO (Figure 12.17) as in the power measurement system, the thermal solution has the resolution to measure temperature with a precision $\ll$ 1°C.

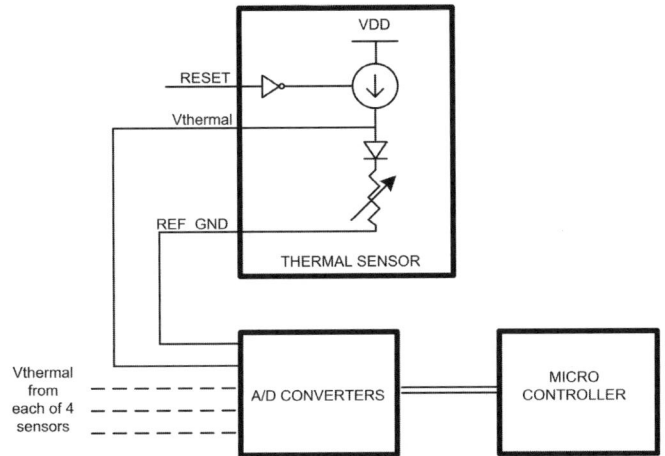

Figure 12.17 Block diagram of thermal measurement. (© IEEE 2006)

However, in order to calibrate the system a known temperature with $\ll$ 1°C of error needs to be supplied by the test environment. The test environment has to test parts with varying power draw, in a short amount of time, and with limited thermal probes. To achieve the desired thermal control in a test environment, the part would need to be submerged in an oil bath. This is not possible while achieving the required test throughput. As a result, the accuracy of the thermal monitoring system is not limited by the processor capabilities, but instead is limited by the capabilities of the test environment.

As more and more adaptive techniques are used to stretch the capabilities of silicon, investments will need to be made in validation and test systems to fully utilize the new capabilities. Adaptive circuit techniques have the ability to reduce processor guard-bands provided the test infrastructure can emulate the use conditions adequately.

12.4 Guard-Band Concerns of Adaptive Power Management

After one considers the correctness of adaptable systems, one must deliver the value that they offer in the product environment. One of the primary manufacturing considerations in designing an adaptive frequency/power control system is performance variability tolerance. A system based on any type of analog measurement will inherently be susceptible to part to part variation as well as environmental variation.

For example, the Montecito system that makes an on-die analog measurement of the power being consumed will be subject to part-to-part variation —no two parts will have exactly the same mix of leakage and dynamic power. This means as voltage is raised or lowered, the power consumed by parts will vary compared to one another. The same is true with temperature variation, which affects the leakage power but not the dynamic power. Also, the ideal voltage versus frequency curve is subject to part-to-part variation, and attempting to optimize this on a per-part basis will introduce additional variability.

This variability can also be a function of more subtle effects such as the aging of components. Voltage regulator outputs may drift as they age, cooling systems may provide less airflow, and even the leakage of the processor itself changes with aging. Thus, it is exceedingly difficult to make a processor that behaves identically from run-to-run and part-to-part throughout its lifetime if it depends on an analog power measurement for the basis of its performance adaptability. Systems that depend on a temperature measurement to adapt performance are subject to similar variability compared to those that measure power directly.

Reducing the number of possible operating conditions from a continuous curve to a series of a few discrete conditions greatly reduces the exposure to variability, as most variation will not be enough to move from one operating condition to the next. However, if absolutely deterministic behavior is required of a design, another approach is to replace analog sensing with architectural event counters.

Using architectural counters [19], specific architectural events can serve as a proxy for power dissipation, by weighting each one according to its expected contribution to the power. Assuming the weighting is not done on a part-by-part basis, all processors will behave identically on identical code streams. This potentially gives up some benefits of the analog schemes, which squeeze out more from the design by using actual power or temperature measurements instead of a proxy. However, this even-based approach guarantees part-to-part and workload-to-workload repeatability—also making benchmarking and design debug much more straightforward.

From a manufacturability standpoint, both analog and architectural designs require similarly sized guard-bands (Adaptive Op. Point, Figure 12.18) to guarantee power stays within limits. Because of issues in testing and operation, this guard-band is larger than the guard-band required at a non-adaptive operating point. From an analog perspective, the design is dependent on the ability to make an accurate current measurement, often in the noisy environment of a running system.

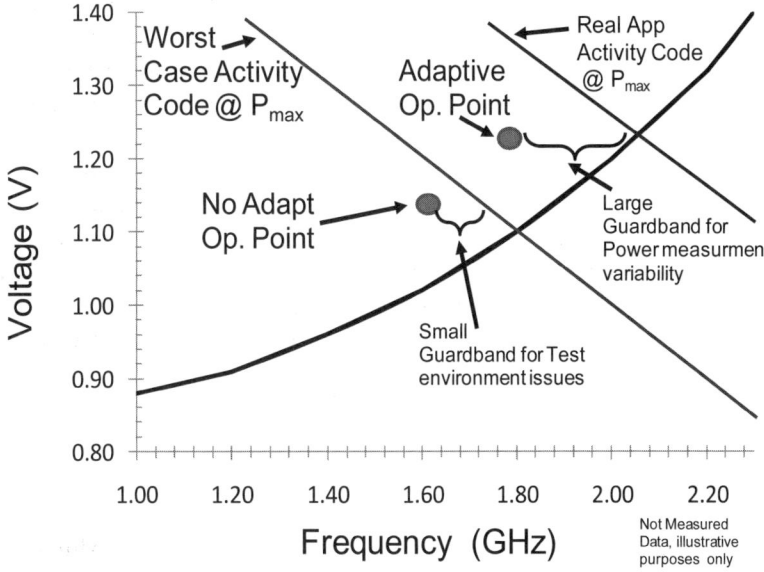

Figure 12.18 Comparison of operating point with and without adaptation.

Architectural counters are not subject to analog noise or accuracy, but they must be placed and weighted carefully in order to provide the best mapping to power. One drawback of the architectural approach is that the worst-case power event needs to be well understood to be detected and the system needs tuning based on silicon-collected data to be accurate. Another drawback is that it is very difficult to cover data-dependent power. That is to say, you can map a certain architectural operation to a given power level, but you cannot easily modify that power level based on the operands or the specific data being manipulated, as this requires too deep a penetration of the architectural monitors.

Determinism and repeatability give architectural power estimates a significant advantage over the analog measurements. Unlike the situation where the analog measurement-based power management must be disabled for almost all production testing, an architectural power-based system will determine steps to maintain a constant power level. While voltage and frequency responses may not be properly emulated on the tester, the measurement system itself will behave in a predictable and testable manner.

12.5 Conclusion

From wafer test to final testing of parts in systems, determinism and repeatability are the cornerstones of bringing a processor design to market. Adaptive techniques used in modern processors like those demonstrated in this chapter make determinism and repeatability difficult to achieve. In some cases, the test infrastructure is not able to keep up with the processor's ability to adapt, and as a result the guard-bands that adaptation is trying to eliminate will remain. Careful planning, along with novel test techniques like the ones described in this chapter, needs to be employed to realize the full potential of adaptive techniques. Additional significant breakthroughs will be required for higher levels of adaptation involving applications, OS, firmware, system components, and the processor to be fully production testable.

References

[1]   Naffziger, S., et al., "The Implementation of a 2-core Multi-Threaded Itanium-Family Processor," IEEE Journal of Solid-State Circuits, Vol. 41, No. 1 pp. 197–209, Jan. 2006

[2]   Thompson, S., et al., "A 90 nm logic technology featuring 50 nm strained silicon channel transistor, 7 layers of Cu interconnects, low k ILD, and 1 μm^2 SRAM cell," Electron Devices Meeting, 2002. IEDM '02. Digest. International, pp. 61–64, Dec. 2002

[3]   Mahoney, P., Fetzer, E., et al., "Clock distribution on a dual-core, multi-threaded Itanium®-family processor," Solid-State Circuits Conference, 2005. Digest of Technical Papers. ISSCC. 2005 IEEE International, Vol. 1, pp. 292–599, 6–10 Feb. 2005

[4]   Anderson, F.E., Wells, J.S., Berta, E.Z., "The core clock system on the next generation Itanium microprocessor," Solid-State Circuits Conference, 2002. Digest of Technical Papers. ISSCC. 2002 IEEE International, Vol. 1, pp. 146–453, 3–7 Feb. 2002

[5]   Geannopoulos, G., Dai, X., "An adaptive digital deskewing circuit for clock distribution networks", Solid-State Circuits Conference, 1998. Digest of Technical Papers. 45th ISSCC 1998 IEEE International, pp. 400–401, 5–7 Feb. 1998

[6]   Peterson, W.W., Weldon, E.J., Jr., Error-Correcting Codes, 2nd editions, MIT Press: Cambridge Mass., 1972

[7]   Ziegler, J. F., Srinivasan, G. R., et al, "Terrestrial cosmic rays and soft errors," IBM Journal of R and D, Vol. 40 No.1 1996

[8]   Ershov, M., Saxena, S., et al., "Dynamic recovery of negative bias temperature instability in p-type metal-oxide-semiconductor field-effect transistors," Applied Physics Letters, , Vol. 83, No. 8, pp. 1647–1649, August 25 2003

[9]   Agostinelli, M., et al., "Erratic fluctuations of SRAM cache Vmin at the 90nm process technology node," Electron Devices Meeting, 2005. IEDM Technical Digest. IEEE International, pp. 655–658, Dec. 5 2005

[10] McGowen, R., Poirier, C., et al., "Power and Temperature Control on a 90-nm Itanium Microprocessor," Solid-State Circuits, IEEE Journal of Vol. 41, No. 1, pp. 229–237, Jan. 2006

[11] Wayne Needham, Cheryl Prunty, Eng Hong Yeoh, "High Volume Microprocessor Test Escapes, An Analysis Of Defects Our Test Are Missing", IEEE International Test Conference, pp. 25–34, 1998.

[12] Mike Mayberry, John Johnson, Navid Shahriari, Mike Trip, "Realizing the Benefits of Structural Test For Intel Microprocessors", IEEE International Test Conference, pp. 456–463, 2002.

[13] Ismet Bayraktaroglu, Jim Hunt, Daniel Watkins, "Cache Resident Functional Microprocessor Testing: Avoiding High Speed IO Issues", IEEE International Test Conference Conference, 2006.

[14] Huston, R., "Microprocessor Functional Test Generation on the Sentry 600", IEEE International Test Conference, 1974.

[15] Praveen Parvathala, Kailas Maneparambil, William Lindsay, " FRITS – A Microprocessor Functional BIST Method", IEEE International Test Conference, pp. 590–598, 2002.

[16] Krantis, N., Xenoulis, G., Paschalis, A., Gizopoulos, D., Zorian, Y., "Application and Analysis of RT-Level Software-Based Self-testing for Embedded Processor Cores", IEEE Intetrnational Test C440.

[17] Wei-Cheng Lai, Kwang-Ting Cheng, "Instruction-Level DFT for Testing Processor and IP Cores in System-on-a-Chip", Design Automation Conference ,pp. 59–64, 2001.

[18] Tsang, J., et. al., "Picosecond imaging circuit analysis", IBM Journal of Research and Development, Vol. 44, No. 4, pp. 583–603, 2000.

[19] Leon, A. S., et al., "A Power-Efficient High-Throughput 32-Thread SPARC Processor," IEEE J. Solid-State Circuits, Vol. 42, No. 1, pp. 7–16, Jan. 2007.

[20] Harry Hsiung, "Manufacturing and test Solutions with EFI", Intel Developers Forum, 2003.

[21] Peter Maxwell, Ismed Hartanto, Lee Bentz, "Comparing Functional and Structural Tests", IEEE International Test Conference, pp. 400–407, 2000.

[22] Satish M. Thatte, Jacob A. Abraham, "Test Generation For Microprocessors", IEEE Transactions On Computers, Vol. 29, No. 6, pp. 429–441.

[23] Advanced Configuration and Power Interface Specification, rev 3.0b, http://www.acpi.info/spec.htm, October 2006

Index

Continued from page ii

Abstraction Refinement for Large Scale Model Checking
Chao Wang, Gary D. Hachtel, and Fabio Somenzi
ISBN 978-0-387-28594-2, 2006

A Practical Introduction to PSL
Cindy Eisner and Dana Fisman
ISBN 978-0-387-35313-5, 2006

Thermal and Power Management of Integrated Systems
Arman Vassighi and Manoj Sachdev
ISBN 978-0-387-25762-4, 2006

Leakage in Nanometer CMOS Technologies
Siva G. Narendra and Anantha Chandrakasan
ISBN 978-0-387-25737-2, 2005

Statistical Analysis and Optimization for VLSI: Timing and Power
Ashish Srivastava, Dennis Sylvester, and David Blaauw
ISBN 978-0-387-26049-9, 2005

Printed in the United States of America